The Right to Food Debates

The Right to Food Debates
Social Protection for Food Security in India

Edited by

Harsh Mander
Ashwin Parulkar
Ankita Aggarwal

Orient BlackSwan

THE RIGHT TO FOOD DEBATES: SOCIAL PROTECTION FOR FOOD SECURITY IN INDIA

ORIENT BLACKSWAN PRIVATE LIMITED

Registered Office
3-6-752 Himayatnagar, Hyderabad 500 029, Telangana, India
e-mail: centraloffice@orientblackswan.com

Other Offices
Bengaluru, Bhopal, Chennai, Guwahati,
Hyderabad, Jaipur, Kolkata, Lucknow, Mumbai,
New Delhi, Noida, Patna, Vijayawada

ISBN 978-93-5287-284-8

027512

Typeset in
Minion Pro 10.5/12.5
by Manmohan Kumar, Delhi

Printed in India at
Glorious Printers, Delhi

Published by
Orient Blackswan Private Limited
3-6-752, Himayatnagar Hyderabad 500 029
e-mail: info@orientblackswan.com

Contents

Acknowledgements

We would like to acknowledge several members of Centre for Equity Studies' S. R. Sankarant Unit on Hunger and Social Exclusion, who contributed research in various stages of preparation for this book: Jeevika Shiv, Agrima Bhasin, Gitanjali Prasad and Anamika Lahiri, and former interns Rukmani Riar, Karl-Axel Lindgren and Anneka Singh.

We are grateful to Dr N. C. Saxena, former Commissioner of the Supreme Court on the right to food case, and colleagues in the Commissioner's office, Biraj Patnaik, Sejal Dand, Dipa Sinha, Sajjad Hassan (who is also a contributor to this volume), Sandeep Chachra, Aditya Shrivastava and Pallavi Nayak.

We would also like to extend our gratitude to members of the Right to Food Campaign: Kavita Srivastava, Jean Dreze, Colin Gonsalves, Anuradha Talwar, Arundhati Dhuru, Balram, Rupesh Kumar, Gurjeet Singh, Sunil Kaul, Sachin Jain, Samir, and Sulakshana Nandi.

A very special thanks to Nina Elinger, as well as to former colleagues of co-editor Harsh Mander's at the National Advisory Council.

The aim of this book is to present salient viewpoints on aspects of food security that were debated by academics, members of civil society and policymakers in the years before and during the drafting of the National Food Security Act (2013). As such, we acknowledge responsibility for any shortcomings in our summary and analysis of these debates. We hope to do justice to all views.

In closing, we are especially grateful to Dr Bhaskar Vira of Cambridge University for graciously facilitating an international peer review of the manuscript before publication. We hope this volume will stimulate and inform debates on state provisioning of food, particularly in the context as part of a human rights-backed social protection system for the poor—in India and abroad.

Harsh Mander
Director, Centre for Equity Studies

Ashwin Parulkar
Senior Researcher, Centre for Policy Research

Ankita Aggarwal
Coordinator, Right to Food Campaign, New Delhi

List of Tables

Abbreviations

AAY	Antyodaya Anna Yojana
AIADMK	All India Anna Dravida Munnetra Kazhagam
APL	above the poverty line
ASHA	accredited social health activist
BJP	Bharatiya Janata Party
BMA	Biscuit Manufacturers Association
BMI	body mass index
BPL	below the poverty line
CCT	conditional cash transfer
CEDAW	Convention on the Elimination of All Forms of Discrimination Against Women
CES	Centre for Equity Studies
CPI	Communist Party of India
CPI(M)	Communist Party of India (Marxist)
DGRO	district grievance redressal officer
DPT	diphtheria, pertussis and tetanus
FAO	Food and Agriculture Organization
FCI	Food Corporation of India
FPS	fair price shop
GDP	gross domestic product
HRD	Human Resource Development
IBFAN	International Baby Food Action Network
ICDS	Integrated Child Development Services
ICESCR	International Covenant on Economic, Social and Cultural Rights
ICMR	Indian Council of Medical Research
IFPRI	International Food Policy Research Institute
IMF	International Monetary Fund
IMS Act	Infant Milk Substitutes, Feeding Bottles and Infant Foods (Regulation of Production, Supply and Distribution) Act, 1992
IT	information technology
JSY	Janani Suraksha Yojana
LPG	liquefied petroleum gas

MDMS	Midday Meals Scheme
MGNREGA	Mahatma Gandhi National Rural Employment Guarantee Act
MHRD	Ministry of Human Resource Development
MP	member of Parliament
MSP	minimum support price
MWCD	Ministry of Women and Child Development
NAC	National Advisory Council
NCHS	National Center for Health Statistics
NCPRI	National Council for People's Right to Information
NDA	National Democratic Alliance
NFHS	National Family Health Survey
NFSA	National Food Security Act, 2013
NFSB	National Food Security Bill
NGO	non-governmental organisation
NMBS	National Maternity Benefit Scheme
NRC	nutrition rehabilitation centre
NRHM	National Rural Health Mission
NSS	National Sample Survey
NSSO	National Sample Survey Office
OBC	Other Backward Castes
OECD	Organisation for Economic Co-operation and Development
PDS	public distribution system
PRIs	panchayati raj institutions
PVTG	Particularly Vulnerable Tribal Groups
PUCL	People's Union for Civil Liberties
RDA	recommended daily allowance
RTE	Right to Education (Act)
RTI	Right to Information (Act)
RUTF	ready-to-use therapeutic food
SAM	severe acute malnutrition
SC	Scheduled Caste
SECC	Socio-economic Caste Census
SNP	Supplementary Nutrition Programme
ST	Scheduled Tribe
TB	tuberculosis
TPDS	targeted public distribution system

UDHR	Universal Declaration of Human Rights
UID	Unique Identification
UN	United Nations
UNICEF	United Nations Children's Fund
UPA	United Progressive Alliance
WHA	World Health Assembly
WHO	World Health Organization
WTO	World Trade Organization

Introduction

State Food Provisioning as Social Protection in India

Ashwin Parulkar and Harsh Mander

India's National Food Security Act (2013) (NFSA) is among the largest government-sponsored food security programmes in the world. The law guarantees 75 per cent of people in rural areas and 50 per cent of people in cities—over 800 million people—a legal right to purchase 5 kilograms of wheat, rice or millets from government fair price shops (FPS) at highly subsidised prices. It also guarantees free lunches for schoolchildren in public primary schools (approximately 174 million children[1]), free meals for pre-school children, pregnant and nursing mothers at nutrition centres registered under the national Integrated Child Development Services (ICDS) programme, and maternity entitlements in the form of cash for female informal economic sector workers upon deliveries. The law also provides for district and state grievance redressal bodies to ensure that rights to these entitlements are restored if or when beneficiaries are denied access to them.

The United Progressive Alliance (UPA) government, which ruled at the centre between 2004 and 2014, proposed this law in 2009, signalling a shift in consensus among policymakers on the kind of social policies required to address hunger. Agricultural production policies through the 1970s, consisting of land reforms and the provision of high-yield seeds to farmers, transformed a once food-deficit India into a country that became and would remain food secure on the national level—a momentous accomplishment, famously called the Green Revolution. Based on National Nutrition Monitoring Bureau surveys, Prema Ramachandran has shown that malnutrition in India declined substantially in the two decades following the Green Revolution's success. From 1975 to 1996, the percentage of underweight female children under the age of 5, for instance, declined from 74.4 per cent to 49.1 per cent. The percentage of female children of the same age group who were stunted[2] fell from 81.1 per cent to 52.6 per cent (Ramachandran 2016). India continued to produce foodgrain surpluses into the twenty-first century and, at the end of the 2000s, the nation's economy was the second fastest growing in the world. Yet, malnutrition rates declined modestly, remaining at levels out of step with India's newfound image as an emerging nation—by 2011–12, according to the National Nutrition Monitoring Bureau, 41.6 per cent of female children and 42.9 per cent of male children under the age of 5 were still underweight (ibid.).

Meanwhile, early this century, social activists from a range of civil society organisations (e.g., the Right to Food Campaign, People's Union for Civil Liberties [PUCL] and Human Rights Law Network) began sustained research-based advocacy campaigns to highlight reasons why poor people, particularly in rural areas, could not avail public work, food and nutrition programmes that had been in existence in varying forms since the 1960s and 1970s. Government corruption and faulty methods of identifying programme beneficiaries (commonly termed 'exclusion errors') were major reasons why people were denied benefits. As we will elaborate later, in 2001, activists associated with PUCL filed a public interest litigation in the Supreme Court on behalf of people who could not avail services and died of hunger. The petitioners demanded that local governments implement these programmes as a matter of poor people's constitutional right to life (Article 21). The Supreme Court decided that access to these programmes was, indeed, poor people's 'right to food' and ordered local governments to provide services. These decisions, and additional directions in interim orders throughout the decade, set the agenda for the right to food in India that eventually culminated in legislation 12 years later.

Apart from maternity benefits, the entitlements currently guaranteed under the NFSA were previously identified by the Supreme Court as 'rights to food' across three existing social security programmes: the public distribution system (PDS), the Midday Meals Scheme (MDMS), and some provisions in the ICDS programme. Other 'rights to food' under the Supreme Court orders, however, are not included in the current law: the Indira Gandhi National Widow Pension Scheme, the Indira Gandhi Old Age Pension Scheme, and Indira Gandhi National Disability Pension Scheme.

The NFSA was conceived as a human rights law meant to ensure people's access to the social protection system. We may, therefore, understand the current status of the right to food in India as the set of social services legally guaranteed by this law. How did lawmakers decide on these entitlements? Does the law reflect the letter and spirit of the right to food as conceived by earlier precedents in national laws and policies and by international human rights mandates? This book explores these questions by examining debates that emerged between academics, policymakers, advisory bodies and civil society actors on how to ensure and sustain food security under human rights law. These discussions began and intensified over a period that approximately covers the years 2001 to 2013, and were held in the context of historical precedents: Indian social policy, the nation's constitutional legacy and international human rights mandates that define the right to food. The discussion and chapters that follow are therefore not an assessment of the implementation of the law, but a survey of ideas and debates on how the right to food should be formalised in legislation to address the multiple dimensions of hunger experienced by poor people in India. We view the debates that unfolded between the court-ordered rights to food and the formalisation of *some* of those provisions in law as a combination and convergence of three distinct, yet interrelated, factors. We summarise these factors here and elaborate on them further in this chapter and throughout the book.

(1) *The emergence this century of a human rights-based social security system for poor people*

Previously, pensions for the elderly, health insurance and maternity benefits were available only for people employed in the formal sector of the economy, which comprised only about 10 per cent of the workforce (Drèze and Khera 2017). State provisioning of food through the PDS became a central component of the emerging social security system for poor people who worked exclusively in the informal sector of the economy, particularly farm labourers in rural areas. Barriers poor people face in accessing the system are often in the realm of human rights violations. In their survey of the PDS and ICDS programmes, Swain and Kumaran (2012) found that people are often denied access to these services due to discrimination on the basis of caste, gender and religion. Service providers (e.g., ration shop dealers), they also found, distribute lower quantities of entitlements specified by programmes, charge higher prices and mislead people with false information (ibid.). Jean Drèze and Reetika Khera (2017) note that in a context in which poor people face difficulties accessing social services, a human rights-based social security system is essential for three reasons. It ensures pathways to legal justice by enabling poor people to defend their rights in court. Second, a rights-based system creates financial obligations and administrative arrangements that make governments accountable for implementing services. Finally, this system of reciprocity influences public perceptions on the context-specific necessity (e.g., access barriers) of implementing social programmes in which the government is accountable for delivering services to poor people who experience further deprivations when denied benefits that are intended for them (ibid.). Drèze and Khera also, however, note the challenges to a rights-based system in a country as large as India: laws that appropriate budgets and administrative arrangements for social programmes are difficult to amend after they are passed, and there is a propensity for excessive central government control, which makes implementation difficult (ibid.).

Other social security measures protected by human rights law include the public works programme offered under the 2005 Mahatma Gandhi National Rural Employment Guarantee Act (MGNREGA). This law guarantees rural residents 100 days of work near their villages at state-specified minimum wages to be paid within 15 days of start dates, and has redressal mechanisms in place in the event that workers are not paid (Drèze and Khera 2017).

A human rights based social security system in India—by which national laws create duties upon the state to implement social programmes and ensure beneficiaries' legal claims to services and pathways for redress in the event of entitlement failures—has been enabled by the enforcement, over time, of social and economic rights located in the Constitution of India, promulgated in 1950, and international human rights mandates: the UN's (United Nations) Universal Declaration of Human Rights (UDHR) adopted by the General Assembly in 1948, and, more specifically, Article 11 on the right to food in the UN's International Covenant on Economic, Social and Cultural Rights (ICESCR) adopted by state parties in 1976. We briefly describe the implications of both.

(2) *The dynamism of the Constitution of India*

The architects of India's Constitution acknowledged that the depth and scale of poverty in which the majority of people after independence in this nation lived posed a barrier to ensuring democracy. 'How to solve the problem of the poor and the starving masses', Jawaharlal Nehru told his colleagues in the Constituent Assembly, was 'the greatest and most important question' facing them. 'The first task of the Assembly', he said, 'is to free India through a new constitution to feed the starving people and clothe the naked masses.'[3]

The aftermath of the 1943 Great Bengal Famine and the deaths of over one million people in riots following Partition in 1947 loomed in the framers' minds. They understood that a Constitution based on civil and political rights—e.g., the rights to vote, the rights to privacy and assembly, and freedom from discrimination—would not alone address unfreedoms rooted in social and economic inequalities that produced hunger, illiteracy and poor health among the masses. The Constitution's main author, Dr Bhimrao Ramji Ambedkar, insisted that democracy in India would not be achievable without legal mechanisms to address these ills. The reason, he claimed, was that Indian society 'was based on the principle of graded inequality', particularly in the areas of caste and gender, which consecrated the permanent 'elevation for some and degradation for others' (Government of Maharashtra 1987, 24–25; see Seshaiah 1997, 229; Bhattacharjea 1997, 228–36). The economy, meanwhile, comprised 'a society in which there are those who have immense wealth as against many who live in abject poverty.' Reversing these inequalities, the framer believed, was needed to both eradicate poverty and secure a pathway to a functional democracy in India. How? They differentiated civil and political rights, deemed 'fundamental rights' (Section III), from social and economic rights, included in the 'directive principles of state policy' (Section IV). While fundamental rights were enforceable in courts, directive principles were ideals intended to guide future social policies throughout the life course of Indian democracy that would eventually require proactive government responses—e.g., state budgetary allocations, administrative arrangements, enforcement and accountability mechanisms—to address structural inequalities pertaining to, for instance, employment and health. As we will explain later, several Supreme Court judgments have, over time, brought social and economic rights located originally in the directive principles under the realm of fundamental rights, such as Article 21 on the right to life. We list here the directive principles relevant to social security provisions for the poor that have over time been considered the remit of the right to food in India:

- Article 47 of the Constitution of India makes it a primary duty of the state to raise the standard of nutrition and the standard of living of its people and to improve public health
- Article 41 of the Constitution of India obliges the state to make effective provision for securing the right to public assistance in cases of unemployment, old age, sickness and disablement, and in other cases of undeserved want
- Article 39(a) of the Constitution of India obliges the state to direct its policy towards ensuring that citizens, men and women, equally have the right to an adequate means of livelihood
- Article 42 of the Constitution of India obliges the state to make provision for securing just and humane conditions of work and for maternity relief

(3) *International human rights mandates*

The UDHR has influenced most national constitutions since 1948, including that of India, which broadly followed the pattern of listing the civil and political rights of the UDHR as fundamental rights that are enforceable in courts. The ethical and political global consensus on the equal intrinsic worth and dignity of every human being is reflected in the UDHR, adopted by the UN General Assembly in 1948. This has led some countries to incorporate the duties of the state to provision food into rights-based legislation, whereas many others are content to retain these as programmatic commitments of the state but not guaranteed by law.

The right to food is one component among UDHR's Article 25.1's 'the right to a standard of living adequate for health and well-being', which includes 'clothing, housing, and medical care' (http://www.un.org/en/universal-declaration-human-rights/). The enforcement of this right is legally binding on the 160 nations who are signatories to the ICESCR, which came into force in 1976. ICESCR's definition of the right to adequate food (located in Article 11, General Comment 12) hinges on concepts of availability and accessibility; availability refers to 'food in a quantity and quality sufficient to satisfy the dietary needs of individuals, free from adverse substances, and acceptable within a given culture' (ICESCR Article 11, General Comment 12, p. 6), and access means 'accessibility of such foods in ways that are sustainable and do not interfere with the enjoyment of other rights' (ibid.; FAO 2011, 4).

State parties, including India, are bound to take administrative and legal steps to progressively (over time) 'respect, protect and fulfill (facilitate and provide)' (ICESCR Article 11, General Comment 12, p. 4) individuals' access to food. In other words, the ICESCR places obligations on states to (*a*) not hinder individual access to food (*respect*); (*b*) take proactive measures—e.g., social policies—to ensure people can avail food (*protect*); (*c*) provide food directly, such as aid in times of emergencies and/or for vulnerable people experiencing acute hunger (*fulfil*) (FAO 2011).

These international mandates provide a logical and legal framework to guide states to realise the otherwise aspirational goal of achieving freedom from hunger.

Conventional wisdom among liberal democratic governments during the Cold War held that states could be morally enjoined to ensure food security for their citizens. Yet there was little consideration given to passing legislation to achieve this, given the complex technical, budgetary and administrative arrangements involved. If one's brother dies because he was tortured in a police station, public officials could be held criminally liable. If, however, one's child dies from starvation because one could not find work with a decent wage, locally available food at affordable costs, or access social services such as health care facilities, public officers cannot be punished under existing laws.

Accordingly, in 2004, the Food and Agriculture Organization (FAO) of the UN adopted by consensus a set of *Voluntary Guidelines to Support the Progressive Realisation of the Right to Adequate Food in the Context of National Food Security* (FAO 2005). The guidelines recommend constitutional and legislative action to achieve the right to food.

Social Security for the Poor in India

Debates on targeted social protection programmes for India's poor in health, nutrition, employment and food security have evolved rapidly in the twenty-first century in response to deprivations afflicting large sections of India's vulnerable populations. India's nutrition and health deprivations are stark indicators of social inequality in a time of rapid overall economic progress. In comparison to five other South Asian countries, for instance, India is the worst performer in child undernutrition and child immunisation rates while having the highest per capita GDP in the region (see Drèze and Sen 2013; UNDP 2013).

Discourse on how to address the vulnerabilities of the country's poorest to ill health, hunger, disease and unnecessary death has catalysed public action. This mobilisation has, in some instances, led to rights-based social policies backed by formal legislation, such as MGNREGA, the Rashtriya Swasthya Bima Yojna (a national health insurance scheme), Unorganised Workers' Social Security Act, Aam Aadmi Bima Yojna (life insurance for the common man) and, most recently, the NFSA.

Thus far, the government's record on implementation of these social protection programmes has proved uneven. Some reasons include restrictions on the coverage of beneficiaries to families identified as below the poverty line (BPL), faulty methods of identification of the poor, lack of communication from the government to poor people regarding eligibility criteria, and discrimination and corruption by officials. Many beneficiaries, particularly residents of the poorest states, do not access the full range of guaranteed entitlements. Many poor people cannot access social protection programmes at all.[4]

Still, discourse on the need for social policy in India to address persistent vulnerabilities to disease, hunger and undernutrition faced by hundreds of millions of Indians does reflect growing agreement between policymakers[5] and scholars on the core economic and social determinants of extreme poverty that mandate a social protection system befitting a democracy seeking recognition as a first-world power. India's poor health and nutrition indicators alongside its high economic growth attest to the existence of social inequality. Scholars have also recently brought to light how a period of high economic growth without an adequate social security system has produced inequality in wages and employment opportunities and insecurity in labour conditions across sectors of the economy. High-growth industry and services sectors are not producing enough jobs with decent wages and social security for workers who are leaving a declining agricultural sector in rural India.

Agriculture still employs about 56 per cent of India's economy and is the means by which most of the labouring poor earn their living

(Anant et al. 2006; Sundaram 2007, cited in Nagaraj 2012). In terms of share of output in India's gross domestic product (GDP) and in comparison to industrial and service sectors, output in agriculture has decreased since economic liberalisation began in the early 1990s. The fact that agriculture's share in the GDP is falling in comparison to modern sectors would not be alarming in itself if enough workers were leaving farms for jobs in thriving sections of the economy which provided liveable wages and the minimum base of social security typically associated with employment conditions in the modern sectors of growing or developed economies. This trend, called structural transformation, emerged in western developed nations as their economies industrialised in the early to mid-twentieth century, as well as in select East Asian countries, namely, South Korea and China, during the latter part of the twentieth century. India's economy has instead acquired a 'dual face', Nagaraj observes. 'It is rural and agrarian with 68 percent of the population living in villages, and with 56 percent of the workforce still subsisting on agriculture (as in 2004–05); yet it is a services economy, contributing 55 per cent of the domestic output' (Nagaraj 2012, 3).

Binswanger-Mkhize (2013) notes two major implications of India's 'stunted structural transformation' for the lives of the poor. First, the combination of low agricultural contribution to GDP with a high share of workers in that sector indicates that earnings of most workers in this country are falling behind incomes earned in the rest of the economy. While the real wages of all workers increased significantly between 1983 and 2010, by 3.2 per cent a year, wage inequalities also increased between regular and casual workers, organised and unorganised workers, and rural and urban workers. '[T]he average daily earnings of a casual worker in 2011–2012 were Rs. 138 in rural and 173 in urban areas, and that of regular workers was Rs. 298 in rural and Rs. 445 in urban areas' (Planning Commission 2008, cited in Institute for Human Development 2013, 4). In 2006–07, while the overall GDP per worker (including both wage earners and the self-employed) was estimated to be Rs 9,000 a month, the average monthly income of a cultivator (own-account farmer) worked out to Rs 2,000 (ibid.). The growing disparity in productivity and incomes between non-agricultural sectors (which employ fewer people) and the agricultural sector (which employs more people and is less stable) means that policymakers cannot consider overall poverty to have declined in the era of economic growth.

Binswanger-Mkhize's second crucial point is that in the absence of better-quality jobs in the formal sectors of the economy, 'workers move primarily from the agricultural sector to the rural non-farm sector, rather than to more secure jobs with pension and health benefits in the urban economy' (2013, 5). The insecurity of a large share of the Indian workforce underscores another major socio-economic determinant of extreme poverty in India. The working population in the informal sector comprises 84 per cent of India's total economy (NCEUS 2006). These people have no access to a guaranteed formal social security system.

The informal sector is largely made up of self-employed small enterprise owners and daily wage workers—brick kiln and construction workers and rickshaw pullers, for example—as well as small or landless farm labourers and domestic workers. Their employment conditions are marked by 'low and irregular wages, low levels of organisation which prevent collective action, no fixed or clear terms of employment, no access to formal credit', and employment relations driven, in some labour markets, by feudal social dynamics such as caste and gender exploitation (Jha 2012, 152). Although income security in agriculture is declining without corresponding opportunities having been made available in other sectors of the economy, informal workers still contribute 45 per cent of the total output of the Indian economy.

Additionally, over half of the 16 per cent of those employed in the formal sector are informal employees whose working conditions, including the lack of social security provisions, mirror those of people working in the informal sector (K. P. Kannan 1992, cited in Papola and Sahu 2012; NCEUS 2009, cited in CES 2014). High economic growth in the 2000s coexisted with a growing number of seasonal migrants. Kannan and Breman (2013) note that while the National Sample Survey (NSS) estimates there are 16 million such labourers who leave their villages for towns and cities across the country in search of work, independent research suggests that the number could be as high as 40 million.

Indeed, a formalised social security system does exist in India. Yet, access to provisions such as guaranteed pension schemes, health benefits and maternity entitlements is largely a privilege restricted to the minority of formal workers employed in the formal sector (a very small percentage, 0.4 per cent, of the informal sector employs formal workers) (Nagaraj 2012). Not all who work in the informal sector are poor,[6] but 'poverty is a key characteristic' of this sector, ' and there appears to be a clear association between the incidence of poverty and participation in the informal sector (Jha 2012, 151). The percentage of informal labour in the formal sector also grew in the period when India achieved its highest economic growth rates, from 46 per cent in 2004–05 to 51 per cent in 2009–10 (K. P. Kannan 1992, cited in Papola and Sahu 2012; NCEUS 2009, cited in CES 2014). The percentage of informal workers as a share of the total Indian economy is higher than in other countries with similar growth and development trajectories, including Mexico (55 to 62 per cent) and South Africa (28 to 34 per cent), and in whole regions of developing countries such as Latin America (51 per cent), Sub-Saharan Africa (72 per cent) and North Africa (51 per cent) (ibid.).[7]

Veena Jha has argued that the continued existence of the informal sector in India in the wake of high economic growth corrects earlier 'assumptions that economic growth in the last twenty years should inevitably trickle down to the poor' (2012, 145). Kannan and Breman (2013) consider a large share of the 425 million people working in the informal sector—about three-fourths—to be living in a situation of persistent insecurity: they are either working poor people, or living in conditions that make poverty inevitable.

A large percentage of the 'ultra-poor' engaged in informal work are people who belong to identified vulnerable groups, such as Scheduled Castes (SCs) and Scheduled Tribes (STs), whose health and nutrition indicators have not improved with high economic growth (as explained in greater detail later). Indeed, these groups have historically registered both higher rates of undernutrition, food insecurity and landlessness, and higher casual workforce participation rates in rural and urban India in comparison to other social groups and all-India averages. Of the total informal labour force, 40.4 per cent of people employed in the informal economy are rural casual labourers[8] or daily wage workers, of whom 59 per cent belong to SC groups and 46.5 per cent to STs.[9] In terms of low-paid agricultural work, the proportion of STs, SCs and Other Backward Castes (OBCs) is much higher than other social groups.[10]

The high percentage of STs who work in low-paying jobs with little to no access to social security provisions in rural India correlates with high rates of landlessness in this community. Increasing landlessness has impacted their food security, because these communities have traditionally maintained a 'self-sufficient economy' from the lands and forests they have inhabited for generations. Landlessness has 'driven many to seek jobs in urban areas or trapped them into forced labour arrangements' (CES 2014, 116). The *India Exclusion Report* shows, on the basis of National Sample Survey Organisation (NSSO) data, that '76.5 per cent of Scheduled Tribe (ST) households in rural areas were either landless or had less than 1 hectare of land,' and that the 'share of tribal households with small and marginal holdings has been steadily increasing over time' (ibid.).

What kind of social protection system is required to address vulnerabilities associated with extreme poverty in India? We briefly recap the major facets of extreme poverty that require a social security protection system:

1. the failure of high-growth sectors of the formal economy to produce enough decent-waged jobs with social security provisions for labourers fleeing a dwindling agricultural sector in a time of overall economic growth;
2. the high shares of workers in the informal sector who work in arduous and, at times, exploitative conditions and who contribute nearly half of India's economic output but do not have assured access to a formal system of social protection; and
3. higher shares of vulnerable social groups—in terms of poor nutrition and health—in informal labour compared to other groups and all-India averages.

Informality in the Indian economy is associated with insecurity, abysmal living, working and health conditions among the poor, particularly those among vulnerable social groups. Economic growth has actually led to higher overall rates of informality. Given this context, we now ask what kind of social protection system is required to address the sources and outcomes of insecurity that poor people in India face.

Scholars have noted that the labouring poor experience two types of insecurity resulting from employment conditions in the informal sector. One concerns random though frequent shocks such as untimely deaths and ill health of household members. The second involves 'structural features of the household which remain constant over a period of time', which are rooted in unequal social dynamics such as 'age, gender, ownership of assets and caste', entailing chronic hunger, undernutrition and low wages (Unni and Rani 2003, 130). Unni and Rani recommend a comprehensive social protection system that addresses both features of poverty. Health and food programmes, they suggest, are necessary to address 'basic social securities', while employment guarantees in compliance with labour standards are necessary to ensure jobs with adequate pay, stability and safety standards.

Since the late 1990s and early 2000s, policymakers and scholars have debated how to implement a range of social protection programmes, including state food provisioning through the public food distribution system, a nationwide public works employment scheme (MGNREGA) and a national health insurance scheme (the Rashtriya Swasthya Bima Yojna). Such programmes are intended to reduce 'long-standing vulnerabilities and deprivation in India to ameliorate consequences of growing informalisation of labour' (Sen and Rajasekhar 2012, 92) by both securing basic needs in income, food, nutrition and health, and addressing 'structural' and chronic features of poverty that economic growth alone has not reversed. An exploration of each facet of the comprehensive social protection system for the poor, which we endorse, is beyond the scope of this book. We will instead discuss the legal, economic and political debates concerning food security. Programmes such as the PDS are a salient component of the overall social protection system that may address basic deprivations and structural features of poverty and inequality in India.

State Food Provisioning as Social Protection in India: Origins and Rationale

India has used state provisioning of food as an instrument of social protection for poor people since the mid-1960s. Policymakers reformed the existing food distribution system established by the British before independence. Originally used to ration food to select cities in times of emergencies, the Indian government 'universalised' the system, or made subsidised food available to all people at all times, to address absolute poverty in India during a period of widespread food insecurity. India's food crisis was marked by successive crop failures and droughts in 1965 and 1966 after years of low agricultural productivity. The nation's farming sector had been struggling since independence, due, in part, to the failure in implementing land redistribution policies for small and landless farm labourers, and partly due to low public investment in agriculture. The nationwide public food distribution system coincided with a policy to import both grains and input technology in seeds, fertilisers and irrigation systems from the United States. Input technology would later usher in the Green Revolution, marked by nationwide food production surpluses from the 1970s onwards.

Indeed, public food provisioning was intended to reverse inequalities that had resulted from the failure of earlier agricultural policies, such as land redistribution. Between 1950 and 1965, the planners of newly independent India placed high importance on the potential of the redistribution of land to poor farmers, who had historically been deprived of rights of ownership, to reverse poverty and inequality. The idea was that laws that defined the terms of contracts, established ceilings on the size of landholdings, and eliminated absentee landlordism and intermediaries, would enable agricultural workers to produce food and other goods with their own means of production (Suryanarayana 2012). Policymakers believed that the success of land reforms would boost overall agricultural output per person and reduce long-standing inequalities in income and ownership of productive assets between historically powerful and dispossessed classes and castes.

Land reforms failed, however, because central and state policymakers did not follow the mandates in the Constitution for states to implement reform laws by a clear, integrated plan of action. Instead of a more equitable distribution of land amongst the nearly 70 per cent of the population employed in farming, 'the share of agricultural labour in the rural workforce increased' (Nagaraj 2012, 36). Cultivators struggled with little support because planners accorded little public investment to agriculture over the first three Five Year Plans. These plans invested in small 'cottage' industries and large manufacturing sector industries such as steel to boost employment and economic growth. Surprisingly, food production did rise by 12 per cent after the First Five Year Plan, but surpluses were invested in the industrial sector. Ayesha Jalal's synopsis is worth quoting in full:

> [By the Second Five Year Plan] agriculture's share of the total investment was slashed by nearly half that of the first five year plan. Within fifteen months of the plan Indian agriculture was in the throes of a serious crisis as food grain production fell well below expectations. By the summer of 1957 a 50 per cent hike in food prices sent the wholesale price index spiralling, forcing the central government to import huge quantities of wheat. (Jalal 1996, 129)

The universalised public food distribution system was at the centre of the shift in the country's public policy regime between 1950–65 and 1966–80. We have mentioned that in the first of these periods, heavy investment was accorded to industry over agriculture and land reform laws were relied upon heavily to redress poverty and inequality. Policymakers in the second period began to acknowledge the necessity of social policies to alleviate chronic poverty hastened by 'shocks and crises' to the overall economy that had occurred in the absence of state-led social protection measures. State provisioning of food was thus conceived as one measure to address the structural causes of hunger, hastened by stagnating farm wages amidst the uneven successes of the Green Revolution. The Green Revolution was 'more of a wheat revolution' and 'led to a regional imbalance in production', since 'wheat is mostly grown in the north-west of the country' (Jalal 1996). While US-based agricultural technology was intended to increase the productive capacities of the nation's farmers, 'the policy upon which the PDS was founded was geared towards attaining food sovereignty and towards ensuring food security for the entire Indian population' (Frodin 2010, 445).

Madhura Swaminathan (2008, 3) identifies three broad phases of the PDS between the 1960s to 2008 (just before the NFSA was conceived). Between 1960 and 1978, the government took a 'holistic approach to food security', and prioritised 'domestic procurement and storage' by setting up key institutions such as the Agricultural Prices Commission and the Food Corporation of India (FCI). From 1978 to 1991, the government expanded the system throughout the country by setting up more procurement and storage facilities. After 1991, a period that coincided with economic liberalisation, the universal system was converted to a targeted one where only people identified as BPL through government surveys would be entitled to rations (ibid.). An analysis of the success of and challenges to the PDS's impact on ensuring food security for the poor must take into account the economic regime changes in 1950–65, the mid-1960s through the 1980s, and the post–economic liberalisation period from 1991: each period has accorded varying degrees of investment to agriculture and state food provisioning. These changes—roughly characterised as the phase of heavy investment in India's nascent industrial sector at the expense of agriculture (the 1950s and early 1960s); concerted social protection responses to persisting deprivations through institutions and coverage (mid-1960s through the 1980s); and economic liberalisation in the 1990s, which further scaled back public investment in agriculture—have impacted food production, procurement and distribution, and, inevitably, access to state-subsidised food for the poor.

While procurement of foodgrains continued to surpass buffer norms from 1989 to 2004, the rate of foodgrain production was lower than the rate of population growth during this period. Second, food availability per head has been declining since 1991, from 485 grams in 1989–91 'to a low of about 445 gram[s] per head per day by 2006, a level not seen since the drought years of the 1970s' (Saxena 2012, 29). Saxena argues that the decline in availability is due to 'both a drop in production and increasing exports' (ibid., 30), marked by the government's export of a record 28 million tons of foodgrains from 2001 to 2005. Since liberalisation, policymakers have attempted to boost agricultural productivity through subsidies in 'power, water and fertilisers' (ibid., 37) and by increasing the minimum support price (MSP) 'rather than through building new capital assets in irrigation, power and infrastructure' (ibid.). Private investment through the provision of specific inputs has grown while 'public investment in agriculture has fallen dramatically since the 1980s' (ibid.). Further, a higher MSP does not ensure gains to farmers overall, because the FCI does not issue an MSP for farmers in many poor and hunger-affected states.

We now discuss the roots of food security as a rights-based measure of social protection, which came to the fore in the early twenty-first century as a result of inefficiencies in several social protection programmes to provide food, nutrition and public works to the numerous poor who were entitled coverage.

State Food Provisioning in India: A Rights-based Approach to Social Protection

India is party to several international agreements that recognise and define the right to food (see Box I.1). The UN's ICESCR, for example, defines the right in Article 11, General Comment 12, in terms of the state's duty to respect, protect and enforce the individual's physical and economic access to food.[11] Yet, as many writers have noted, much of the progress on the right to food in India has been made possible by human rights principles available in this country's own Constitution. Indeed, the Indian Constitution, as we will explore further in this chapter, provided the basis in 2001 for the Supreme Court to interpret the right to food in the context of the right to life in Article 21, and to issue orders that enshrined food, work and nutrition programmes as legal entitlements of the poor.

Box I.1 International agreements India has signed that recognise the right to food

Universal Declaration of Human Rights (UDHR): Article 25 recognises the right to an adequate standard of living, including food.

UN International Covenant on Economic, Social and Cultural Rights (ICESCR): Article 11 General Comment 12 defines the right to food in terms of the state's duty to respect, protect and enforce the individual's physical and economic access to food.

Convention on the Elimination of All Forms of Discrimination Against Women (CEDAW): Article 12 recognises the right of pregnant and lactating women to special protection with regard to adequate nutrition; Article 14 recognises the right of rural women to equal access to land, water, credit and other services, social security and adequate living conditions.

Convention on the Rights of the Child: Article 25 recognises the right to the highest attainable standard of health; Article 27 recognises the right to an adequate standard of living, which includes food and nutrition.

For now, it is important to say that since independence, India has developed a range of strategies for state food provisioning as social protection to address widespread hunger and poverty, which evolved into the right to food law passed by India's Parliament in September 2013. The push to formalise the right in a comprehensive legislation came from various organisations working with food-insecure populations around the country, including people's movements, human rights organisations, non-governmental organisations (NGOs) as well as scholars, journalists and activists. In 2001, the landmark right to food case was made possible by a petition filed by PUCL, one of India's leading human rights organisations, in the Supreme Court.

Starvation deaths that June in the north-western state of Rajasthan catalysed the PUCL's demand that the Supreme Court address India's hunger crisis in the context of human rights, since the state government's failure to release surplus foodgrains or open relief works, as mandated by its own Famine Codes, had raised questions regarding states' duties to enforce social protection programmes, particularly in times of high rural unemployment or drought. The deaths in Rajasthan were not an isolated case. Journalists and human rights groups had also reported hunger deaths and defunct social programmes in five other Indian states that year. The deaths were symbolic of a nationwide hunger crisis, characterised by the persistent struggles of poor people in India to secure enough food to avoid starvation despite the government's accumulation of over 50 million tons of grain in public silos and the existence of numerous social protection programmes, including the PDS. National health surveys also reported nearly 50 per cent of children in Rajasthan as malnourished and almost half of the state's rural population as BPL, pointing to conditions of chronic hunger for millions of people.

Through a public interest litigation, the petitioners sought justice for the families of victims who had starved to death in Rajasthan.[12] The major criticism at the time was that the existence of a social protection system did not ensure its implementation. The petitioners argued that local officials' failure to implement the food, public works and nutrition programmes had a bearing on the deaths of poor people suffering chronic hunger in these already impoverished regions. They demanded redress, citing a violation of the right to life enshrined in Article 21 of the Constitution. They asked: 'Article 21 …

protects for every citizen a right to live with human dignity. Would the very existence of life for those families which are below poverty line not come under danger for want of appropriate schemes and implementation thereof, to provide requisite aid to such families?' (Human Rights Law Network 2008).

In response, the court ordered access to these programmes—PDS, ICDS, the National Old Age Pension Scheme, the National Family Health Benefit Scheme and MGNREGA—as poor people's 'rights to food'. The decision engendered vital debates in India on why a human rights backed social protection system was needed to combat hunger. Starvation deaths proved that the existence of a social protection system for the poor did not ensure delivery of services to them. Second, their access to the system proved a matter of life and death. In this regard, state accountability in delivering social services to the poor became a matter of protecting the constitutional right to life.

India's case of the conversion of access to public programmes into rights is as an example of what Gita Sen and D. Rajasekhar have called 'the public goods instrumental rationale for collective action':

> while entitlements are programmatic (even if derived from laws), rights are more basic, in that they can be intrinsic, inalienable and reflected in foundational social agreements such as constitutions or the Universal Declaration of Human Rights. Entitlements can be modified or even done away with through programmatic changes, but rights, once acknowledged, are more difficult to change or annul. In this sense, the linking of social protection and social policy, more generally to a rationale of rights, provides them with the strongest foundation for action and demands for accountability. (Sen and Rajasekhar 2012, 92)

The push to enforce a legal right to food was also a response to the legitimate frustration over the failure of economic growth to reduce high rates of malnutrition. A model of governance that relied on human outreach over market forces was imperative for development. In the last 25 years, absolute poverty has nearly halved from 50–60 per cent to 27.6 per cent, but nearly half of all Indian children still remain underweight and about half suffer from anaemia (Nagaraj 2012; see also Deaton and Drèze 2009). The reforms in the 1990s are often credited for India's emergence as a future global economic power, but in the last 20 years the government has invested very little in agriculture (about 6 per cent of GDP in 2009–10), which is the sector in which most people in the country find employment. Most of India's poor are small and landless farmers, and the relative neglect of agriculture over time has resulted in these people earning comparatively less than they did 20 years ago (Parulkar 2012). In the absence of a well-functioning social protection system that includes a functional food distribution system, poor agricultural labourers have been hurt by rising food prices. Between 2008 and 2010, for instance, economist C. P. Chandrasekhar showed that rice prices rose by nearly half in northern cities and more than half in southern cities, while government surveys on food consumption showed that from 1993 to 2010, the poor spent more of their monthly expenses on food, but consumed less of it (Chandrasekhar 2012).

A human rights backed social protection system for the poor has therefore been conceived as a necessary step in addressing persistent deprivations in health, nutrition and food consumption among marginalised communities, which continue to exist as the country prospers. The case of STs in India is illustrative. A 2012 World Bank study shows that a decline of 31 per cent in the overall poverty rate of this group between 1983–84 and 2004–05 is still behind the all-India decline of 40 per cent; other marginal groups, such as SCs, experienced a 35 per cent decline (World Bank 2012). The starkest marker of ST deprivation, however, is the high rate of child mortality: 96 tribal children out of every 1,000 die before their fifth birthday, compared to 74 per 1,000 children across India. The World Bank study shows that risk factors for child deaths include access to health services, the population's geographic isolation, and its high risks of malnutrition. Tribal children fare worse on all nutrition indicators in comparison to the already abysmal all-India averages, including the number of children that are underweight (55 per cent). A 2005 study cited by the World Bank that surveyed tribal populations in two states reported that '99 per cent of tribal households faced chronic hunger, one-quarter had faced semi-starvation during the previous week' (ibid., 58).

In 2009, the federal UPA government promised to pass the National Food Security Bill (NFSB) upon re-election. Politicians, activists and policymakers across party lines could no longer ignore the glaring paradox framing the context of hunger in India: despite decades-long achievements in surplus food production, as well as record economic growth in recent years, India was still home to over 200 million hungry people.[13] On the basis of this understanding, policymakers could at least agree that India's hunger problem was one of access to food, rather than production alone. The road to such understanding was set in motion by empirical evidence presented to the courts during the 2001 case on the failure of implementation of social protection programmes at a time of increasing rural unemployment amidst grain surpluses and exports.[14] The petitioner's demand that the courts address the rights of the hungry and malnourished led, in the years that followed, to the conception of the NFSB as a human rights law meant to protect people's rights to such access.

The petitioners were able to legitimately question whether the right to food should be considered part of the constitutional right to life based on precedents decided in earlier Court decisions that were based on human rights principles located in the Indian Constitution. Judges in the 1981 case of Francis Coralie Mullin vs. the Administrator decided[15] 'the right to live with human dignity and all that goes with it' includes 'the bare necessaries of life such as adequate nutrition'. Given the failure of the public food distribution system to offload grains to people in need, the petitioners argued that this case 'implies that the State has a duty to provide food to people who are affected by drought and are not in a position to purchase food', and that since 'the right to life did include a right to food, state and central governments were therefore duty bound to start relief works and distribute grain'.[16]

This and other court decisions[17] had a lasting impact in expanding the scope of the right to life in Article 21 to eventually mean the state's duty to protect a person's right to live *with human dignity*—a far cry from the original conception of the right to life as the state's duty not to interfere with a person's liberty. Durga Das Basu (2009, 380) argues that Article 21 was originally conceived as a 'negative duty not to interfere with the life or liberty of an individual without the sanction of law'. In the 1980s, however, conditions of *locus standi* in public interest litigations were relaxed to allow petitioners to approach the courts on behalf of vulnerable people whose human rights had been violated but who were too poor or disadvantaged to seek legal counsel or remedy from the courts themselves.[18] Over time, the court's orders on a range of human rights cases have changed the definition of Article 21 to mean a 'positive obligation upon the State to take steps for ensuring to the individual a better enjoyment of his life and dignity' (ibid.).

These two views of the 'right to life' are radically different. The court's decision to declare access to existing social protection programmes as an enforceable right to food is in line with the latter interpretation of Article 21 as the right to live with human dignity, and creates a positive duty upon the state to ensure that people have the means to do so.[19] The court-ordered 'right to food' in India is evidence of the continuing evolution over the last three decades of the meaning and application of Article 21 as the right to live with human dignity, made possible by Article 32 of the Constitution which permits the courts unique powers to make human rights laws or to expand the application of the range of human rights located in the fundamental rights section of the document. This has been made possible by a progressive interpretation of human rights principles located in the fundamental and directive principles of state policy sections of the Constitution, precedents from international human rights treaties that India has signed, and, in some cases, decisions made by courts in other democracies that have had far-reaching consequences for the expansion of human rights in those countries.

Supreme court judges have, in previous court decisions related to hunger and social exclusion, brought human rights principles into the realm of Article 21[20]: on starvation deaths,[21] rights of children to a full development,[22] and to decent clothing and shelter,[23] to name a few.[24] Judges were compelled in these cases to question and expand the realm of the 'right to life' to comply with the continuing evolution of the meaning of human dignity. In these cases, judges were compelled to question and expand the meaning of the 'right to life' in compliance with previous decisions that had progressively defined that right in terms of human dignity.

The courts have specifically mentioned the right to food in the context of the right to life in several earlier cases,[25] before and after the 2001 right to food case. These decisions were made in consideration of a number of economic and social contexts which forced poor people into hunger. Former Justice M. B. Shah details these cases in his 2006 article, 'The Indian Supreme Court Acknowledges the Right to Food as a Human Right'. It is worth quoting him in full in his review of salient cases that addressed the right to food in the context of the right to life on different points and in different contexts in the past three decades:

> In *Olga Tellis & Ors vs. Bombay Municipal Corporation & Ors* (1985; 3 SCC 545) the petitioners were living on pavement/slums in the vicinity of their place of work and they were forcibly evicted and their dwellings were demolished by officers of the Bombay Municipal Corporation. They challenged the act of the Corporation on the ground of violation of their rights under Articles 19 and 21. In that context, the Court held that the sweep of the right to life conferred by Article 21 is wide and far-reaching. If the right to livelihood is not treated as a part of the constitutional right to life, the easiest way of depriving a person of his right to life would be to deprive him of his means of livelihood to the point of abrogation. The Court also observed that person have to eat to live. (Shah 2006, 25)
>
> […] In *C.E.S.C. Ltd vs. Subhash Chandra Bose* (1992; SCC 441 at p. 462, para 30), the question decided was whether employees appointed by contractors whose wages were being paid through such contractors would fall within Section 2 of the 1948 Act. In that context, the Court observed that the right to livelihood springs from the right to life guaranteed under Article 21. The Court referred to the Universal Declaration of Human Rights, the International Convention on Economic, Social and Cultural Rights, which include the right to food, clothing, housing, education, the right to work, leisure, fair wages, decent working conditions, social security, the right to physical or mental health, protection of their families, as an integral part of the right to life. The Court also held that the Preamble and Part IV reinforce them compendiously as socio-economic justice, bedrock to an egalitarian social order. (Ibid., 25)
>
> In *Kapila Hingorani vs. State of Bihar* (2003; 6 SCC1), hundreds of employees of state-owned corporations, public undertakings or other statutory bodies in the State of Bihar died due to starvation, or committed suicide owing to acute financial crises resulting from non-payment of remuneration for a long time. When this fact was brought to the notice of the Supreme Court, it held that

> the State cannot escape its liability when human rights problems of such a magnitude involve starvation deaths. While deciding the case, the Court referred to Article 1 of the International Covenant on Economic, Social and Cultural Rights, 1966 and the '12 Misconceptions About the Right to Food (FIAN)' and approved that human beings have *a right to food and that hunger is a violation of human rights*. The Court held that lack of access to food is in violation of the human right to food, and issued various directives to ensure that no starvation deaths occur.... The Court also observed, The States of India are welfare States. Having regard to the constitutional provisions adumbrated in the Constitution of India, and in particular Part IV thereof laying down the Directive Principles of the State Policy, and Part IV-A laying down the Fundamental Duties, they are bound to preserve the practice of maintaining dignity. (Ibid., 26)

These cases highlight how people are often forced into hunger as the result of various social and economic injustices: forced evictions, the exploitation of manual labourers who work under unfair contracts, and the inability of some people to secure food because their employers refused to pay wages. They also show that the court's move to declare the right to food in the context of the fundamental right to life was an attempt to protect various groups of vulnerable people, since different groups of people have needs suited specifically to the equally diverse social and economic constraints that force them into hunger. In later chapters, we note precedents that exist in Indian law to protect the right to food for various groups of people. We argue that grounds exist in law for the state to protect the right to food with regard to the needs of different groups of vulnerable people such as children and those facing starvation, whose conditions of hunger are determined by diverse and pernicious social and economic dynamics.

The Indian Constitution does not explicitly recognise a fundamental right to food. What is important here is that the Court's expansion of the fundamental right to life in response to cases dealing with hunger and malnutrition led to an acknowledgement of the right to food as part of the right to life with human dignity and thus, eventually, made possible the grounds for legislation. However, the protection of citizens' right to food has always existed in Articles 39(a) and 47 of the directive principles of state policy,[26] located in Section IV of the Indian Constitution. Though this section details social and economic rights which are non-enforceable, many writers have noted that the distinction between the non-enforceable rights of Section IV (directive principles) and the enforceable rights of Section III (fundamental rights) was meant to be temporary.[27] In the last three decades, the courts have used their powers in Article 32 to expand Article 21 by incorporating socio-economic rights from Section IV into the broader meaning of the right to life with human dignity. In earlier cases,[28] for example, the courts made the right to free and compulsory education (Article 45) and the right to education (RTE) within the economic means of the state (Article 41), located in the directive principles, enforceable under Article 21.[29]

The directive principles of state policy of the Indian Constitution lay the ground for the continual expansion of democracy in India through the state's recognition of socio-economic rights as fundamental rights. Even though the Indian Constitution did not explicitly recognise a fundamental right to food, the court—through discretionary powers in Article 32—expanded the interpretation of the right to life under Article 21 of the Constitution to imply the right to life with dignity, and to include all that is required to enable life with dignity; this includes the right to food. In this sense, the expansion of the fundamental right to life made possible a fundamental, thus enforceable, right to food (Gonsalves et al. 2005). Article 32[30] grants the Supreme Court power to enforce fundamental rights over disputes that arise between the federal government and states or between two states. The PUCL's public interest litigation on the right to food satisfied these conditions, and the court considered the petition 'not as adversarial litigation between two parties but as government injury to the public interest and therefore a concern for all' (Birchfield and Corsi 2010, 699).

In over 100 interim orders from 2001 until the passage of the NFSA in 2013, the Supreme Court gradually 'unpacked' the right to food into a series of tangible and enforceable rights.[31] They also created an oversight body, the Supreme Court Commissioners, who until just after the NFSA monitored the implementation of 'right to food' programmes and advised the court (Mander 2012). The Right to Food Campaign organised people and educated people on these rights and remained vigilant when violations occurred.

The UPA coalition reaffirmed its pledge to pass food security law upon its re-election in 2009. The prime minister reconstituted the National Advisory Council (NAC), chaired by Congress president Sonia Gandhi, to advise the union government on social policy and legislation. The National Advisory Council, a group of social and economic policy experts appointed by then Prime Minister Manmohan Singh, undertook the drafting of the National Food Security Bill (NFSB) as its first mandate.

The NFSB was extensively debated for around a year within the NAC, in the union and state governments, India's Planning Commission, the Prime Minister's Economic Council, the ruling and opposition parties, as well as the Right to Food Campaign, academic writing and the press. The union cabinet diluted many provisions that had been agreed upon in the NAC draft, and introduced their own version of the bill to Parliament. After the Parliament's Standing Committee on Food, Consumer Affairs and Public Distribution gave its recommendations on the direction it felt was best for the NFSB, the bill was made an ordinance due to the political gridlock caused by tactics of the opposition parties. The latter did not want to oppose a rights-based bill for fear of being castigated as acting against the people's wishes, but they did not want the bill to pass either due to the political leverage it would potentially give the ruling parties. After being passed as an ordinance, the bill was debated in Parliament to agree on amendments. Although no major amendments were passed, the bill was forced through Parliament, and ultimately enacted as the National Food Security Act in September 2013.

In a way, the chapters of this book present a comprehensive, although not exhaustive, list of major debates that took place between the Supreme Court orders and the passage of the NFSA. These debates cover state's duties and possible strategies for food provisioning as part of a human rights based social protection system. Discourse on these issues arose in India in the context of a national food security law, but readers may find many of the debates relevant for countries that also assume such duties in the absence of framework law. Although these debates occurred in India in the context of a national food security law, many of these debates would be equally relevant even where states assume these duties—as in Brazil and Ethiopia—but without resorting to a binding and enforceable framework of law.

The passage of the NFSA is not the end of the right to food story in India. The Supreme Court orders were seminal. They legitimated the idea that social services could be the tangible equivalent of a human right. Still, legislation was required to ensure the presence of institutions, grievance redressal mechanisms and officials. These gaps rendered the orders principally administrative in nature.

Since the passage of the act, questions have been raised on how best to enforce this right in line with the human rights principles located in the Indian Constitution and numerous international treaties that give credence to the need and possibility of legislating the right to food. While there is mention of the state's duty to ensure adequate nutritional standards in the directive principles of the Indian Constitution, India is also signatory to several international human rights treaties which detail the state's obligation to respect, protect and fulfil each citizen's right to physical and economic access to food. While the NFSA promises to 'provide for food and nutritional security in [a] human life cycle approach', critics such as academics and activists associated with the Right to Food Campaign contend that law does not adhere to a human rights based approach to food security.

On the one hand, some critics of the current act and the processes that led to its passage, such as activists and academics aligned with the Right to Food Campaign, believe that its contents are not rooted in a human rights based approach to food security. A number of the court orders in the right to food case—on old age pensions, widow pensions and cash support to poor families who have suffered the death of an income earner—are not included in the act, and while the law does increase the number of people who will have the right to buy subsidised foodgrains under the PDS from 65 million households to 180 million households, the government does not specify how the states will identify the additional people to be covered under the PDS.

While the Supreme Court passed landmark orders which made food- and work-related programmes tangible rights, many have argued that the NFSA does not reflect the conception and formulation of the right to food based on human rights principles in the Indian Constitution and international covenants that India has signed. We intend to raise questions regarding whether the reasons and methods used to pass the NFSA will provide the chance of realistically achieving the goal of food security for all in the country. Critics from the Right to

Food Campaign, several academics and activists have argued that the current version of the act would significantly dilute the meaning of a legal right to food. This gives us a reason to ask how a clear formulation of the right in law can facilitate the full realisation of the right to food.

India's Right to Food Story in Conversation with Other Countries

Many countries are pursuing diverse pathways to secure sufficient food and nutrition to all their people, and thus there is a great deal to learn from their experiences. Some have developed elaborate programmes for state food provisioning without making access a right in either constitutional or legal statutes. Some governments have made explicit reference to the right to food for all persons in their constitutions (South Africa, Bolivia, Guyana, Haiti), or for specific categories of persons such as children (Brazil, Colombia, Cuba, among others), and for prisoners and detainees (South Africa) (FAO 2009, 35). This is in line with the FAO's assertion that 'the protection of human rights through constitutions is the strongest form of legal protection as constitutions are considered the fundamental or supreme law of the land.... every law in a country must conform to the constitutional provisions and, in cases of conflict, the constitutional norm will always prevail' (FAO 2005, 33). The constitutions of Nigeria and Bangladesh recognise the right to food in their non-binding directive principles (FAO 2009, 40).

The inclusion of the right to food as a fundamental right, even in a progressive constitution such as South Africa's, does not guarantee that people who live with hunger can exercise that right. The realisation of the right requires governments to ensure institutional arrangements, enforcement and grievance redressal mechanisms when violations of the right occur. Policymakers will have to 'unpack' the right to food not only through institutions and monitoring and accountability systems, but also with respect to the needs of various groups of vulnerable people who face distinct barriers to food access based on differences in age, gender, economic resources and productive assets, social standing, political power and wages. A right to food law should, therefore, indicate which vulnerable groups are protected, as well how they are protected.

A number of countries have legislated or are in the process of drafting legislations for the right to food. These include Argentina, Brazil, Bolivia, Ecuador, Guatemala, Honduras, Indonesia, Malawi, Mali, Mexico, Mozambique, Nicaragua, Peru, South Africa, Uganda and Venezuela (FAO 2004, 5). India passed the NFSA in Parliament in September 2013,[32] at a time where many other countries were considering the drafting of such legislations.

India has long grappled with effective implementation of a wide range of state food provisioning programmes—such as subsidised cereals, school meals, infant feeding programmes and pensions for the aged. More recently, nudged in part by an activist Supreme Court, it has pursued legislation for a legal and enforceable right to food—a journey that has taken more than a decade, having involved extensive civil society organisation, political manifestos and debates at the highest levels of government and in India's Parliament.

India's story may provide insight to both countries that have recognised the right to food in their constitutions in various forms (De Schutter 2010; Knuth and Vidar 2011), and those who wish to expand their networks of social protection to include much more extensive state food provisioning. Like India, other countries such as Bolivia, Malawi, Mozambique and South Africa have both recognised the right to food in their constitutions and are drafting, or have drafted, framework laws for national food security. Brazil, meanwhile, has in place a comprehensive programme called Fome Zero (Zero Hunger) that consists of 31 programmes with local-level grievance redressal officers to take up individual violations of the right to food. Ecuador passed a framework law in 2009 that includes provisions for small-scale farmers, such as the promotion of access to capital, resources and inputs required to produce food.

A study by Knuth and Vidar (2011) found that 56 countries throughout the world have recognised the right to food in their constitutions. Of these, 23 countries explicitly recognise the right to food in various forms: as an independent right, as the right of a specific segment of the population (e.g., children), or as part of other human rights such as the right to development. South Africa's Constitution, for instance, recognises the right to food as an independent right in Article 27, which states: 'Everyone has the right to have access to … sufficient food and water, and social security, including, if they are unable to support themselves and their dependents, appropriate social assistance.'

Colombia recognises the right to food for children. Article 44 of the Colombian Constitution states: 'Children have fundamental rights to: life, integrity, health and social security, and adequate food.' Article 21 of Belarus's Constitution recognises the right to food as a part of other human rights: 'Every individual shall exercise the right to a dignified standard of living, including appropriate food, clothing, housing and likewise a continuous improvement of necessary living conditions.'

In 33 countries, the right to food is recognised as a part of other rights, such as the right to an adequate standard of living (see Tables I.1 and I.2). Thirteen other countries including Bangladesh, Nigeria, Sierra Leone and Uganda incorporate the right to food in their directive principles of state policy.

Table I.1. Explicit recognition of the Right to Food in National Constitutions (23 total)

Countries that recognise RTF as independent right	Countries that recognise RTF for a specific segment of the population (i.e. children)	Countries that recognise TRF as part of other human rights (i.e. standard of living, quality of life, development
1. Bolivia (Art. 16)	1. Brazil (Art. 227)	1. Belarus (Art. 21)
2. Brazil (Art. 6)	2. Colombia (Art. 44)	2. Democratic Republic of the Congo (Art. 34.1)
3. Ecuador (Art. 13)	3. Cuba (Art. 44)	3. Malawi (30.2)
4. Guyana (Art. 40)	4. Guatemala (Art. 51)	4. Moldova (art 47.1)
5. Haiti (Art. 22)	5. Honduras (Art. 123)	5. Ukraine (Art. 48)
6. Kenya (Art. 43)	6. Mexico (Art. 4)	
7. South Africa (Art. 27.1)	7. Panama (Art. 52)	
	8. Paraguay (Art. 54)	
	9. South Africa (Art. 38.1.c)	
	10. Costa Rica (Art. 82)	

Source: Based on Knuth and Vidar (2011).

Table I.2. Implicit Recognition of the Right to Food in National Constitutions (33 total)

RTF implied in the right to a decent standard of living	RTF implied in the right to well-being	RTF implied in right to a standard of living not blow subsistence level	RTF implied in the right to a dignified life	RTF implied in the Right to Development
1. Armenia (Art. 34) 2. Bolivia (Art. 158) 3. Cambodia (Art. 63) 4. Costa Rica (Art. 50) 5. Czech Republic (Art. 30) 6. Ethiopia (Art. 43) 7. Guatemala (Art. 119) 8. Pakistan (Art. 38a) 9. Romania (Art. 47.1) 10. Turkey (Art. 61)	1. Azerbaijan (Art. 6) 2. El Salvador (Art. 1) 3. Equatorial Guinea (item 25) 4. Eritrea (Art. 21) 5. Guinea (Art. 15) 6. Peru (Art. 2)	1. Georgia (Art. 32) 2. Germany (arts. 1, 20, and 28) 3. Kyrgyzstan (Art. 27) 4. Netherlands	1. Belgium (Art. 23) 2. Cyprus (Art. 9) 3. El Salvador (Art. 101) 4. Finland (Art. 19) 5. Ghana (Art. 36) 6. Switzerland (Art.12) 7. Thailand (Art. 79) 8. Venezuela (Art. 299)	Burundi (Art. 33) Congo (Art. 58) Ecuador (Art. 59) Eritrea (Art. 10) Malawi (Art. 30)

Source: Based on Knuth and Vidar (2011).

The design of India's legal and policy framework for state provisioning of food is unique and instructive because it shares common ground with all of the above-mentioned categories of countries with large food provisioning programmes, as well as those with the right to food contained in their constitutions and statutes. India has experimented with many forms of food provisioning as an essential part of social protection, and it has both explicitly (through Article 21 and numerous Supreme Court decisions) and implicitly (through

Articles 39[a] and 47) recognised the right to food in its Constitution along with a national framework law to address food security. This book will raise questions on the extent to which India's food security law will actually address both food security programmes and laws in line with constitutional mandates, international norms and current precedents set down by the courts, which have established the base for a rights-based approach to food security.

Organisation of the Book

The aim of this volume is to record major debates that emerged before the passage of the NFSA in 2013. We hope this recent historical record will engage with literature and discourse in India on food security and human rights based social security systems for poor people in the country. A shortcoming of this book is the absence of discussion on food security in the context of urban poverty—in terms both of expanding city spaces and urban migration.

The India case study illustrates that developing a strong food protection programme and drafting a right to food law are very complex tasks. The law needs to be clear about what programmes and rights precisely are being created, for whom, who is responsible to ensure the realisation of those programmes and rights and by what institutional and schematic mechanisms, what precise recourse individuals have if their programmes and rights are being violated, to which institutions such recourse is available, and what will be the consequences, both for the victim and the perpetrator, of any violations.

In our discussion of the India case of legislating food security by formalising a right to the existing social protection system for the poor, we have organised the debates into four main themes.

Part 1: Concepts, Definitions, Scope, and the International Context

1. The Duty of the State to Provision Food as Social Protection

In this chapter, Harsh Mander discusses major debates that emerged in the media on the efficacy and need for food security legislation.

How would the government finance the world's largest food security programme? Was it economically sensible—or sustainable—for the state to subsidise food rather than stimulate economic growth? While advocates of the food security law critiqued the bill's shortcomings on human rights mandates, they also viewed legislation as an essential step towards addressing hunger in India.

Dissidents saw the bill as a way for the UPA government to 'buy' votes using taxpayers' money to secure the next election. Mander argues that subsidies for the poor are often referred to as 'doles' or 'handouts'. Subsidies captured by the rich, on fuel or electricity, are not described the same way. Costs for the food bill are estimated to lie between Rs 1.25 lakh crore and 3 lakh crore[33]—supporters argue that real additional costs would amount to Rs 25,000 crore. This could be mitigated by increasing taxation on the rich and cutting down on corporate tax holidays that amount to Rs 5 lakh crore annually. The costs of not implementing some form of food security are also very high, given existing rates of poverty and malnutrition. Lastly, supporters argue that it is morally untenable for a government to let a situation continue where over 800 million of the country's population are starving.

2. Coverage of the National Food Security Bill

International covenants and declarations on the protection of the rights of vulnerable non-citizens recommend that state governments devise legal frameworks to protect these groups to offset the discrimination they face. In this chapter, Saba Sharma explains that the right to food and freedom from hunger is recognised as an integral part of ending discrimination, and should extend to all individuals based on their personhood, irrespective of national identity. Most states that provision food, with or without a rights-based framework, limit entitlements to citizens (a notable exception is South Africa). India does not have a legal framework to address the status of non-citizens with respect to public services. The Right to Food Campaign demanded that the right be extended to all residents of the country; various drafts presented in Parliament refer to those covered under the act as 'people' or 'residents'. However, the PDS requires proof of identity from beneficiaries, which automatically restricts entitlements to citizens. Seemingly, those debating the bill in Parliament assumed it was directed only at citizens. The final act maintains the condition of providing identity proof for the PDS, but provisions for vulnerable groups, such as women, children, 'destitute persons' and so on, can potentially be extended to non-citizens as they do not state such conditions.

3. Scope of the National Food Security Bill

In this chapter, Ashwin Parulkar explores debates around a central question that emerged during the drafting of food security legislation: should the law restrict legal guarantees only to the provision of food, or include access to other necessities required to ensure health and nutrition, such as sanitation, clean drinking water and healthcare. Relatedly, should the law ensure provisions for farmers to produce food, relevant in India, as poverty and food insecurity among rural Indians are intimately connected with a crisis in agriculture. Parulkar presents debates that emerged on the implications of a law that would cover various aspect of food security, including nutritional security, public provisioning of food and food production.

Part 2: Universalism or Targeting, Food or Cash Transfers, and Conditionalities

4. Universal or Targeted Entitlements

In this chapter, Ankita Aggarwal analyses debates on how to secure food entitlements through social programmes. She focuses on two views on this subject that emerged in India at this time. The universal approach, she explains, sought to provide the right to purchase grains through the PDS to all people in India. Wealthier people, proponents believed, would opt out of this system and purchase food in the open market. Proponents of the targeted approach held that access should be restricted to people officially identified as BPL.

The question of whether to identify specific beneficiaries for access to public services ('targeting') or to provide access to everyone ('universalising') is arguably the most heated debate regarding implementation of India's food schemes. Public infrastructure—e.g., storage warehouses and grain shops—has existed since independence and the system has been reformed at critical points in successive decades to meet the needs of the poor. At a time of food production deficits in the 1960s, the system was open to all people. Reforms in the late 1990s after India adopted economic liberalisation restricted access to beneficiaries identified by state officials.

Aggarwal discusses debates on whether the PDS should be universal and uniform, universal and differentiated, or available only to people deemed food insecure. Her chapter also explores debates on targeting beneficiaries in other food-related schemes.

5. Selection for Targeted Entitlements

In this chapter, Ankita Aggarwal explains the numerous ways beneficiaries can be identified for social programmes: location-based targeting, community-based targeting and means testing methods. She explains that location-based methods are useful when poverty is linked to geographical features, such as isolation, topography or fertility, but fail to account for intra-region inequities. Community-based measures rely on local knowledge of poverty, such as the impact of local elite capture of power. These qualitative methods are at times at odds with quantitative estimates of poverty. Means testing is an assessment of household-level income or assets, intended to capture the quality of life at the family level.

In India, households qualify for social assistance if they meet criteria used to identify families who are BPL. The Planning Commission had also periodically fixed a consumption-based poverty line. Aggarwal discusses the implications of a key problem involving both approaches to understanding poverty in India. State governments identify individual target households through socio-economic surveys, but are required to limit the number of BPL households to match Planning Commission estimates.

As part of the debates on selecting beneficiaries for the PDS, Aggarwal discusses revised methods to determine BPL households and the critiques by civil society organisations, scholars and policymakers.

6. Conditionalities

A conditional cash transfer (CCT) is social security assistance in the form of cash provided to beneficiaries who fulfil stipulated requirements. They are commonly used in education and health programmes. In this chapter, Ankita Aggarwal surveys various kinds of CCTs in existence both in other developing countries and in India. She discusses how beneficiaries are targeted, conditions they have to meet, safeguards in the event that beneficiaries lack sufficient income, and whether such transfers are deemed rights within the social security system. She surveys debates in the international literature on the perceived causes of poverty that serve as the rationale for various types of CCTs, the potential for excluding poor people from CCT programmes, supply-side constraints, and unforeseen outcomes of this social security measure studied in other countries such as Nicaragua.

In this context, Aggarwal discusses two CCT schemes in India. The National Maternity Benefit Scheme (NMBS), in existence since 1995, provides a cash entitlement of Rs 500 to women above 19 years of age who are BPL 8–12 weeks before delivery to support food and nutrition requirements during pregnancy. The Janani Suraksha Yojana (JSY) is a CCT scheme introduced by the central government, and provides an incentive to poor women to give birth at hospitals and health institutions. The conditions beneficiaries have to meet vary between and within (rural and urban areas) states. 'In low-performing states (Assam, Bihar, Chhattisgarh, Jammu and Kashmir, Jharkhand, Madhya Pradesh, Odisha, Rajasthan, Uttar Pradesh and Uttaranchal)', Aggarwal writes, 'all pregnant women delivering in government or accredited private institutions are entitled to Rs 1,400 in rural areas and Rs 1,000 in urban areas, after delivery for up to two live births. In the other states, only BPL and SC/ST women are eligible for cash assistance of Rs 700 and Rs 600 in rural and urban areas respectively.' Based on the differences among and within programmes in India, Aggarwal provides insights on the impacts on women who avail of such programmes, the impacts on those who do not, and a set of conclusions that present the issue of access and conditions in the context of human rights.

7. Cash Transfers or Transfers in Kind

Cash transfers are a form of welfare subsidy in which a government provides money to beneficiaries for a specific social purpose, such as to assist with a child's education, a family's health care, or to prevent destitution after the death of the head of household. Before NFSA, cash transfer schemes existed in the form of pensions and maternity benefits. In this chapter, Ashwin Parulkar captures debates that arose during policy discussion on NFSA when the government considered replacing food entitlements under the PDS with cash transfers.

Proponents argued that cash payments would help poor people circumvent the problem of corruption in the food system, in which local ration shop dealers sold grains on the black market instead of to beneficiaries. On these grounds, cash transfers were hailed as a policy move that would enable consumer choice. However, critics argued that India did not have the banking infrastructure required to help poor people set up and use accounts required to disburse and access money for this purpose. Critics also highlighted failures in other social programmes that involved cash payments, such as MGNREGA, in which labourers' wages were routinely unpaid. Lastly, critics argued that cash transfers did not protect poor people from price fluctuations and inflation.

Parulkar surveys these positions and concludes that reforms to increase access to food through the PDS would better address hunger than direct cash payments. First, making the PDS available to all people (universalising) would cut costs associated with targeting BPL families. Universalisation would also improve monitoring and implementation by increasing the number of people who have a stake in the success of the system. Second, increasing salaries of local fair price shops (FPS) and reducing the price of food would address corruption. Lastly, door to door delivery systems would increase access.

Part 3: Food Entitlements and Special Groups

8. Ways by Which States Can Ensure Food Security

What methods of food distribution should the state adopt to satisfy mandates of a food security law? In this chapter, Ankita Aggarwal explores three ways by which the state ensures food security: direct food provisioning through subsidised rations made available by the PDS; the provision of cooked meals, enabling people to access food, say, through public employment initiatives; and facilitating poor people, many of whom are agricultural workers, to produce their own food. The chapter discusses government capacities and responsibilities to successfully implement each of these methods in the context of the FAO's international guidelines that recommend that states ratify right to food laws that ensure people have access to adequate food, as well as social programmes and historical precedents in India that have been designed to enable food security. The chapter captures debates that arose during the drafting of the NFSA on the extent to which activists, policymakers and academics believed the proposed law either met or fell short of international recommendations and existing social programmes in India intended to address hunger.

9. Starvation

Hunger deaths in rural north India led to public action that resulted in the initial court orders in the right to food. However, in this chapter, Ashwin Parulkar finds that policymakers gave little attention to how to define and respond to starvation throughout the drafting stages of NFSA. The NAC's draft legislation defined starvation as 'involuntary deprivation of food'. Their bill contained a 'Rights of Persons Living in Starvation' section, which included redressal systems to respond to persons living with starvation and families of victims who had died of hunger. Policymakers rejected these recommendations. Given this gap in law, Parulkar surveys the history of starvation relief policies: from nineteenth-century colonial famine codes to post-independence statewide relief policies—e.g., in Odisha, Rajasthan and Andhra Pradesh. State relief codes were patterned on colonial famine policies and shared deficiencies that rendered them ineffective: the lack of proper definitions of starvation and accurate indicators to respond to conditions of extreme hunger. In recent years, activists and academics devised a starvation protocol, which was recommended by the Commissioners to the Supreme Court in the right to food case. The protocol called for a 'community diagnosis' of starvation, by which multiple stakeholders, e.g., local residents and public officials, identify people whose family members have died from starvation and those living with starvation to provide them immediate access to needed social programmes.

10. Children

Children's food and nutritional security was a paramount concern for civil society, academic and policymaking bodies, and was addressed in stages prior to and during the drafting of food security legislation in India. In Chapter 10, Ankita Aggarwal details international guidelines, recommendations from academics on the basis of national surveys, and government programmes intended to secure institutional and service delivery arrangements for children's food security through critical stages of development—chronicling, for example, the role of *anganwadi* centres (childcare centres) in providing supplementary nutrition for children from birth to 6 years of age, midday meals in government schools for children between the ages of 6 and 14, and recommendations regarding provisions for children of marginalised SC and ST communities in these age groups based on their specific vulnerabilities to malnutrition. Aggarwal concludes with a summary of comparisons between entitlements recommended by the FAO for right to food legislation and those present in the NFSA.

11. Malnutrition

A central concern among academics, policymakers and civil society organisations in the run-up to the NFSA was the lack of any significant reduction in malnutrition rates amid India's high economic growth period during the last decade. Debates emerged on how law and policy should ensure access to nutrition programmes, since merely the assurance of subsidised grains from the PDS would have little bearing on the persistence of underweightedness, stuntedness and anaemia. In Chapter 11, Ankita Aggarwal, Ashwin Parulkar and Harsh Mander survey these and other debates that called for policy to address the structural causes of undernutrition, including social inequities and the lack of access of poor people to sanitation and health services. In this context, the authors survey the programmes in place intended to address malnutrition among vulnerable groups, with special attention to the ICDS programme as well institutional arrangements across the country for severe acute malnutrition (SAM) among children. The chapter then surveys critiques of the NFSA with respect to provisions on malnutrition, particularly in the context of court-ordered guarantees.

12. Cooked Food versus Ready-to-Eat Meals and Commercial Interests

A few years before policymakers began drafting the NFSB, government officials in the Ministry of Women and Child Development (MWCD) proposed introducing micronutrient-fortified ready-to-eat meals, generally supplied by private contractors, to address child malnutrition. This idea was rejected by multiple stakeholders, including politicians, human rights groups, academics and policy advising bodies. But a debate soon emerged on whether to consider ready-to-eat packages over cooked meals, which continued through the drafting stages of legislation. In this chapter, Ankita Aggarwal discusses these debates among civil society groups and policymakers on the impact that ready-to-eat meals were predicted to have on existing programmes, such as the MDMS in government schools, and the Supplementary Nutrition Programme (SNP) of ICDS. 'As ready-to-eat food is usually provided by private contractors', Aggarwal notes, 'this option also leads to the debate on the desirability of introducing the private sector in the provision of food for children's feeding programmes.' Aggarwal surveys the history of government-sponsored nutrition programmes, examines why the MWCD supported ready-to-eat food for children before legislation was conceived, and summarises arguments in favour of cooked meals that emerged as a response to the perceived threat of private food suppliers in the public food distribution and delivery system in India. She then summarises the relevant provisions in various drafts of the NFSB.

13. Gender-Just Food Entitlements

In this chapter, Ankita Aggarwal discusses specific vulnerabilities that women in India face with regard to nutritional and food insecurity, calling particular attention to pregnant, nursing and single women in poverty. She draws on international conventions established by CEDAW and the FAO, provisions suggested by policymakers in various drafts of the NFSA, and extant social programmes in India intended to address women's nutritional and hunger deprivations. In this context, she explains how the act falls short in not ensuring social security pensions for single women and widows.

14. Vulnerable Groups

In an extension of the debates on universal and targeted programmes, Saba Sharma considers the rationale and need for guaranteeing special provisions to vulnerable populations who experience access barriers to food and social programmes based on discrimination and lack of employment and wages. Sharma draws on international human rights mandates established by the ICESCR that acknowledge the relationship between discrimination and poverty and the barriers faced by vulnerable groups, such as agricultural workers and refugees, in access to food. She also explores debates between lawmakers in parliamentary discussions on relevant provisions to address the needs of such groups through legislation. In this context, the chapter presents a typology of vulnerable groups based on age, gender, infirmity and employment that food security law should acknowledge so as to adhere to both the aims of international mandates and the intentions of existing national programmes sanctioned by the government.

15. Reforming Delivery

While much of this book examines the implications of food security law for poor people's access to food within a human rights framework, Sajjad Hassan in Chapter 15 explores the potential for such legislation to reform administrative, technical and monitoring gaps responsible for problems in India's public food delivery systems: from farms to public warehouses to ration shops and, finally, to beneficiaries. Past failures of these systems, Hassan notes, were responsible for hunger deaths that occurred in Rajasthan in 2001, which catalysed the 'right to food' social movement that led to the Supreme Court orders and the NFSA. Prior to legislation, Hassan notes, individual states such as Chhattisgarh, Madhya Pradesh, Maharashtra and Tamil Nadu had already reformed their own food delivery systems. Chhattisgarh's 2012 state food security law, for example, created a management information system and permitted citizens to monitor local delivery systems. Based on examples from the states, Hassan provides stepwise policy recommendations: generating awareness (through rights-based entitlements or 'citizens' charters'), assigning clear roles and responsibilities to federal, state and local stakeholders within the supply chain (with 'clear job descriptions of all positions' and an agreed-upon organisational structure), operational guidelines for 'last-mile delivery' of food, upgrading technology (such as 'end-to-end computerisation'), budgeting (earmarking funds for specific food security programmes), decentralising responsibilities across all levels of government (such as in an emergency feeding programme in Odisha in which stocks of emergency grain are monitored by village councils), and participant monitoring as in the Chhattisgarh case.

Part 4: Enforcement and Transparency Systems

16. Enforcement and Grievance Redressal

In this chapter, Amod Shah examines the practical challenges and requirements associated with ensuring legal grievance redressal systems under a right to food law. Redressal systems in this context are official government bodies responsible for adjudicating and prosecuting remedies for people in the event that their access to food and social services under legislation is denied or disrupted. Shah notes that precedents for redressal systems can be found in examples of national legislation that protect other socio-economic rights, including education (RTE, 2009) and livelihood (MGNREGA, 2005). He draws attention to redressal frameworks in these and other laws to highlight the practical requirements of a food security law to restore rights to food when they are denied. These include formulating grievance redressal bodies with appropriate officials, assigning requisite powers and responsibilities to such officials across local, state and federal governments, and clear separation of powers between the executive, legislative and judicial government bodies.

Based on these three points, Shah critiques the structure of grievance redressal in the NFSA, comprising district grievance redressal officers (DGROs), state food commissions and a National Food Commission. He argues that the NFSA's provisions for staffing, roles and responsibilities, and separation of powers across these three bodies are unclear, and thus insufficient. While the NAC 'empowered the National Food Commission to monitor … starvation and penalise errant officials', Shah notes, 'the [NFSA] excludes all of these provisions, and … allows the central government to decide on the powers of the commission.' The law also does not delineate a separation of 'prosecutorial and adjudicatory powers', since DGROs are responsible for both duties, while a lack of clarity exists on these roles between state and national commissions.

17. Victim Support

In this chapter, Saba Sharma and Warisha Farasat argue that a strong victim support system is imperative for any successful human rights legislation. The inability to access justice is a major cause of the chronic deprivations marginalised communities experience. The authors analyse the NFSA's provisions for victim support in the context of international mandates pertaining to both socio-economic and civil-political rights. Concomitantly, they draw on examples in draft legislation and formal law in India that have specific frameworks for victim support. The 2011 Draft Grievance Redressal Bill and the Protection of Women from Domestic Violence Act, 2005, they explain, assign a protection officer who is responsible for connecting female victims of domestic abuse to immediate medical care and shelter homes as well as reporting incidents to local magistrates. In this context, the authors observe that the NFSA 'provides no specific guidelines on how a victim whose right to food has been violated may make a complaint.' In a critique similar to Shah's in the previous chapter, Sharma and Farasat observe that the mere provision of DGROs (without specific guidelines for responding to individual complaints) was criticised by NAC members 'for not taking remedial measures closer to the aggrieved persons at the block or panchayat level.' While the NAC recommended provisions to enable individual complaints to be filed 'through telephone, electronically, in writing or in person' and to have grievance redressal mechanisms made public at local (e.g., block-level) offices, the authors argue that the NAC's recommendations did not include formal or institutional victim support measures such as legal aid. They warn that gaps in these important provisions 'will undoubtedly hinder access to … grievance redressal mechanisms', since it is 'naïve' to expect people living with hunger and starvation to access such systems on their own.

18. Fines, Penalties and Compensation

In this chapter, Farasat argues that the Constitution, Supreme Court and high court judgements and relevant parliamentary legislation have, over time, concretised the enforceable nature of socio-economic rights. As a result, individuals whose rights under relevant laws (e.g., MGNREGA) are violated can demand redress in a court of law. She focuses on how financial compensation is a form of redress in such cases. This discussion draws on UN conventions (Basic Principles on the Right to Remedy and Reparations), international cases (e.g., South Africa) and domestic cases (e.g., MGNREGA).

In this context, Farasat details policy discussions on compensation, which she explains revolved around 'who should pay the compensation' in the event of the denial of food, and further, the amount of compensation granted to victims. According to Farasat, lawmakers agreed that punitive measures should be civil in nature, and grappled with how to 'affix responsibility on the individuals responsible'. In this context, she discusses the differences between parliamentary discussions and the recommendations on these issues by the NAC.

19. Accountability and Transparency

In this chapter, Sajjad Hassan argues that the law should assign specific responsibilities to officials and services providers ('agents') to deliver entitlements to beneficiaries ('rights holders' or 'principals'). Hassan explains that the NFSA makes explicit supply-side arrangements through 'end-to-end computerisation' and demand-side provisions by permitting local-level social audits. He argues, however, that greater acknowledgement of citizen participation is required to link demand-side needs with supply-side changes in personnel and systems. 'Demand-side measures are important', Hassan notes, 'in poor countries where education levels are low; where there is a large section of the marginalised;... and where public systems mostly need to be forced into action.'

Hassan notes that the government accepted NAC recommendations for social audits through the requirement of vigilance committees to monitor delivery systems. He states that supply-side corrections will only be made in real time if community evaluations are systematised. He offers a robust set of standards for: (*a*) the planning, implementation, monitoring and evaluation cycle; (*b*) proactive disclosures from local implementing agencies to make information available to the law's intended beneficiaries; (*c*) models for vigilance committees; and (*d*) a model for social audit that he believes could enable synergy between demand- and supply-side corrections to fulfil human rights obligations under a right to food law.

20. Duty Bearers

In this chapter, Saba Sharma addresses how the NFSA assigns duties to officials. Individual states are responsible for implementing the law. The task of assigning enforcement duties to officials is complex. A range of tasks is divided between state and federal bodies. Sharma explains that 'the state government has the predominant duty to set the eligibility criteria' for selecting the law's beneficiaries, while the federal government is responsible for procuring grains and ensuring adequate stockpiles. The division of roles is a result of fierce political contestation between regional parties vying for greater control over local resources and governance. In this context, Sharma provides a panoramic view of multiple stakeholders impacted by administrative and legal duties assigned by the law: individuals and families, private sector organisations, and civil society and religious organisations.

21. Financing the Right to Food

In this chapter, Amod Shah highlights debates on financing the world's largest food security programme—the NFSA—that arose while the law was being drafted. He surveys the history of government spending on social sectors in comparison to other countries. He evaluates the costs of the NFSA and summarises debates on the use of subsidies that were proposed to facilitate public social spending. Finally, Shah places debates on financing the NFSA in the context of other fiscal concerns, such as taxation, ethics and the plausible economic impact of attempting to publicly finance food security through law.

Notes

1. Information on the number of government and government-aided schools as well as the number of children enrolled in primary and upper primary schools is available at GoI (2014).

2. Stunting is defined by the World Health Organization (WHO) as a child's 'height that is more than two standard deviations below the World Health Organization (WHO) Child Growth Standards median.' See 'WHA Global Nutrition Targets 2025: Stunting Policy Brief', http://www.who.int/nutrition/topics/globaltargets_stunting_policybrief.pdf (accessed 8 March 2018) and 'The WHO Child Growth Standards', http://www.who.int/childgrowth/en/ (accessed 8 March 2018).

3. *Constituent Assembly Debates*, vol. VII, pp. 317, 316 (4)(a); *Constituent Assembly Debates*, vol. IX, p. 979; see Seshaiah (1997, 228), and Bhattacharjea (1997, 228–36).

4. For a comprehensive list of percentages of poor people who have access to a range of social programmes, including old age, disability and widow pensions, public housing, the public works programme (MGNREGA), and the national health insurance scheme, see Table 22, 'Access to Government Programmes (across States and UTs)' in the 2013–14 Centre for Equity Studies *India Exclusion Report* (CES 2014, 253). The case of MGNREGA is illustrative. While 96.9 per cent of households in India that demanded employment obtained some form of work through the programme, only 10.4 per cent obtained the guaranteed 100 days, as stipulated by law (ibid.).

5. Numerous government commissions were appointed between 2004 and 2014 to address the vulnerabilities and deprivations of the poor in India through social laws and policies, such as the National Commission for Enterprises in the Unorganised Sector, the Hasim Committee and the National Advisory Council (NAC), among others.

6. Studies on the self-employed have shown that some are 'found to have average earnings several times the minimum wage, leading to the possible inference of a lower likelihood of poverty among them' (Nopo and Valenzuela 2007).

7. Veena Jha points out that 'some countries include informal employment in agriculture in their estimates. This significantly increases the proportion of informal employment: from 83 per cent of non-agricultural employment

to 93 per cent of total employment in India; from 55 to 62 per cent in Mexico and from 28 to 43 per cent in South Africa' (2012, 150).

8. A casual labourer is defined by the Government of India as 'a person who was casually engaged in others' farm or non-farm enterprises (both household and non-household) and, in return, received wages according to the terms of the daily or periodic work contract, as a casual wage labourer' (NSSO 2012, 15).

9. NSSO (2012), calculated by Kompier et al., 'Labour Markets: Exclusion from "Decent Work"', Chapter 4 in CES (2014).

10. NSSO (2013), calculated by Institute for Human Development (2013).

11. Article 11, General Comment 12, defines the accessibility of food in terms of economic and physical accessibility. Because the covenant's definition of economic accessibility is relevant to later discussions in this chapter, it is worth noting in full here: 'Economic accessibility implies that personal or household financial costs associated with the acquisition of food for an adequate diet should be at a level such that the attainment and satisfaction of other basic needs are not threatened or compromised. Economic accessibility applies to any acquisition pattern or entitlement through which people procure their food and is a measure of the extent to which it is satisfactory for the enjoyment of the right to adequate food. Socially vulnerable groups such as landless persons and other particularly impoverished segments of the population may need attention through special programmes' (UN Economic and Social Council 1999).

12. 'Originally brought against the Government of India, the Food Corporation of India (FCI), and six state governments on the claim that these bodies had ineffectively managed the public distribution of foodgrains, the litigation was expanded to apply to all state governments and to address larger, more complex issues of hunger, unemployment and food security.' Further, 'the Petitioners requested that the court issue a writ of mandamus or any other appropriate order to: (1) direct the state and central governments to enforce the Famine Code; (2) direct the Government of India and the FCI to release surplus food grains lying in storage for relief to drought affected areas; and (3) direct all Respondents to revisit the PDS and "frame a fresh scheme of public distribution for scientific and reasonable distribution of grains"' (Birchfield and Corsi 2010, 696, 717).

13. See the International Food Policy Research Institute's (IFPRI) Global Hunger Index. On the basis of key hunger indicators such as the percentage of a population that is undernourished, the proportion of children who are underweight, and the mortality rate of children under 5, IFPRI's hunger index has identified India as being home to 231 million hungry people, placing India 67th out of 81 surveyed countries. Distressingly, Supreme Court Commissioner on the Right to Food Case Dr N. C. Saxena has observed that India is only

one of three developing countries surveyed that did not improve in hunger-related indicators between 1996 and 2011. Based on his reading of the report, he states that several countries that actually faced economic strife and political conflict during that time did, in fact, reduce hunger. A partial list includes Pakistan, Nepal, Bangladesh, Vietnam, Kenya, Nigeria, Myanmar, Uganda, Zimbabwe and Malawi.

14. On the one hand, high rates of rural unemployment in 2001 meant that many rural people could not buy grains from the PDS. The inability to distribute foodgrains through the subsidy system also resulted, in part, in the government's controversial decision to sell wheat stocks to private traders for export to alleviate the surplus burden. In June 2001, Cargill purchased 20,000 tons of wheat at BPL prices from the FCI for export to the Philippines and emerged as the largest buyer of subsidised wheat in India at that time. See Damodaran (2001); see also Shiva (2004, 2007).

15. See *Francis Coralie Mullin v. the Administrator* (1981) 2 S.C.R. 517, 518 (India); see also *People's Union for Civil Liberties v. Union of India*, Writ Petition (Civil) no. 196 of 2001. While in *Francis Coralie* the courts made explicit mention of 'adequate nutrition' as one of the 'bare necessaries' to 'live life with human dignity', the precedent for viewing Article 21 in terms of 'human dignity' and all that was required to realise this goal was laid down three years earlier in the case of *Maneka Gandhi v. Union of India* (1978). The judges in this case wrote, 'Right to life enshrined in Article 21 means something more than animal instinct and includes the right to live with human dignity, it would include all these aspects which would make life meaningful, complete and living.'

16. *People's Union for Civil Liberties v. Union of India*, Writ Petition (Civil) no. 196 of 2001, para 50; Birchfield and Corsi (2010, 698).

17. *Vincent Panikurlangara v. Union of India*, AIR 1987 SC 990: (1987) SC 990: (1987) 2 SCC 165; *J. P. Unnikrishnan v. State of A.P.*, AIR 1993 SC 2178: (1993) 1 SCC 645; *M. C. Mehta v. Union of India* (1987) 4 SCC 463: AIR 1988 SC 1037; *Rural Litigation and Entitlement Kendra v. State of U.P.*, AIR 1987 SC 359: (1986) Supp. SCC 517; *Subhash Kumar v. State of Bihar*, AIR 1991 SC 420 (para 7): (1991) 1 SCC 598; *State of H.P. v. Umed Ram Sharma*, AIR 1986 SC 847: (1986) 2 SCC 68; *Bandhua Mukti Morcha v. Union of India*, AIR 1984 SC 802 (para 10): (1984) 3 SCC 161; *Sher Singh v. State of Punjab*, AIR 1983 SC 465 (para 11): (1983) 2 SCC 344; *Javed Ahmed Abdul Hamid Pawala v. State of Maharashtra*, AIR 1985 SC 231 (para 4): (1985) 1 SCC 275; *Sheela Barse v. Union of India* (1986) 3 SCC 596: AIR 1986 SC 1773; *Upendra Baxi (Dr.) v. State of U.P.* (1986) 4 SCC 106: AIR 1987 SC 191; *Buffalo Traders Welfare Association v. Maneka Gandhi* (1994) Supp. (3) SCC 448: 1994 (3) Scale 1.; *State of J.&K. v. H.C. Bar Assocn.* (1994) Supp. (3) SCC 708 (para 4): 1995 (2) Scale 239; *Nilabati Behera v. State of Orissa*, AIR 1993 SC 1960: (1993) 2 SCC

746; *Sheela Barse v. Union of India* (1993) 4 SCC 204: 1993 (4) JT 558.; *B. L. Wadehra (Dr.) v. Union of India* (1996) 2 SCC 594 (para 22): AIR 1996 SC 2969; *Papaiah v. State of Karnataka* (1996) 10 SCC 533 (para 8); *M. C. Mehta v. Union of India* (1997) 8 SCC 770 (paras 13 and 14)

18. In *S. P. Gupta v. Union of India*, Justice Bhagwati stated why it was necessary to make it possible for petitioners to approach the courts through a public interest litigation in cases when human rights violations have impacted the society at large: 'Where a legal wrong or a legal injury is caused to a person or to a determinate class of persons by reason of violation of any constitutional or legal right or any burden is imposed in contravention of any constitutional or legal provision or without authority of law or any such legal wrong or legal injury or illegal burden is threatened and such a person or determinate class of persons is by reasons of poverty, helplessness or disability or socially or economically disadvantaged position, unable to approach the Court for relief, any member of the public can maintain an application for an appropriate direction, order or writ in the High Court under Article 226 and in case of breach of any fundamental right of such person or determinate class of persons, in this Court under Article 32 seeking judicial redress for the legal wrong or injury caused to such a person or determinate class of persons.' *S. P. Gupta v. Union of India*, AIR 1982 SC 149.

19. This refers to the 'positive' obligation upon the state to uphold the expanded interpretation of Article 21 as the right to live with dignity, as decided in *Vincent Panikulangara v. Union of India*, AIR 1987 SC 990: (1987) 2 SCC 165; and *J. P. Unnikrishnan v. State of A.P.*, AIR 1993 SC 2178: (1993) 1 SCC 645.

20. Legal scholars Birchfield and Corsi (2010, 717) argue that 'India was ultimately established as a social welfare state' on the basis that 'its Constitution has defined and recognized justices as social, economic, and political.' The authors cite the preamble which 'designates India as a 'socialist … republic'. They build their argument on P. L. Mehta and Neena Verma's (1999, 46) assertion that the 'elimination of inequality of income opportunities and status in securing a just social order, is the philosophical foundation of Part IV, embodying the concept of the welfare state'. Meanwhile, Article 38 of the directive principles explicitly states: 'the State shall strive to promote the welfare of the people by securing and protecting as effectively as it may a social order in which justice—social, economic and political—shall inform all institutions of national life.'

21. *Kishen v. State of Orissa*, AIR 1989 SC 67; *Francis Coralie Mullin v. Union Territory*, AIR 1981 SC 746 (753): *Bundhua Mukti Morcha v. Union of India*, AIR 1984 SC 802 (811–12): (1984) 3 SCC 161—all cases cited in Basu (2009).

22. *Shantistar Builders v. Narayan Khimalal Totame* (1990) 2 SCJ 10 (paras 13–14): AIR 1990 SC 630: (1990) 1 SCC 520; cf. *Lakshmi Kant Pandey v. Union*

of India (1984) 2 SCC 244 (para 6): AIR 1984 469; *Vishal v. Union of India* (1990) 3 SCC 318 (para 14): AIR 1990 SC 1412.

23. *Shantistar Builders v. Narayan Khimalal Totame* (1990) 2 SCJ 10 (paras 12–14): AIR 1990 SC 630: (1990) 1 SCC 520.

24. Others include *C.E.S.C. Ltd v. Subhash Chandra Bose* (1992) 1 SCC 441.

25. Cases before the 2001 right to food case that made explicit mention of the state's duty to protect the right to food: *Bandhua Mukti Morcha v. Union of India* (1997) 10 SCC 549; *P. G. Gupta v. State of Gujarat*, 1995 Supp (2) SCC 182; *Olga Tellis v. Bombay Municipal Corporation*; *Shantistar Builders v. Narayan Khimalal Totame* (1990) 1 SCC 520; *C.E.S.C. Ltd v. Subhash Chandra Bose* (1992) 1 SCC 441; *Chameli Singh v. State of U.P.* (1996) 2 SCC 549; *All India Statutory Corporation v. United Labour Union* (1997) 9 SCC 377.

26. Article 47 details the duty of the state to raise the level of nutrition and the standard of living of its people as a primary responsibility.

27. Birchfield and Corsi (2010, 717) write: 'The drafting history of the Constitution strongly suggests that the division into judicially and non-judicially enforceable sections was meant to be a temporary deference to India's status as a newly independent state while still suffering under the weight of colonialism.' The authors cite Mehta's argument as the basis for this claim: 'The non-enforceability clause only provides that the infant state shall not be immediately called upon to account for not fulfilling the new obligations laid down upon it. A state just awakened to freedom with its many pre-occupations might be crushed under the burden unless it was free to decide the order, the place and the mode of fulfilling them' (see Mehta and Verma 1999).

28. See *Mohini Jain v. State of Karnataka* (AIR 1992 SC 858), and *J. P. Unnikrishnan v. State of A.P.* (AIR 1993 SC 2178).

29. According to Kundu and Jain (2004, 17), 'In Mohini Jain v. State of Karnataka [AIR 1992 SC 1858], ... the Court ... came to recognise the right to education as "a right to call upon the State to provide education facilities ... within the limits of its economic capacity and development." It argued that "by saying so, we are not transferring Article 41 from Part IV to Part III—we are merely relying upon Article 41 to illustrate the content flowing from Article 21." We cannot believe that any State would say that it need not provide education to its people even within the limits of its economic capacity and development. It goes without saying that the limits of economic capacity are, ordinarily speaking, matters within the subjective satisfaction of the State.'

30. Dr Babasaheb Ambedkar, the principal writer of the Constitution, once said, 'if I am asked which is the most important provision of the Indian Constitution, without which the Constitution would not survive, I would point to none other than article 32 which is the soul of the Indian Constitution.' See Ostwal (2012).

31. In some cases these rights are 'universal'. For example, every child in India enrolled in a public school has the 'right' to the nationwide MDMS, which provides free hot cooked lunch in public schools across the country. In some cases, the rights are issued on a 'target' basis. For example, all families identified as BPL are guaranteed the right to buy subsidised grains from their local public grain shop.

32. Press Information Bureau, Government of India, 12 September 2013, http://pib.nic.in/newsite/erelease.aspx?relid=99309 (accessed 17 August 2017).

33. 1 lakh crore = 1 trillion.

References

Anant, T. C. C., R. Hasan, P. Mohapatra, R. Nagaraj and S. K. Sasikumar. 2006. 'Labour Markets in India: Issues and Perspectives', in Jesus Felipe and Rana Hasan (eds), *Labour Markets in Asia: Issues and Perspectives*. Manila: Palgrave Macmillan, for Asian Development Bank, pp. 205–96.

Basu, Durga Das. 2009. *Shorter Constitution of India* (eds A. R. Lakshmanan, V. R. Manohar and B. P. Banerjee), 14th edn. New Delhi: LexisNexis.

Bhattacharjea, A. (ed.). 1997. *Social Justice and the Constitution*. Shimla: Indian Institute of Advanced Study.

Binswanger-Mkhize, Hans P. 2013. 'The Stunted Structural Transformation of the Indian Economy: Agriculture, Manufacturing and the Rural Non-farm Sector'. *Economic and Political Weekly*, 48(26–27): 5–13.

Birchfield, Lauren, and Jessica Corsi. 2010. 'Between Starvation and Globalization: Realizing the Right to Food in India'. *Michigan Journal of International Law*, 31(4): 691–764.

CES (Centre for Equity Studies). 2014. *India Exclusion Report*. New Delhi: Books for Change.

Chandrasekhar, C. P. 2012. 'Food Price Levels and Volatility: Sources, Impact and Implications', Special Issue: Standing on the Threshold: Food Justice in India. *IDS Bulletin*, 43(1): 74–83.

Damodaran, Harish. 2001. 'Cargill India bags Manila's 20,000-tonne wheat order'. *Business Line*, 15 June. http://www.thehindubusinessline.com/2001/06/16/stories/14160303.htm (accessed 16 August 2017).

Deaton, Angus, and Jean Drèze. 2009. 'Food and Nutrition in India: Facts and Interpretations'. *Economic and Political Weekly*, 44(7): 42–65.

De Schutter, Olivier. 2010. 'Countries Tackling Hunger with a Right to Food Approach'. Briefing Note 01, May. http://www2.ohchr.org/english/issues/food/docs/Briefing_Note_01_May_2010_EN.pdf (accessed 17 August 2017).

Drèze, Jean, and Amartya Sen. 2013. *An Uncertain Glory: India and Its Contradictions*. London: Allen Lane.

Drèze, Jean, and Reetika Khera. 2017. 'Recent Social Security Initiatives in India'. *World Development*, 98: 555–72.

FAO (Food and Agriculture Organization). 2005. *Voluntary Guidelines to Support the Progressive Realization of the Right to Adequate Food in the Context of National Food Security*. Rome: FAO.

———. 2009. *Guide on Legislating for the Right to Food*. Rome: FAO.

———. 2011. *Right to Food: Making it Happen*. Rome: FAO.

Frodin, Olle. 2010. 'Feeding India. The Spatial Parameters of Food Policy'. *European Journal of Development Research*, 22(3): 445–46.

Gonsalves, Colin, P. Ramesh Kumar and Anup Kumar Srivastava (eds). 2005. *Right to Food: Commissioners' Reports, Supreme Court Orders, NHRC Reports, Articles*. New Delhi: Human Rights Law Network.

GoI (Government of India). 2014. *Statistics of School Education 2011–12*. Ministry of Human Resource Development, Bureau of Planning, Monitoring & Statistics, New Delhi. http://mhrd.gov.in/sites/upload_files/mhrd/files/statistics/SSE1112.pdf (accessed 8 March 2018).

Government of Maharashtra. 1987. *Dr. Ambedkar's Writing and Speeches*, vol. 3. Mumbai: Government of Maharashtra.

Human Rights Law Network. 2008. *Right to Food*, 3rd edn. New Delhi: Human Rights Law Network.

Institute for Human Development. 2013. *India Labour and Employment Report 2014: Workers in the Era of Globalisation*. New Delhi: Academic Foundation and Institute for Human Development.

Jalal, Ayesha. 1996. *Democracy and Authoritarianism in South Asia*. New Delhi: Cambridge University Press.

Jha, Veena. 2012. *India Emerging: The Reality Checks*. New Delhi: Academic Foundation and Ottawa: International Development Research Centre.

Kannan, K. P. 1992. 'Social Security in a Globalizing World', in T. S. Papola and Gerry Rodgers (eds), *Labour Institutions and Economic Development in India*, Research Series 97. Geneva: International Institute of Labour Studies.

Kannan, K. P., and Jan Breman (eds). 2013. *The Long Road to Social Security: Assessing the Implementation of National Social Security Initiatives for the Working Poor in India*. New Delhi: Oxford University Press.

Knuth, Lidija, and Margret Vidar. 2011. *Constitutional and Legal Protection of the Right to Food in the World*. Rome: FAO.

Kundu, Amitabh, and Satish Jain. 2004. 'Right to Food Case Study: India', Intergovernmental Working Group for the Elaboration of a Set of Voluntary Guidelines to Support the Progressive Realization of the Right to Adequate Food in the Context of National Food Security, FAO, Rome.

Mander, Harsh. 2012. 'Food from the Courts: The Indian Experience', Special Issue: Standing on the Threshold: Food Justice in India. *IDS Bulletin*, 43(s1): 15–24.

Mehta, P. L., and Neena Verma. 1999. *Human Rights under the Indian Constitution: The Philosophy and Judicial Gerrymandering*. New Delhi: Deep & Deep Publications.

Nagaraj, R. 2012. 'Introduction', in R. Nagaraj (ed.), *Growth, Inequality and Social Development in India: Is Inclusive Growth Possible?* New York: Palgrave Macmillan, pp. 25–62.

NCEUS (National Commission for Enterprises in the Unorganized Sector). 2006. *Report on Social Security for Unorganized Workers*. New Delhi: Government of India.

———. 2009. *The Challenge of Employment in India: An Informal Economy Perspective*. New Delhi: Government of India.

Nopo, Hugo, and Patricio Valenzuela. 2007. 'Becoming an Entrepreneur'. Discussion Paper no. 2716, Institute for the Study of Labor, Bonn.

NSSO (National Sample Survey Office). 2012. *Employment and Unemployment Situation among Social Groups in India*. NSS 68th Round, July 2011–June 2012. New Delhi: Ministry of Statistics and Programme Implementation, Government of India.

———. 2013. *Employment and Unemployment Situation among Major Religious Groups in India*. New Delhi: Ministry of Statistics and Programme Implementation, Government of India.

Ostwal, Abhaykumar Dilip. 2012. 'Judicial Interpretation of Article 21 of the Indian Constitution', 19 June. http://guide2lawyers.blogspot.in/2012/06/judicial-interpretation-of-article-21.html (accessed 16 August 2017).

Papola, Trilok S., and Partha P. Sahu. 2012. *Growth and Structure of Employment in India: Long-Term and Post-reform Performance and the Emerging Challenge*. New Delhi: Institute for Studies in Industrial Development.

Parulkar, Ashwin. 2012. 'The Big Letdown of the Food Security Bill'. *Wall Street Journal* (*India Real Time* blog), 18 October.

Planning Commission. 2008. *Eleventh Five Year Plan 2007–2012*, vol. 3. New Delhi: Government of India.

Ramachandran, P. 2016. 'The Assessment of Nutritional Status in India', in S. Desai, L. Haddad, D. Chopra and A. Thorat (eds), *Undernutrition and Public Policy in India*. New Delhi: Routledge, pp. 19–45.

Saxena, N. C. 2012. 'Hunger, Under-nutrition and Food Security in India'. Working Paper, Centre for Equity Studies, New Delhi.

Sen, Gita, and D. Rajasekhar. 2012. 'Social Protection Policies, Experiences and Challenges', in R. Nagaraj (ed.), *Growth, Inequality and Social Development in India: Is Inclusive Growth Possible?* New York: Palgrave Macmillan.

Seshaiah, S. 1997. 'Ambedkar's Economic Thought', in A. Bhattacharjea (ed.), *Social Justice and the Constitution*. Shimla: Indian Institute of Advanced Study.

Shah, M. B. 2006. 'The Indian Supreme Court Acknowledges the Right to Food as a Human Right'. *Agriculture and Rural Development*, 2: 24–26.

Shiva, Vandana. 2004. 'The Future of Food: Countering Globalisation and Recolonisation of Indian Agriculture'. *Futures*, 36(6–7): 715–32.

———. 2007. *Cargill: The Corporate Hijack of India's Food and Agriculture*. New Delhi: Navdanya/RFSTE.

Sundaram, K. 2007. 'Employment and Poverty in India, 2000–2005'. *Economic and Political Weekly*, 42(25): 2129–40.

Suryanarayana, M. H. 2012. 'Economic Development and Inequalities', in R. Nagaraj (ed.), *Growth, Inequality and Social Development in India: Is Inclusive Growth Possible?* New York: Palgrave Macmillan.

Swain, B., and M. Kumaran. 2012. 'Why Do ICDS and PDS Exclude and What Can Be Done to Change This?', Special Issue: Standing on the Threshold: Food Justice in India. *IDS Bulletin*, 43(s1): 32–39.

Swaminathan, Madhura. 2008. 'Programmes to Protect the Hungry: Lessons from India'. DESA Working Paper no. 70, Department of Economic and Social Affairs, UN, New York.

UNDP (United Nations Development Programme). 2013. *Human Development Report 2013: The Rise of the South—Human Progress in a Diverse World*. New York: UNDP.

UN Economic and Social Council. 1999. 'Substantive Issues Arising in the Implementation of the International Covenant on Economic, Social and Cultural Rights', General Comment 12, Right to Adequate Food (Art. 11), E/C.12/1999/5. Committee on Economic, Social and Cultural Rights, 20th Session, Geneva, 12 May. https://documents-dds-ny.un.org/doc/UNDOC/GEN/G99/420/12/PDF/G9942012.pdf?OpenElement (accessed 6 August 2017).

Unni, Jeemol, and Uma Rani. 2003. 'Social Protection for Informal Workers in India: Insecurities, Instruments and Institutional Mechanisms'. *Development and Change* 34(1): 127–61.

World Bank. 2012. *Poverty and Social Exclusion in India*. Washington, D.C.: World Bank, and New Delhi: Oxford University Press.

Part I

Concepts, Definitions, Scope, and the International Context

1

The Duty of the State to Provision Food as Social Protection

Harsh Mander

In the autumn of 2013, the Indian Parliament passed a law which legally binds the Indian government to provision very cheap grain to two-thirds of the country's households, school meals to all children in government schools, infant feeding, and near-universal maternity entitlements. The passage of the NFSB became a moment of national debate both inside and outside Parliament about whether indeed the state should provision food at all as a segment of its duties of social protection.

In Parliament, as the NFSB was debated, the clearest enunciation of the case for a law which defines and makes enforceable the duties of the state to provision food was made from the government's side by Mrs Sonia Gandhi, chairperson of the ruling alliance, the UPA. She described the food law as part of an 'empowerment revolution' and reached out to the opposition, which had flagged a series of shortcomings with the scheme. 'This legislation is only the beginning,' she appealed to them. 'We will be open to constructive suggestions, we will learn from experience, but we must put aside our differences.' She also articulated her conviction that costs could not be a barrier to the passage of the law. 'The question is not whether we have enough resources or not and whether it would benefit the farmers or not. We have to arrange resources for it. We have to do it' (Dhar 2013).

The experience of the long journey of the food bill in Parliament was interesting. Between 2011 and 2013, when the Parliamentary Standing Committee was considering the provisions of the bill, a great deal of enthusiasm to make the bill stronger or more comprehensive was not visible. In fact, the Standing Committee further diluted a bill which was already a much weaker version of the NAC draft. It removed most provisions dealing with vulnerable populations, such as community kitchens for the destitute and homeless, and all the provisions related to starvation and state responsibility. But once the arc lights of the national public came to rest on this measure, opposition parties (mainly the Bharatiya Janata Party [BJP], the Communist Party of India [Marxist] [CPI(M)], the Communist Party of India [CPI], the All India Anna Dravida Munnetra Kazhagam [AIADMK] and the Akali Dal) introduced a total of 237 amendments to the official draft. Many of these amendments, if accepted, would have further strengthened the provisions of the bill; however, none was accepted. Nevertheless, the bill was passed in Parliament. While voicing the BJP's support for the bill, Sushma Swaraj pointed out that they were doing so despite the bill being 'weak and shaky' (*Economic Times* 2013b). It is noteworthy that not a single political party—from the right to the left—opposed the principle that the state should provision food to those who were food deprived or food insecure.

But the story outside Parliament was quite the opposite. There was widespread middle-class opposition to, even rage against, the law, reflected in television studio debates, articles in newspapers, and so on. The basic opposition was to the core idea that the state would use large public resources to feed the poor.

First, the language used by commentators to describe the subsidised food assured by the new food law is telling. Based on where they stood on the political, economic and indeed ethical spectrum, they variously described it as a 'freebie', a 'dole', a 'handout', 'populist', a subsidy, an investment in human capital or a moral obligation. Industrialist Rahul Bajaj commented that the food bill was a populist measure at the time of elections (*Business Today* 2013). Sadanand Dhume, in a particularly withering article in the *Wall Street Journal*, compared the 'hare-brained' law to 'something cooked up by a Soviet planner on a bad day', derived from 'the limitless imagination' of 'a gaggle of activists and do-gooders' and 'leftist intellectuals who have never encountered a government hand-out they didn't like'. He went on to reject other rights-based welfare programmes such as MGNREGA: 'Other programs from the same stable of ideas include an expansive rural employment guarantee scheme (effectively a "right to dig ditches"), and a "right to education" policy that has sunk India's already abysmal math and reading scores (teachers aren't allowed to fail students). But even by these exalted standards, the food security bill marks a new low' (Dhume 2013). Gurcharan Das (2013) described it as a strategy to trap the poor 'into a permanent vote bank', suggesting that 'giving away virtually free food and power, waiving loans to farmers, creating "make work" jobs, as the UPA government has been doing, has undermined the social contract' and 'contributed to inflation in our country'. Yashwant Sinha, finance minister in the former National Democratic Alliance (NDA) government, dubbed it 'senseless welfarism' (Sinha 2013). The basic premise of these often visceral critiques was that a morally and intellectually bankrupt government was trying desperately to use taxpayers' money to bribe impoverished voters by offering them freebies.

Economist Pranab Bardhan was more reasonable when he protested this talk of the 'politics of dole' to describe public money spent on anti-poverty programmes. 'Most of the subsidies of the Indian government', he pointed out,

> are actually to the business class and middle and upper classes, but that is not regarded as dole. Giving help to the poor for education, health, food or employment is called dole and to me, this is really tendentious. This serves to distract from the huge amount of subsidies and hand-outs of the government to better-off people in the form of petroleum subsidies, diesel, fertiliser, LPG [liquefied petroleum gas] and several other subsidies. There are estimates that suggest that the amount of subsidies for better-off people is about three to four times more than the money the government spends on anti-poverty programmes. (*Telegraph* 2013)

International Monetary Fund (IMF) research (Anand et al. 2013) shows that fuel subsidies (on diesel, petrol, LPG and kerosene) alone increased from 0.6 per cent of GDP in 2004–05 to 1.9 per cent in 2007–08. After a decrease in subsequent years, it rose again to about 1.9 per cent in 2011–12.[1] An analysis of household consumption expenditure shows that, on an average, the top two income deciles allocated 6 per cent of total household expenditure to fuel consumption, compared to 1.6 per cent in the bottom two deciles. The richest households spent nearly 20 times per capita what the poorest spent, in effect capturing these subsidies (ibid.). Tax exemptions in 2012–13 totalled over five trillion rupees, or 5.7 per cent of the GDP (Bandyopadhyay 2013). On the other hand, an Asian Development Bank report estimated that India spent only 1.7 per cent of GDP on social protection schemes for the poor in 2009, 38 per cent of which was because of the Mahatma

Gandhi National Rural Employment Guarantee Act (MGNREGA). The additional cost of the food bill was Rs 25,000 crore.[2] In the year 2012–13, the petroleum subsidy was estimated at almost Rs 96,880 crore, and subsidy for fertilisers at Rs 65,974 crore. The bulk of the subsidised fertiliser is picked up by rich farmers, because small and marginal farmers account for just 37 per cent of farmland, much of which is rain-fed land where farmers cannot afford to use chemical fertilisers.

Those commentators who were profoundly dismayed by the law did not deny that poor people in India deserved a better deal. But they were convinced that a food law requiring state food provisioning on a large scale was disastrously ill conceived: wasteful, populist, profligate, unaffordable and non-implementable. Most of these criticisms focused on the PDS segment of the law that guaranteed cheap rice, wheat and millets to 66 per cent of the population. They did not comment on many equally significant elements of this law, such as near-universal maternity entitlements, universal feeding for young children, and school meals. They also did not consider the critiques of those of us who cautiously welcomed the legislation as a long-delayed but potentially beneficial statute, as it legally guaranteed for the first time to all the people in the country food to enable healthy and dignified survival, but who were concerned that the law did not go far enough: it is not universal, neglects agriculture, ignores the starving and destitute, and fails to guarantee clean water, sanitation and healthcare which are imperative for nutritional security.

Those who worry that the law is wasteful and populist suggest that this scale of public money would be far better invested in public infrastructure, education and healthcare, rather than in distributing cheap food to large populations. Critics like Jagdish Bhagwati (who launched a very visible and excessively personalised public spat with Amartya Sen for his advocacy of the food law; see, for instance, Bhattacharya [2013]) believe that the answer to hunger is not for the state to feed people, but to stimulate economic growth. If large public resources are expended on food subsidies, it will impact investment and growth, and damage the pace of growth, jobs, and people's capacity to earn enough to buy (or grow) their own food.

The answer to mass hunger is not that the state should feed people in perpetuity; far from it, we need a range of measures to tackle the causes of poverty, and these include stimulating economic growth.

Ending hunger requires a range of public measures to accelerate sustainable agricultural growth, sanitation, clean water, healthcare, social and gender equity, and decent and assured employment, alongside much else. But while all of this unfolds, it can be pointed out that it is economically (and morally) unacceptable for people to have to live with hunger and its consequences, and therefore the state must provision food as long as it remains necessary. Dismayed at the opposition's stalling of Parliament which led to the delay of consideration of the NSFB, Amartya Sen spoke at a press conference of how every day's delay was resulting in the avoidable deaths of hungry and malnourished children (Parsai 2013). It has also been observed that economic growth has not generated as much employment as expected,[3] and senior policymakers in India do not feel assured of the end of even minimally defined poverty for many decades.

It is argued by supporters of the law that one should see this not as merely a dole for the undeserving, but rather as an investment in ensuring that the working people of this country are well fed, which is critical both for their productivity and their morale. According to the child rights organisation CRY (2013), every second child in India is malnourished, which means that the brains and bodies of every second young adult are not developed to their full potential. There is no disagreement that for poverty to end, far more needs to be done than simply feeding people. However, while all of this is being done, the child's brain and body are forming today, and it is the duty of a caring state, especially in a rapidly growing economy, to ensure that while more lasting solutions are crafted and implemented, people today do not suffer from preventable hunger.

Those who regard the law as profligate populism point also to the high cost of the food law. The measure is attacked for being 'unaffordable', what the *Economic Times* describes as 'money-guzzling': an estimated annual burden of Rs 1.25 lakh crore, they argue, is profligate, and will inflate deficits and fuel inflation (*Economic Times* 2013a). These figures, suggest supporters of the law, need to be taken with a pinch of salt. First, what is relevant is not the total but the marginal increase in public expenditure that the food bill entails. This amounts to Rs 25,000 crore, which seems a more reasonable expansion if we are convinced that what the law offers is a useful public investment. Second, we can manage public deficits if we are willing to tax more. India's tax to GDP ratio is lower than that of most industrialised market economies (Alam and Das 2014). Further, the country is excessively reliant on indirect instead of direct taxes (*Economic Times* 2014), something that burdens the poor disproportionately. India also gives tax holidays amounting to Rs 5 lakh crore every year to the corporate sector, and this is justified as necessary for wealth and job creation (Bidwai 2013). The overwhelming evidence from the high-growth years is that this has been a period of virtually jobless growth, which underlines that there is no substitute for public investment to enhance livelihoods. There is need also to enhance the integrity of India's tax efforts. All of this suggests that there is considerable scope for taxing the rich to ensure investment in the nutrition, health and education of the working poor.

Oxford economist Sabina Alkire (2013) in an article offered a telling global comparison. She pointed out that India 'has a higher proportion of stunted children than nearly any other country on earth, yet spends half the proportion of GDP that lower middle income Asian countries spend on social protection and less than one-fifth of what high income countries in Asia spend.' In lower-middle-income countries, these expenses are 3.4 per cent of GDP. India's expenditure is a mere half of that at 1.7 per cent, and even this low level is reached largely because of the rural jobs guarantee programme that ensures 100 days of paid work to all poor households in villages. The average for upper-middle-income countries is 4 per cent of GDP and 10.2 per cent for high-income countries. Japan spends 19.2 per cent and China, 5.4 per cent. Even Singapore spends more than twice as much on social protection as India, at 3.5 per cent of GDP.

It is further argued that it is not right to assume that the pot of public revenues is fixed and given, therefore if we spend more on food, we will either have to pull back on other important expenditures or raise deficits. A third option is to raise more taxes, and with India's low tax to GDP ratio, there is considerable scope for higher tax efforts. Also, too much of our taxation is indirect, burdening the poor unfairly. Further, we need to demand greater integrity in India's tax effort, rather than a moratorium on more public spending for the poor. And finally, we need to weigh the costs of *not* making these investments—the enormous costs of hunger, preventable diseases and deaths and their impact on the morale and productivity of several hundred million working people and growing children.

The third cluster of criticisms of the food security law is that it is not implementable, that state administrations demonstrably lack the capacity to actually deliver on the promises of the law, as evidenced by even official studies which confirm enormous leakages of subsidised PDS grains into the black market. Again, supporters of the law argue that there is no doubt that runaway corruption and inefficiencies greatly cripple the state's capacity. But it is important to remember that both corruption and its debilitating impacts on governance are not restricted to programmes for the poor. They apply also to defence deals, for instance, or to urban infrastructure development, or to the allocation of natural resources like coal mines. But it is no one's case among the same middle-class commentators that because of corruption, we should place an embargo on defence purchases, or on developing our cities, or on mining coal. Then why should we selectively apply corruption as a reason to veto investing public resources only in programmes for the poor?

It is hoped that the pressures created by the food security law, and democratic mobilisation around it, will compel incremental improvements in the PDS over time in all states, in the way that the MGNREGA, RTE and RTI laws have done. It can be no one's case that the states are fully geared up to deliver the rights to work, education or transparent governance, but the creation of legal rights by statutes has generated unprecedented pressure on central and state administrations to greatly improve their delivery systems over time. The same can reasonably be expected to happen with regard to the right to food. The experience of states as diverse as Chhattisgarh and Tamil Nadu demonstrates that, given political will, the leaky and creaky PDS can be credibly fixed.

Gurcharan Das (2013) felt that not only would cheap food disincentivise work, but that impoverished people would work less if they were able to access cheaper cereals. Such a view fails to acknowledge both the hard toil that characterises the lives of millions of India's poor, and the fact that like all of us, they aim for much more than a full stomach. Poor people are not another species who, merely if they have rotis or rice in their stomachs, would stop working because they have no dreams. Like the rest of us, they want a better life for their children, and a better material life for themselves.

Moreover, spending on food is not a populist dole but an investment in India's greatest economic resource—its vast young population in the productive age group—which is imperative for consolidating the gains of India's demographic dividend. A well-nourished and well-educated workforce would be more productive and have higher morale, the best guarantee to sustain higher economic growth. The savings on the cost of cereals would place more disposable income in their hands, and this enhanced spending by millions would stimulate growth from below. An expanded national market could stimulate agriculture, which still employs the majority of India's food-deprived people.

Some supporters also often also make a moral case for the food bill. In an article in the *Economic and Political Weekly*, Mander argued:

> In a country which for too long is scarred by its absence of outrage about and suffering with desperate inequality, the greatest imperative for a right to food law is to breach our collective indifference. There is a great gaping hole in our collective souls, which we must mend. The people of this ancient land must push into history the enormous silent hopeless agony of generations, over centuries, of the inability to feed one's loved ones and oneself. (2011, 50)

Siddhartha Singh (2013) specifically countered this moral position in an article in the *Mint*, where he spoke of the fragility of moral reasoning, arguing that rationality imposes strict limits on morality. 'While the government will have acted in a moral manner, how moral is it for officials to steal food meant for the poor? If one examines the entire distribution chain, one "moral" act by the government will unleash immoral behaviour by many' (ibid.). In defence, it may be stated that the law will not impose a perfect morality on public actors; far from it. But it will establish both a legal and moral benchmark for the unacceptability of avoidable human suffering associated with hunger. The enormous burden of continued, preventable suffering among millions of the poor should persuade even those convinced about the efficacy of markets in ultimately ending poverty that, at least in the medium term, there is no substitute for large state investments to ensure that all people have work, food, education and healthcare. The point is to consider not only the costs of the NFSB, but also the costs of not investing in the better nourishment of millions of our people. Growth cannot be sustained much longer on the thin shoulders of hungry people.

Notes

1. In order to reduce the fuel subsidy, the government subsequently took measures such as capping the supply of subsidised LPG cylinders at six per family per year, and also raising the price of diesel by Rs 5.62 per litre. Despite this, the finance minister, P. Chidambaram, estimated that major subsidies (on fuel, fertiliser and food) for 2012–13 would amount to 2.4 per cent of the GDP (*Indian Express* 2012).

2. 1 crore = 10 million.

3. In the decade 1999–2000 to 2009–10, while GDP growth accelerated to 7.52 per cent per annum, employment growth was just 1.5 per cent, below the long-term employment growth of 2 per cent per annum over the four decades since 1972–73. Only 2.7 million jobs were added in the period from 2004 to 2010, compared to over 60 million during the previous five-year period (Kompier et al. 2014).

References

Alam, J., and S. Das. 2014. 'Exclusion in Planning and Budgetary Processes', in Centre for Equity Studies, *India Exclusion Report 2013–14*. New Delhi: Books for Change, pp. 165–81.

Alkire, Sabina. 2013. 'This bill won't eat your money'. *Hindu*, 29 July.

Anand, R., David Coady, Adil Mohommad, Vimal Thakoor and James P. Walsh. 2013. 'The Fiscal and Welfare Impacts of Reforming Fuel Subsidies in India'. IMF Working Paper WP/13/128. http://www.imf.org/external/pubs/ft/wp/2013/wp13128.pdf (accessed 19 August 2017).

Bandyopadhyay, Sankanath. 2013. 'Tax Exemptions in India—Issues and Challenges: A Discussion Paper', Centre for Budget and Governance Accountability, March. http://www.keralabudgetwatch.org/downloads/tax_exemptions_in_india.pdf (accessed 19 August 2017).

Bhattacharya, P. 2013. 'Everything you wanted to know about the Sen-Bhagwati debate'. *Mint*, 20 July. http://www.livemint.com/Politics/zvxkjvP9KNfarGagLd5wmK/Everything-you-wanted-to-know-about-SenBhagwati-debate.html (accessed 19 August 2017).

Bidwai, P. 2013. 'Invest in the poor, tax the rich'. Rediff.com India News, 22 March. http://www.rediff.com/news/column/column-invest-in-the-poor-tax-the-rich/20130322.htm (accessed 19 August 2017).

Business Today. 2013. 'Volatile rupee hurts business: Rahul Bajaj', 6 July. http://businesstoday.intoday.in/story/volatile-rupee-hurts-business-rahul-bajaj/1/196508.html (accessed 19 August 2017).

CRY (Child Rights and You). 2013. *Status Report on Child Rights in India: An Overview of the Past Decade in India*. http://www.cry.org/resources/pdf/Status-Report-on-Child-Rights-2013.pdf (accessed 19 August 2017).

Das, Gurcharan. 2013. 'Food security bill: Corruption by another name'. *The Times of India* Blogs, 31 March. http://blogs.timesofindia.indiatimes.com/men-and-ideas/food-security-bill-corruption-by-another-name/ (accessed 19 August 2017).

Dhar, A. 2013. 'Let's sink differences to pass "historic bill": Sonia'. *The Hindu*, 26 August. http://www.thehindu.com/news/national/lets-sink-differences-to-pass-historic-bill-sonia/article5061381.ece (accessed 19 August 2017).

Dhume, Sadanand. 2013. 'New Delhi's hunger games'. *Wall Street Journal*, 20 June.

The Economic Times. 2013a. 'Fiscal deficit to slip up by 50 bps on new food scheme: DBS', 4 July.

———. 2013b. 'Lok Sabha approves $20 bn food security bill', 26 August.

———. 2014. 'Economic Survey 2014: Tax revenue to GDP ratio in India appallingly low, says Jahangir Aziz of JPMorgan', 9 July.

The Indian Express. 2012. 'Subsidy bill may touch 2.4% of GDP: FM', 15 September. http://archive.indianexpress.com/news/subsidy-bill-may-touch-2.4--of-gdp-fm/1003089/ (accessed 19 August 2017).

Kompier, C., Archana Prasad, Sajjad Hassan, Smita Premchander, Sudhir Katiyar et al. 2014. 'Labour Markets: Exclusion from "Decent Work"', in Centre for Equity Studies, *India Exclusion Report 2013–14*. New Delhi: Books for Change, pp. 109–39.

Mander, Harsh. 2011. 'Ending Indifference: A Law to Exile Hunger?' *Economic and Political Weekly*, 46(25): 45–50. http://www.righttofoodindia.org/data/right_to_food_act_data/June_2011_ending_indifference_law_to_exile_hunger_harsh_mander.pdf (accessed 19 August 2017).

Parsai, G. 2013. 'Pass food bill even without amendments: Sen'. *The Hindu*, 7 May. http://www.thehindu.com/news/national/pass-food-bill-even-without-amendments-sen/article4689314.ece (accessed 19 August 2017).

Singh, Siddhartha. 2013. 'Morality and the food law'. *Mint*, 16 July. http://www.livemint.com/Opinion/XOugwZuTrKNgzsfNTFX0UO/Morality-and-the-food-law.html (accessed 19 August 2017).

Sinha, Yashwant. 2013. 'Food Security Bill is proof that PM is happy to go along with Sonia Gandhi's senseless welfarism'. *The Economic Times*, 9 July. http://articles.economictimes.indiatimes.com/2013-07-09/news/40469285_1_congress-party-finance-minister-fiscal-deficits (accessed 19 August 2017).

The Telegraph. 2013. 'Dirty D-word & unspoken subsidy', 30 December. https://www.telegraphindia.com/1131230/jsp/nation/story_17733538.jsp (accessed 19 August 2017).

2

Coverage of the National Food Security Bill

SABA SHARMA

INCLUSION OF NON-CITIZENS

In drafting the NFSB, an issue of debate was of coverage—should the right to food be extended only to citizens of the country, or to all residents including non-citizens as well? At any point of time, India is home to a large number of immigrants, documented and undocumented, legal and illegal.[1] Refugees, asylum seekers and stateless persons together form a broad category of resident non-citizens with fluid identities in India. A large majority of them share some types of ties with India—historical, religious, ethnic or linguistic. The legal basis for their stay varies, based on treaties or agreements India has signed with these countries. For instance, because of the Indo-Nepal Treaty of Peace and Friendship, 1950, Nepali nationals are allowed unrestricted passage into the country, can own property and carry out economic activities (as can people of Indian origin in Nepal). These privileges are not extended to nationals from other neighbouring countries, such as Bangladesh.

There is no explicit legal framework to determine the rights of non-citizen residents to various public services. The majority of these residents are economic refugees, escaping extreme poverty in their homelands and seeking employment in the relatively more prosperous and dynamic economies of urban India. They are rarely documented, and usually drift into low-end jobs such as waste picking, domestic work, rickshaw pulling, head loading,[2] and sex work. There is considerable paranoia related to illegal immigrants, especially Muslims from Bangladesh and Pakistan, among political parties such as the BJP.

HUMAN RIGHTS INSTRUMENTS RELATING TO THE RIGHT TO FOOD

The ICESCR, which came into force in 1976, is the human rights instrument that deals most explicitly with the right to food. Article

11 of this covenant recognises the 'right of everyone to an adequate standard of living for himself and his family, including adequate *food*, clothing and housing, and to the continuous improvement of living conditions' (UN 1966; emphasis added). It also stipulates that state parties must 'recognise the right of everyone to be free from hunger' (ibid.), and take adequate steps to ensure those rights. Article 2.1 of the covenant invokes the principle of non-discrimination, which is at the core of many human rights instruments. According to this article, state parties have to guarantee these rights 'without discrimination of any kind as to race, color, sex, language, religion, political or other opinion, *national* or social origin, property, birth or other status' (ibid.; emphasis added). However, the following subsection of Article 2 clarifies that 'developing countries, with due regard to human rights and their national economy, may determine to what extent they would guarantee the economic rights recognised in the present Covenant to non-nationals' (ibid.).

The principle of universality, and consequently of non-discrimination, is crucial to the ethical and philosophical foundations of human rights. In addition, the framework of human rights also provides for corrective measures to reverse the process of discrimination. Moreover, as specified in General Comment 12 of the Committee on Economic, Social and Cultural Rights, food aid cannot be conditional on economic or political issues, and cannot be used as a tool of political pressure. The committee adds that particularly in the case of refugees and internally displaced persons, states have a duty to provide humanitarian assistance (UN Economic and Social Council 1999). If the legal right to food seeks only to guarantee access to food, and not ensure the adequate production of food, then eliminating discrimination is even more crucial to legislating the right to food.

Furthermore, Article 25 of the UDHR states that everyone has the right to an adequate standard of living (which encompasses the right to food and health) without any qualifications of nationality. The International Convention on the Protection of the Rights of All Migrant Workers and Members of Their Families calls for all state parties to ensure the human rights of all migrant workers and their families within their territories, without discrimination on the basis of sex, race, colour, language, religion or conviction, political or other opinion, national, ethnic or social origin, nationality, age, economic position, property, marital status, birth or other status. Article 27 of the convention specifically states that migrant workers and their families have a right to the same extent of social security that is provided to nationals (UN 1990).

The advisory committee on the right to food in the Human Rights Council estimates that many of the one billion people who suffered from undernourishment in 2009 did so because of underlying patterns of discrimination, clarifying that many migrants are refugees because of hunger. When famine strikes a region, people have no choice but to flee; in such a situation, there is a need to protect such refugees from the immediate threats to their lives. However, in the international legal context, fleeing from hunger does not qualify a person as a refugee, and, instead of having their rights protected, such refugees are often held in detention centres and usually forcibly repatriated to their countries (Human Rights Council 2010).

Millions of people living in Africa, especially Sub-Saharan Africa, regularly face famine and destitution. The close links between hunger and conflict are often strengthened when food and famine are also used as weapons of war against certain groups or communities, as in the case of many African countries. Currently, refugees from hunger are seen as 'economic migrants', thus leaving it up to the discretion of the concerned countries whether they are to be allowed in or not. This definition, however, neglects to distinguish between someone leaving to seek a better life versus a refugee who faces a life-threatening situation (in this case, hunger or famine) (Human Rights Council 2008). A seminal report by the Special Rapporteur on the Right to Food urged all states to consider the creation of a new legal instrument to protect people fleeing hunger and violations of the right to food. At the very minimum, legal provisions must decriminalise hunger refugees and not send them back to dangerous situations of hunger, or expel them, as this would constitute a violation of their basic human right to food. The Special Rapporteur argued that the increasing criminalisation of migration would simply lead to further violations of the right to life and the right to food (Human Rights Council 2007).

International Perspectives on the Right to Food: Entitlements of Non-citizens

Among the various countries that have legislated the right to food, some have extended this right to all who reside within the country, while others restrict it to only their citizens. Some countries, like the Democratic Republic of the Congo, Haiti and Nepal, mention 'citizens' in the context of the right to food. Nicaragua has a provision for 'Nicaraguans'. Others, like Moldova, Suriname, Costa Rica and Brazil, speak of the obligations of the state, but without specifying what the ambit of such a duty is, or who exactly is covered under it (FAO 2011).

Brazil is known to have one of the most progressive constitutions in the world, as well as one of the strongest right to food legislations. International treaties to which Brazil is party are immediately recognised as a part of the country's national law, and are legally enforceable. Thus, by extension, the right to food is justiciable in Brazil, and violations can be brought to a Brazilian court of law. In practice, however, this applies very rarely, and economic, social and cultural rights are not seen as justiciable in the Brazilian courts (Commission on Human Rights 2003). In February 2010, Brazil made the right to food an explicit human right in its Constitution, among other social rights that had already been included, with the law clarifying that it is *citizens* who may address and hold authorities accountable for their actions and omissions (FAO 2010b). The stated objective of a subsequent policy on food and nutrition security is 'to guarantee the human right to adequate food on the entire national territory' (FAO 2010a), which too is ambiguous and open to interpretation.

Article 27(b) of South Africa's Bill of Rights states that 'everyone has the right to have access to ... sufficient food and water' (FAO 2011), thus making food security a constitutional right. In a case in South Africa, the court directed that the term 'everyone' in South Africa's Bill of Rights be interpreted inclusively and read as though the words 'or permanent resident' appeared after 'citizen' in the relevant legislative provisions.[3]

In Switzerland, following a case filed by three Czech refugees, the Federal Tribunal recognised the rights of every individual on Swiss soil to certain minimal conditions of existence, including 'food, clothing and shelter', to ensure that individuals are not reduced to an undignified existence (FAO 2004). Under Article 21 of the Constitution (adopted in 1999), 'Right to Aid in Distress', this was recognised as a basic right,[4] even for non-citizens. (However, the rights of non-citizens are restricted through other measures, such as curtailing the rights of asylum seekers, detention of illegal immigrants, strict naturalisation laws, etc.)

The Indian Context

Shanti Bhushan, activist and senior advocate of the Supreme Court of India, evoked Article 21 of the Indian Constitution, which guarantees the right to 'life and personal liberty', in reference to the right to basic sustenance, security and protection from 'unwarranted State intervention, by way of arbitrary arrests, search and seizure etc.' (Bhushan 2009, 1). According to Bhushan, in a rights discourse, certain basic rights, among them sustenance, are to be made available to everyone; to paraphrase Henry Shue, the framework of rights may be built upon two 'basic and axiomatic rights'—the right to security and the right to sustenance (ibid.). Bhushan points out two approaches to the rights of undocumented migrants in this regard: first, that the state has the duty to treat undocumented migrants in the same way as other nationals, and second, that the treatment of aliens is to be subjected to a higher, international 'minimum standard'. Using both approaches, it is argued that the right to food can and should be extended to undocumented migrants as well. Under the first approach, one which is upheld by the Indian Constitution, fundamental rights are to be provided to everyone, but those rights that do not fall under the characteristic of 'basic' are restricted to citizens (ibid.).

Thus, the right to sustenance does not require membership of a political community, and ration cards, for instance, could be issued to undocumented migrants without needing to provide the right to vote. Under the second approach, for which Bhushan uses the US legal system as an example, certain minimum standards are upheld by all nations, and there are constitutional rights that are 'beyond the immigration regulatory context … attached to the undocumented alien by virtue of their personhood alone' (Bhushan 2009, xv).

The problem with restricting socio-economic rights to citizens is not just its impact on non-citizens. There is a large pool of extremely poor citizens, like rural migrants to cities and tribal forest 'encroachers', who typically lack the documents that are required to prove their citizenship. Restricting the right to food to 'citizens' may have the unintended consequence of denying the right to a large number of the extremely poor who are most vulnerable to hunger. Among other socio-economic legislations in India, the RTE Act, 2009, prescribes that '*every child of the age of six to fourteen years* shall have the right to free and compulsory education' (GoI 2009; emphases added). Thus, the act does not make a specific reference to citizenship as a prerequisite to access a socio-economic right.

The Right to Food Campaign (2009) demanded that '*all residents of the country* ... must be covered by the Public Distribution System.' The NAC draft of the NFSB also specified that 'every person shall have physical, economic and social access, at all times, either directly or by means of financial purchases, to quantitatively and qualitatively adequate, sufficient and safe food, which ensures an active and healthy life' (NAC 2011). In its preamble, the NAC draft invokes the UDHR, which 'recognises the right of everyone to adequate food', and also General Comment 12 of the Committee on Economic, Social and Cultural Rights, which further elaborates the 'responsibilities of all State Parties to recognise the right of everyone to be free from hunger' (ibid.). With particular regard to children, the draft specifies that no child may be denied the right to food. Furthermore, special provisions, such as provisions for the destitute, homeless, disabled or those living in starvation, are all defined in terms of personhood, i.e., 'destitute person', 'persons living in starvation', and so forth. As these groups of vulnerable people extend beyond those who are citizens, this draft can be interpreted to apply to all these categories of vulnerable residents.

The government draft of the NFSB does not specify whether the law will extend only to citizens or to non-citizen residents as well. However, the entitlements it provides under the targeted public distribution system (TPDS) require the possession of a ration card, which in theory can only be provided to citizens, thereby effectively restricting these entitlements only to citizens. Internal migrants are at a disadvantage as their ration card or BPL card is valid only within their 'home' state; issuing authorities use the condition of proving domicile to refuse cards to such migrants. However, as in the NAC draft, the government draft also makes special provisions for women and children without any requirements of qualifications or documentation. Similarly, for vulnerable populations, the draft uses terms such as 'destitute persons', 'homeless persons', etc., which again can be interpreted to extend beyond the purview of citizenship.

The final NFSA still requires the possession of a ration card, and contains clear eligibility criteria for inclusion, meaning that entitlements under the PDS will be restricted to citizens. Special provisions for women and children remain universal, however, with every pregnant woman or lactating mother and every child from 6 months to 14 years being eligible. The clauses dealing explicitly with vulnerable groups have been excised from the NFSA, however, and this removes all ambiguity. In the Standing Committee recommendations, vulnerable groups would have been part of a broader 'included' category, without specific reference to citizenship. The implicit assumption among politicians, reflected in the parliamentary debates surrounding the NFSB, was that the bill was exclusively for citizens, and no substantive debate was held about the rights of non-citizens.

Notes

1. For instance, according to the 2001 Census of India, 5.1 million persons immigrated from outside India in the last decade, of which about 3 million were from Bangladesh. See Census of India (2001).

2. Head loaders, whose primary job is to transport material on the head, usually work in construction or other sectors that rely on manual labour.

3. *Khosa & Ors v. Minister of Social Development & Ors*, 2004(6) BCLR 569 (CC) (2004), 2009, https://www.escr-net.org/sites/default/files/judgement.pdf (accessed 8 March 2018).

4. 'Persons in need and unable to provide for themselves have the right to assistance and care, and to the financial means required for a decent standard of living.' See the translation of the Swiss Constitution available at http://www.admin.ch/opc/en/classified-compilation/19995395/index.html#a12 (accessed 20 August 2017).

References

Bhushan, Shanti. 2009. 'Right to Food of Undocumented Immigrants: A Paradox?', 18 November. https://papers.ssrn.com/sol3/papers.cfm?abstract_id=1598725 (accessed 8 March 2018).

Census of India. 2001. 'Data Highlights: Migration Tables'. http://www.censusindia.gov.in/Data_Products/Data_Highlights/Data_Highlights_link/data_highlights_D1D2D3.pdf (accessed 20 August 2017).

Commission on Human Rights. 2003. 'Economic, Social and Cultural Rights: The Right to Food—Mission to Brazil'. Report by Special Rapporteur on the Right to Food, E/CN.4/2003/54/Add.1.

FAO (Food and Agricultural Organization). 2004. 'Recognition of the Right to Food at the National Level'. Intergovernmental Working Group for the Elaboration of a Set of Voluntary Guidelines to Support the Progressive Realization of the Right to Adequate Food in the Context of National Food Security, Information Paper, FAO, Rome. http://www.fao.org/docrep/MEETING/007/J0574E.HTM#P166_26666 (accessed 20 August 2017).

———. 2010a. 'Brazil Adopts a New Policy on Food and Security and Nutrition'. FAO Newsroom. http://www.fao.org/righttofood/news-and-events/news-detail/en/c/44945/ (accessed 22 August 2017).

———. 2010b. 'The Right to Food is Now in the Constitution in Brazil'. FAO Newsroom. http://www.fao.org/righttofood/news-and-events/news-detail/en/c/157360/ (accessed 22 August 2017).

———. 2011. 'Constitutional and Legal Protection of the Right to Food around the World'. http://www.fao.org/docrep/016/ap554e/ap554e.pdf (accessed 22 August 2017).

GoI (Government of India). 2009. 'The Right of Children to Free and Compulsory Education Act, 2009'. Ministry of Law and Justice, Legislative Department, *Gazette of India*, 27 August. http://eoc.du.ac.in/RTE%20-%20notified.pdf (accessed 8 March 2018).

Human Rights Council. 2007. 'Implementation of General Assembly Resolution 60/251 of 15 March 2006 Entitled "Human Rights Council"'. Report of the Special Rapporteur on the Right to Food, Jean Ziegler, A/HRC/4/30.

———. 2008. 'Promotion and Protection of All Human Rights, Civil, Political, Economic, Social and Cultural Rights, including the Right to Food'. Report of the Special Rapporteur on the Right to Food, Jean Ziegler, A/HRC/7/5.

———. 2010. 'Study on Discrimination in the Context of Right to Food'. Prepared by the Drafting Group of the Advisory Committee on the Right to Food, Sixth Session, A/HRC/AC/6/CRP.1.

NAC (National Advisory Council). 2011. 'National Food Security Bill', July 2011. http://www.insightsonindia.com/wp-content/uploads/2011/07/national-food-security-bill-2011.pdf (accessed 8 March 2018).

Right to Food Campaign. 2009. 'Right to Food: Essential Demands'. http://www.righttofoodindia.org/data/rtf_act_essential_demands_of_the_rtf_campaign_220709.pdf (accessed 20 August 2017).

UN (United Nations). 1966. 'International Covenant on Economic, Social and Cultural Rights'. General Assembly Resolution 2200A (XXI) of 16 December 1966. http://www.ohchr.org/EN/ProfessionalInterest/Pages/CESCR.aspx (accessed 20 August 2017).

———. 1990. 'International Convention on the Protection of the Rights of All Migrant Workers and Members of Their Families'. General Assembly resolution 45/158 of 18 December 1990. http://www.ohchr.org/EN/ProfessionalInterest/Pages/CMW.aspx (accessed 20 August 2017).

UN Economic and Social Council. 1999. 'Substantive Issues Arising in the Implementation of the International Covenant on Economic, Social and Cultural Rights'. General Comment 12, Right to Adequate Food (Art. 11), E/C.12/1999/5. Committee on Economic, Social and Cultural Rights, 20th Session, Geneva, 12 May. https://documents-dds-ny.un.org/doc/UNDOC/GEN/G99/420/12/PDF/G9942012.pdf?OpenElement (accessed 6 August 2017).

3

Scope of the National Food Security Bill

ASHWIN PARULKAR

FOOD SECURITY OR NUTRITIONAL SECURITY

A crucial debate in the context of the NFSB regarded its scope, that is, the question of whether the bill should be restricted to ensuring food security, or if it should also aim at achieving nutritional security. Nutrition is a broader concept than food security: apart from access to adequate quantities of nutritious food, it also requires adequate absorption of the food. This in turn requires good health and the absence of frequent bouts of infection, for which access to clean drinking water, sanitation, healthcare and education are essential.

The Right to Food Campaign was of the view that the NFSB should not just be restricted to creating legal guarantees to food, but should also ensure good nutrition. This recalls Article 47 of the directive principles of state policy of the Indian Constitution, which mandates that the state shall regard the raising of the level of nutrition and the standard of living of its people as being among its primary duties (Right to Food Campaign 2012). The campaign argued that given the extremely poor nutritional status of such a large proportion of the country's population, nutrition could not be seen separately from food security. This argument drew from the FAO (2009) guide, which advocates that a right to food law should ensure not just food security, but also good nutrition.

The draft of the NFSB prepared by the NAC also referred to Article 47 in its preamble as well as the Supreme Court's interpretation of the right to life as including the right to food and nutrition. However, the legally enforceable guarantees of this draft fell far short of ensuring good nutrition for all. The draft did include legal entitlements for treating malnutrition in children, as well as entitlements to nutritious food for children, lactating women, pregnant women and destitute persons, but the non-food components of nutrition were only mentioned in the 'progressive realisation of food security' section of the draft, which is not legally enforceable. The draft did provide legal entitlement to the non-food services of the ICDS, including immunisation, health check-ups, referral services, growth monitoring and promotion and pre-school education to all children below 6 years of age. The NAC took the view that since the Supreme Court had already given these services the status of legal rights, they should also become a part of the law.[1] Although pensions of adequate amounts are crucial for ensuring the food security of extremely disadvantaged groups such as the elderly, people with disabilities and single women, the NAC draft did not provide for pensions as a binding obligation of the state, and these too were relegated to the section on 'progressive realisation of food security'.

The government draft of the NFSB mentioned nutritional security as one of its objectives, but failed even more than the NAC draft in fulfilling this mandate. While it did not add to the NAC draft in any respect, it diluted the entitlements for children with malnutrition and rendered the non-food services of the ICDS legally non-binding on the state.

The NFSA similarly places provisions for pensions, healthcare, sanitation, drinking water, and support for adolescent girls in the legally non-binding section, meaning that they do not have the legal protection that other aspects of the act do. The Standing Committee recommended that a clause be inserted ensuring proper implementation of welfare schemes to combat the underlying causes of malnutrition, such as the Rajiv Gandhi Drinking Water Mission, the Mahatma Gandhi Total Sanitation Programme, and the Rural and Urban Health Missions, but this was not done. Despite this, provisions for women and children remain, with cooked meals to be provided through anganwadis and schools, as well as increased food rations for children between 6 months and 6 years who suffer from malnutrition. Furthermore, the act has no place for non-grain food subsidisation, despite the Standing Committee's recommendation on the inclusion of pulses, oils, fortified flour and sugar as well as vocal and widespread calls in Parliament, particularly by the opposition, in favour of including pulses, oils and salt.

Law to Guarantee Only Public Provisioning of Food

In a later chapter on duty bearers, we observe why a law to guarantee the right to be free from hunger would primarily create duties for the state. This duty would be to ensure that all persons have sufficient food at all times. In its most basic interpretation, this would require the government to supply food to all persons who are food insecure, or in danger of becoming food insecure. In other words, this interpretation of the scope of a bill for ensuring food security would primarily emphasise the public provisioning of food.

In order to fulfil these duties, the state could resort to a range of instruments. How best it can fulfil these duties is the subject of later chapters. For instance, should the entitlements for ensuring food security be targeted to the poor and food-vulnerable populations only, or should they be universal? Should the state supply food directly to vulnerable populations, or should it offer cash to procure food, and should these cash transfers be conditional or unconditional? Should the food be cooked or in the form of dry rations? What should be the quantum of the entitlements? Should they be provided free of cost, or at a charge, and if charged, at what rate?

The lack of consensus on the meaning of 'hunger' poses a challenge for the substantive provisions under the right to freedom from hunger. As prolonged hunger usually leads to undernutrition, which is a less ambiguous concept, the FAO guide (2009, 14) suggests that the minimum food entitlements should be defined in terms of what is required to achieve freedom from hunger and undernutrition.[2] The FAO recommends entitlements of a minimum amount of food according to the person's age, gender, health status and occupation for every person suffering from hunger or undernutrition or at risk of suffering from hunger or undernutrition. The law should accordingly define the minimum amount of food entitlement that the state is required to provide to ensure the right of every person to be free from hunger. The actual content of the 'minimum amount of food' and the details of eligibility requirements and procedures for delivery should be indicated either in the main law or in subsidiary legislation or rules. The goal would be for the law to set down the key elements underlying the minimum food entitlements as well as the basic conditions for implementation; these include designating a competent public authority and establishing its legal responsibility for regular, reliable and timely delivery of the 'minimum amount of food' to any person who is suffering from hunger or undernutrition or is at risk of suffering from these conditions.

The FAO guide also suggests that the law should specify the scope of the 'minimum amount of food' entitlement. The entitlement to a minimum amount of food, which could also include a certain amount of water, should be defined in terms of freedom from hunger and undernutrition. This means that the minimum amount of food should cover the minimum dietary necessities of an individual allowing him or her to live with dignity and freedom from hunger and undernutrition. In this regard, the law should explicitly state the dietary necessities that are required for a person to function physically and mentally and maintain a normal resistance to illness, as well as creating optimal conditions for children to grow.

The guide also suggests that the minimum amount of food should be defined on an individual basis. This means that each *individual* member of the community, and not each household taken as a whole, or its head, is entitled to receive the assistance. It should indicate the minimum amount of food determined by the exact quantity of calories, proteins and micronutrients according to the age, sex, health status and occupation of a person. These categories may include, in particular, children (according to age group: up to 6 months of age, between 6 months and 3 years, 3 to 6 years, and school-age children), adolescents (boys/girls), adults (men/women), pregnant and lactating women, persons with disabilities, persons with illnesses, etc. (FAO 2009).

As discussed earlier, the NAC draft, the government draft and the final act restricted the scope of the law to the public provisioning of food, and in doing so did not address the structural causes of hunger or undernutrition. Resultantly, they did not create legal obligations for the state to ensure that farmers have adequate control over resources such as land, water, forests and seeds that are required for producing sufficient quantities of food.

These limitations of the NFSA can be compared with the restricted entitlements under another recent progressive rights-based legislation in India, MGNREGA. This law guarantees 100 days of manual employment a year to every rural household which seeks it. The MGNREGA programme aims at reducing rural unemployment rather than eliminating it entirely by addressing the structural causes of unemployment, such as unequal distribution of resources like land and the absence of affordable and quality education and healthcare for all. These resources increase the capability of people to gainfully utilise or create employment opportunities for themselves.

The NFSA can be viewed as the first of a series of rights-based laws that have the potential to not only alleviate hunger but also address malnourishment and the structural causes of poverty and inequality. Arguments on the necessary provisions of a food security law hinge on five major axes.[3] The first of these is scarcity in food production, wherein scholars argue that failures in agricultural policies in terms of investment in agriculture, the use of appropriate technology, land redistribution, subsidies to rich farmers, etc., have created a crisis in agriculture, and massive corrections are required. A related strand of discourse is that on food sovereignty, which stresses local self-sufficiency and control over food production and distribution systems and advocates ecologically sustainable technologies. Others privilege non-food prerequisites of health and nutrition, such as water, sanitation and healthcare. Inequalities in terms of religion, caste and gender which prevent access to livelihoods, wages, food and water, and health and sanitation form the focus of another body of scholars. And finally there are conversations about policies of social protection and the right to food, and the need to create legal and enforceable duties for the state to publicly provision food through programmes such as subsidised grain transfers, cooked meals in schools, childcare centres, soup kitchens and feeding centres for the destitute, unconditional cash transfers to the aged, maternity benefits and a legal right to work. It is essential to recognise that these are not mutually exclusive or dispensable when addressing the structural causes of poverty and hunger.

Law to Guarantee Food Production for Food Security

It is interesting to note that internationally, notions of government obligations to prevent hunger have swung from considerations of production, to distribution, and now back to production. In the 1960s and 1970s, the idea of food security largely developed with a focus on the 'supply side'—'of assuring the availability and to some degree the price stability of basic foodstuffs at the international and national levels' (FAO 2009, 16). In the decades that followed, the focus shifted from production to distribution—to the accessing of entitlements and not merely ensuring food availability—as food security became increasingly viewed and conceptualised through the lens of human rights. This was influenced by studies such as that by Jean Drèze and Amartya Sen (1989), which argued that famines occur even when food is plentiful but people do not have the means to access this food. Yet now, the clock is swinging back again—especially in the light of the crisis in agriculture—to issues of production and the availability of food.[4]

Defining the scope of a substantive food rights law therefore entails defining terms pertinent to food security—e.g., food, food security, nutritional security and vulnerable groups—and, on this basis, ensuring that the provisions that follow are binding on each branch of government and relevant administrative bodies across local, state and federal levels (FAO 2009). The NFSA defines food security as 5 kilograms of foodgrains per person, or the grains quota that poor people are entitled to under the law. This is a much narrower understanding of the term 'food security' than the definition that India and other UN member states agreed to at the 1996 World Food Summit Declaration—as the condition 'when all people at all times, have physical and economic access to sufficient, safe and nutritious food to meet their dietary needs and food preferences for an active and healthy life.' The framework to operationalise this definition was then based on four pillars: availability (adequate quantities of quality food); access (at the individual level); utilisation (in terms of diet and sanitation); and stability (consistently secure stocks of food at the family or individual level at all times) (ibid.). The NFSA does not define nutritional security or vulnerable groups.

In India, the Right to Food Campaign and others pointed out that the NAC and government drafts of the bill, as well as the final act, were restricted to the distribution of food by the state, mainly by expanding the coverage of the PDS[5] and obligating the state to supply cooked food to certain sections of the population. They argued that in doing so, lawmakers failed to abide by the spirit of the directive principles of state policy of the Indian Constitution (which calls for the continual expansion of democracy in India through the state's recognition of socio-economic rights as fundamental rights[6]) as well as the Supreme Court's historic interpretation of the right to food in the context of the constitutional right to life. The Right to Food Campaign believed that the Government of India was acting in an inconsistent way. In international negotiations at the World Trade Organization (WTO) to promote food security, the Indian delegation invoked the FAO's definition of that term, according to which food security exists when 'the physical and economic access for all people at all times to enough food for an active, healthy life with no risk of losing such access as such is directly connected with livelihood in developing countries' (WTO 2001).[7]

The Right to Food Campaign pointed out that food producers such as small farmers and agricultural labourers whose access to food is directly connected with their livelihood are among those most afflicted by hunger. Agriculture employs nearly 55 per cent of India's total workforce (GoI 2014a), and generates more than 55 per cent of rural income (GoI 2014b). Despite this, its share in the country's GDP fell to 15 per cent in 2011–12, in contrast to industry, which employs less than 25 per cent of the workforce, and the services sector which employs 26.8 per cent of the workforce. The latter two sectors account for 33 per cent and 57 per cent of the GDP, respectively (ibid.). While India is a food-surplus nation (FAO 2011), agricultural production began to fall short of population growth rates in the early 2000s. Production lags have recently recovered, but the sustainability of agriculture in India remains in question. Since the 1991 economic reforms, farmers have experienced declines in farm income, consumption,[8] employment and credit availability. Additionally, high incidences of farmers suicides (see Bello 2007)[9] and the below-subsistence level of food expenditure of farm households across India,[10] has forced even the Government of India to admit that the country faces an agricultural crisis.[11]

Most of India's 231 million hungry people live in rural areas and depend on some form of agricultural work to survive. As net consumers of food, their livelihood security and purchasing power are at the mercy of policies that impact agricultural production and food prices, which have been rising since the initial 2007–08 food price hikes.[12] India's crisis in agriculture is therefore linked to its crisis of hunger, which has resulted in the widespread consensus that the revival of the agrarian sector is key to economic growth as well as food security (Himanshu 2012). Specifically, critics argue that to achieve food security,[13] it is imperative to assure farmers inviolable and equitable access to land, water and affordable inputs required to meet their food requirement. This goal will remain elusive if farmers continue to suffer from unemployment or declining wages in agriculture in comparison to other sectors of the economy, or from displacement, landlessness, and chronic hunger. Each of these problems has increased in the post-reform era, partly as a result of decreased public investment in agriculture and corporate incursion into all aspects of the food supply chain (Pal and Ghosh 2007).[14]

Jayati Ghosh (2005) argues that agriculture in India was hurt by trade policies that favoured large farmers and overall economic policies that scaled back the state, but also because no specific agricultural reforms accompanied the dramatic policy shifts of the early 1990s. The lack of such reforms negatively impacted not only agricultural production but also, for instance, the protection of resources, which are increasingly under the control of corporations and private interests, and overall rural employment. Ghosh notes that the direct relationship between trade, liberalisation and agriculture is complex because farmers bear the costs and benefits of a range of factors impacting both international and domestic markets. These factors are not easy to predict or identify, as crop price fluctuations in world markets are volatile, but the most relevant factors are those that create possibilities for agricultural growth and sustainability of cultivation, as well as those impacting employment and food security.

Ghosh claims that policymakers assumed that freeing agricultural markets and liberalising external trade in farm commodities would provide price incentives, lead to greater investment and output in agriculture overall, and shift the terms of trade between sectors in India in favour of agriculture. Yet that did not happen, as the government reduced public spending at the same time as it adopted fiscal policies that benefited the corporate sector. Economic reforms not only resulted in trade imbalances in agricultural goods, but also affected the conditions and patterns of cultivation of the small farmer (Ghosh 2005).

In the Introduction to this book, we discussed the dimensions of poverty in rural India that are attributable to the inability of the organised sector to absorb the large numbers of agricultural workers forced out of farming. Pravin Jha contends that after World War II, South Korea implemented policies of land redistribution and macro-economic reforms such as public support to agriculture, marketing of farm products, and greater emphasis on education. While the percentage of land redistribution from 1950 to 2005 was about 30–40 per cent in South Korea, the corresponding figure is a mere 1 per cent in India (70 per cent of which occurred in just three states). In 1950, the per capita incomes of India and South Korea were relatively equal. Today, however, South Korea far surpasses India in per capita income and all indicators of basic well-being.[15]

Vandana Shiva (2004) argues that post-liberalisation reforms which amended the state land reforms acts and the Urban Land Ceiling and Regulation Act of 1976 have reversed the gains of these progressive land-based legislations. She adds that liberalisation policies also initiated corporate control of production and input systems. Sukanya Pillay (2009) also argues that economic reforms have violated India's right to food obligations in numerous ways, including: (*a*) reversing investment in agriculture; (*b*) failing to protect small farmers and rural landowners against land acquisition by corporate agribusiness resulting in the displacement and dismantling of small-scale farming; and (*c*) allowing reforms in credit and intellectual property laws, which remove resources from farmers and subject them to unfair laws.

Notes

1. Even though supplementary nutrition for adolescent girls is one of the services provided under ICDS, it was not a legal entitlement under the NAC draft and was only mentioned in the list of 'enabling provisions' along with some other services for adolescent girls.

2. The guide also explains that 'freedom from hunger is the only right qualified as "fundamental" by the International Covenant on Economic Social and Cultural Rights.'

3. For a detailed discussion, see Mander and Parulkar (2016).

4. For a review of the literature on agricultural production, see for instance Dethier and Effenberger (2011).

5. The PDS provides wheat, rice, sugar and kerosene to Indian families that are classified as BPL at highly subsidised rates. Before 1997, the system was available to all citizens in India. In 1997, the universal system was converted to a targeted PDS, where only families identified by government surveys as BPL were eligible for cheap grains. In 2001, the Supreme Court converted these

benefits into legal entitlements as part of the interim orders in the PUCL right to food case. The PDS was the major focus of the NFSA passed by Parliament in 2013, which assures 67 per cent of Indians—about 820 million people—access to subsidised grains.

6. On 25 November 1949, the night before India signed the Constitution into law, Dr B. R. Ambedkar, the principal writer of that document, addressed the Constituent Assembly. Called to defend the document, he instead addressed the challenges India would face in transforming democracy in form (a working Constitution) into one of practice (institutions and federal and state bureaucracies that could implement laws intended to ensure freedom and equality). The barriers to that goal, he argued, were the iniquities governing Indian social life—economic inequality, poverty and human indignity. He claimed that if ignored by man and law, those ills would paralyse the country on its path to democracy. 'We are going to enter into a life of contradiction,' he warned. 'In politics we will have equality and in social and economic life we will have inequality. In politics we will be recognizing the principle of one man one vote and one vote one value. In our social and economic life, we shall, by reason of our social and economic structure, continue to deny the principle of one man one value.' The aspirational dimensions of the directive principles may be viewed as a way for the state to ensure 'one man, one value' as the country grew in terms of both resources and the political will required to ensure a democracy in practice within the federal structure of perhaps one of the most socially, culturally and linguistically diverse countries in the world.

7. India is signatory to numerous international agreements which define the right to food and food security from a human rights perspective. Among them are the UDHR, the ICESCR (1966), the Convention on the Rights of the Child (Articles 27.1 and 27.3) and the Convention on the Elimination of All Forms of Discrimination against Women (Article 12).

8. Pal and Ghosh (2007, 15) show through NSSO data that 'per capita food-grain consumption declined from 476 grams per day in 1990 to 418 grams per day in 2001, while aggregate calorific consumption per capita declined from just over 2,200 calories per day in 1987–1988 to around 2,150 in 1999–2000.'

9. Bello discusses how WTO trade liberalisation, characterised by the removal of tariffs and quantitative restrictions in India, have resulted in what Utsa Patnaik (2004, 1) calls 'a collapse of rural livelihoods and incomes'. According to reports of the National Crime Records Bureau, the total number of farmer suicides in India crossed 270,000 in the period between 1995 to 2011. Also see Sainath (2012).

10. The NSSO (59th Round) report indicates that the average monthly per capita consumption expenditure of farm households was Rs 503 in 2003 (NSSO 2003).

11. On 15 August 2006, Prime Minister Manmohan Singh said in his speech at the Red Fort: 'I see that our farmers in many parts are in a crisis, not managing to eke out a decent living from their land.... The agricultural crisis that is forcing them to take the desperate step of committing suicide needs to be resolved.'

12. According to the ILO's (2011) report, wholesale food prices in India increased by 40 per cent more than between the mid-1990s and 2004. The report also noted that for the NFSA to be effective, the storage and distribution systems had to ensure universal access to basic food items at an affordable price. Additionally, Utsa Patnaik argues that the primary cause of price hikes beginning in 2007 have been cuts in government expenditure in rural areas that caused unemployment to rise and income inequality to increase, 'with real incomes (money adjusted for price change) for much of the population declining, forcing people to cut back on food' (Patnaik 2009).

13. From an interview with Dipa Sinha, Right to Food Campaign, August 2011. Ms Sinha states that in the early drafting period of the bill, the campaign had fought to include access to land and natural resources as well as production and procurement reforms through a guaranteed MSP.

14. Pal and Ghosh conclude that economic inequality increased in rural and urban India in the post-reform period due to 'the stagnation of employment generation in both rural and urban areas across states'. Employment growth in the 1990s plummeted in both rural and urban areas. With regard to factors relating to agriculture, the rural employment growth rate was 0.67 per cent. This was directly attributable to the stagnation of agricultural employment in this decade. The authors site NSSO data, which indicate that total employment in the agricultural sector increased from 190.72 million in 1993–94 to 190.94 million in 1999–2000, an increase of 0.02 per cent, as opposed to an increase of 2.57 per cent from 1987–88 to 1993–94. The authors also state that the decline in self-employment in agriculture—especially sharp for women—may have been related to changes in production patterns, which forced some peasants out of direct cultivation (Pal and Ghosh 2007, 1–28).

15. Interview with Pravin Jha, Professor of Economics, Jawaharlal Nehru University, New Delhi, 25 February 2012.

References

Bello, Walden. 2007. 'Why Asia's Farmers Deserve Protection from Free Trade'. *Global Asia*, April.

Dethier, J., and A. Effenberger. 2011. 'Agriculture and Development: A Brief Review of the Literature'. Policy Research Working Paper no. 5553. http://www-wds.worldbank.org/external/default/WDSContentServer/IW3P/IB/2011/01/31/000158349_20110131131635/Rendered/PDF/WPS5553.pdf (accessed 29 August 2017).

Drèze, Jean, and Amartya Sen. 1989. *Hunger and Public Action*. New Delhi: Oxford University Press.

FAO (Food and Agriculture Organization). 2009. *Guide on Legislating for the Right to Food*. Rome: FAO.

———. 2011. 'India and FAO: Achievements and Success Stories'. FAO Representation in India, March. http://www.fao.org/3/a-at007e.pdf (accessed 30 August 2017).

Ghosh, Jayati. 2005. 'Trade Liberalization in Agriculture: An Examination of Impact and Policy Strategies with Special Reference to India'. *Human Development Report* Occasional Paper 2005/12.

GoI (Government of India). 2014a. 'Agriculture and Food Management'. *Economic Survey*, Chapter 8. http://indiabudget.nic.in/es2013-14/echap-08.pdf (accessed 30 August 2017).

———. 2014b. 'Services Sector'. *Economic Survey*, Chapter 10. http://indiabudget.nic.in/es2013-14/echap-10.pdf (accessed 30 August 2017).

Himanshu. 2012. 'To feed a billion mouths'. *Forbes India Magazine*, 4(3), 3 February.

ILO (International Labour Organization). 2011. *World of Work Report: Making Markets Work for Jobs*. Geneva: ILO.

Mander, Harsh, and Ashwin Parulkar. 2016. 'Elephant in the Dark: Finding Ways to End India's Hunger and Malnutrition', in Sonalde Desai, Lawrence Haddad, Deepta Chopra and Amit Thorat (eds), *Undernutrition and Public Policy in India: Investing in the Future*. New Delhi: Routledge and NCAER-IDS, pp. 49–68.

NSSO (National Sample Survey Organisation). 2003. 51st Round, January–December 2003, NSS Report no. 503. Ministry of Statistics and Programme Implementation, Government of India, New Delhi.

Pal, Parthapratim, and Jayati Ghosh. 2007. 'Inequality in India: A Survey of Recent Trends', Working Paper no. 45, UN Department of Economic and Social Affairs, July.

Patnaik, Utsa. 2004. 'External Trade, Domestic Employment, and Food Security: Recent Outcomes of Trade Liberalization and Neo-Liberal Economic Reforms in India'. Paper presented at the International Workshop on Policies against Hunger III, Berlin, 20–22 October.

———. 2009. 'Origins of the Food Crisis in India and Developing Countries'. *Monthly Review*, 61(3).

Pillay, Sukanya. 2009. 'India Sinking: Threats to the Right to Food, Food Security and Development in an Era of Economic Growth'. *Windsor Yearbook of Access to Justice*, 27(1): 127–69.

Right to Food Campaign. 2012. 'Right to Food Campaign's Critique of the National Food Security Bill, Part II: Clause by Clause Critique of the National Food Security Bill, 2011'. http://www.righttofoodindia.org/data/right_to_food_act_data/events/March_2012_clause_by_clause_critique_final_18_february_2012.pdf (accessed 28 August 2017).

Sainath, P. 2012. 'Farm suicide rise in Maharashtra, state still leads the list'. *The Hindu*, 3 July.

Shiva, Vandana. 2004. 'The Future of Food: Countering Globalisation and Recolonisation of Indian Agriculture'. *Futures*, 36(6–7): 715–32.

WTO (World Trade Organization). 2001. 'Negotiations on WTO Agreement on Agriculture, Proposals by India'. Committee on Agriculture, Special Session, G/AG/NG/W/102, 15 January. wto.org/english/tratop_e/agric_e/ngw102_e.doc (accessed 30 August 2017).

PART II

UNIVERSALISM OR TARGETING, FOOD OR CASH TRANSFERS, AND CONDITIONALITIES

4

Universal or Targeted Entitlements

ANKITA AGGARWAL

BACKGROUND TO PDS AND TPDS

The PDS is an avatar of a rationing system introduced by the British during World War II at a time when famine struck Bengal. This system was abolished at the end of the war, and reintroduced in 1950 to curb inflation (Nawani 1994) and manage the supply of food rations (Sharma et al. 2000). The PDS was also intended to contribute to household food security by providing foodgrains and other essential commodities such as sugar and edible oil at below market prices (ibid.; Nawani 1994). Initially, the PDS served only urban areas, but in the First Five Year Plan (1951–56), policymakers extended access to rural areas afflicted by chronic food shortages.

Between 1966 and 1992, everyone in India could buy grains from ration shops. The first period of targeting was introduced in June 1992. On the grounds of corruption and inefficiency, policymakers decided to restrict access to cheap grains in 1,775 blocks of the country in tribal, arid, hilly and remote areas (Khera 2011c). In 1997, the second period of targeting commenced. The targeted public distribution system (TPDS) would provide grains to households classified as above the poverty line (APL) or BPL based on surveys conducted by the Planning Commission in 1993–94 (Kochar 2005). The next section will discuss how targeting in the PDS was part of the structural adjustment policy of the World Bank and the IMF, which required as loan conditions cuts in subsidies to reduce the fiscal deficits of developing countries that borrowed from international financial institutions (Bunsha 2002). The conversion to the TPDS in India was aided by the recommendations of various studies in the 1990s (Parikh 1994; Radhakrishna and Subbarao 1997) that concluded that the universal food distribution system was inefficient and did not reach the poor, primarily because the poorer states of north India received less subsidy than the richer states of the south.

In 2001, an important element of affirmative action was introduced into the PDS in the form of the Antyodaya Anna Yojana (AAY).[1] The

AAY programme provides food-based assistance to destitute families, identified as the most vulnerable families. Ten million households identified for this scheme were provided Antyodaya cards, which entitled them to 25 kilograms of foodgrains every month at Rs 2 per kilogram for wheat and Rs 3 per kilogram for rice. The monthly entitlement of Antyodaya households was increased to 35 kilograms in April 2002. On 2 May 2003, the Supreme Court of India ordered that all households belonging to the following categories would be entitled to Antyodaya cards:

1. Aged, infirm, disabled, destitute men and women, pregnant and lactating women, destitute women
2. Widows and other single women with no regular support
3. Old persons (aged 60 years or above) with no regular support and no assured means of subsistence
4. Households with a disabled adult and no assured means of subsistence
5. Households where due to old age, lack of physical or mental fitness, social customs, the need to care for a disabled person, or other reason, no adult member is available to engage in gainful employment outside the house
6. Primitive tribes

Because of the low official poverty estimates that exclude many food-insecure households, many state governments have moved towards a more inclusive, if not universal, PDS. While Tamil Nadu and Himachal Pradesh have instituted a universal PDS, Andhra Pradesh and Chhattisgarh have a 'quasi-universal' PDS which covers almost 80 per cent of their populations (Khera 2011b).

Arguments Supporting TPDS

The main objective of food subsidy programmes is to provide fixed quantities of food to consumers at fixed prices. Such programmes aim at transferring the impact of price changes from consumers to the government, thereby reducing both chronic and transitory food insecurity among the poor (Subbarao et al. 1995). The ICESCR states that although the full realisation of the right to food may take time, state parties are required to take 'steps' towards this end within a reasonably short time after ratification of the covenant. These steps should be deliberate, concrete and *targeted* (FAO 2009a). The FAO's *Right to Food Methodological Toolbox* (see FAO 2009c) briefly discusses the debate on whether food entitlements should be universal or targeted, but makes repeated reference to the need to target policies and programmes to the most needy and vulnerable. However, neither the ICESCR nor the Methodological Toolbox specifies whether targeting is recommended for all or for some specific types of programmes and policies aimed at achieving food security.

High Costs and Low Efficiency of Universal Programmes

The limited budgets of poverty reduction programmes are believed by some to create the necessity of targeting to ensure that the precious resources deployed in these programmes are distributed only to those who are poor, and are not wasted on those who are not poor. Hence, improving the efficiency of poverty reduction programmes requires that those who are most in need are targeted with the highest priority (Krishna 2007). This in turn requires the identification of the neediest and understanding the reasons for their deprivation (FAO 2009b). The various bases on which targeting can be done include income, age, gender, disability and geography. It is acknowledged that targeting involves trade-offs between the desire to cover all poor people and the risk of wasting resources if the benefits of the programme leak to those who are not poor. The extent of coverage in targeted programmes is largely dictated by the budgets of the programmes (Subbarao et al. 1995).

P. S. George (1994) cautions that in order to prevent subsidised food distribution from becoming a severe burden on the exchequer, it should be strictly targeted. Subbarao et al. (1995) criticise the untargeted nature of the subsidised food programmes of several Asian countries for being very costly and fiscally unsustainable. Bangladesh's PDS, apart from providing some non-monetary benefits to the poor, led to massive leakages of benefits to the non-poor due to being untargeted. It is also one of the most expensive programmes in the region. The cost of transferring 1 taka of income to an eligible household is six times the income transfer. Pakistan's partially targeted food ration programme has also been criticised for being inefficient. Analysis of the targeted food stamp programme launched in Sri Lanka in 1979 reveals that the programme was well targeted and the food stamp receipts were progressive. Simulations also suggest that poverty with food stamps decreases more than it does under untargeted transfers with the same gross budget (Datt and Gunewardena 1997).

On the other hand, universal subsidised food programmes in many countries have been criticised for their negligible impact on the poor. In Nepal, the benefits of the subsidised food provided by the National Food Corporation are reaped by civil servants as well. Although food is flown to remote areas at highly subsidised transportation costs, distribution of this food has not had a significant impact on the poor. In the Philippines, due to the untargeted food subsidy and its urban bias, the poorest income groups and regions are unable to benefit from this programme. For instance, from 1992 to 1995, the National Capital Region and the Cagayan Valley, which account for only 2.8 per cent of the country's poor, received 34.5 per cent of the total subsidised rice (Subbarao et al. 1995). A World Bank discussion paper claims that universal programmes in many countries have not only been fiscally unsustainable, but also significantly distortionary in their impacts on the agricultural sector, which has resulted in several countries restructuring their food programmes (Radhakrishna and Subbarao 1997).

In India, Jha and Srinivasan (2001) conclude that the TPDS has a higher benefit–cost ratio[2] than the universal PDS. Universal PDS has also been criticised for being cost-ineffective (Dev 1995). This is illustrated by Radhakrishna and Subbarao (1997) with the example of Andhra Pradesh's rice scheme, in which Rs 3,492.5 million was spent in 1986–87 to transfer Rs 550 million to the poor. Radhakrishna and Subbarao argue that under conditions of 'perfect' targeting, not more than Rs 1,200 million (inclusive of costs incurred in targeting) would have to be spent to transfer the same amount to the poor. Supporters of TPDS argue that in the face of fiscal constraints, it is not prudent to spend food subsidies on those who are not poor (Nawani 1994).

Imperfect Targeting Better at Reaching the Poor Than Universal Programmes

Some of the studies which recommend 'proper' targeting of the PDS do not actually suggest how targeting should be done so that the poor do not get excluded and the rich do not get included in the list of beneficiaries. It is interesting to note that in their World Bank discussion paper, Radhakrishna and Subbarao (1997) refer repeatedly to 'perfect' targeting, although they ultimately conclude that no method of targeting has been found to be absolutely accurate. However, they argue that even imperfectly targeted programmes are better at reaching the poor and reducing costs than universal programmes.

Marginal Impact of Universal PDS on Income Transfers to Poor Households

It has been argued that universal PDS has a marginal impact on income transfer to poor households (Dev 1995). This shortcoming is said to stem from its universal nature, which forces the benefits of the scheme to be spread thinly over a large number of beneficiaries who are both poor as well as rich (Parikh 1994). Based on an analysis of 1986–87 household data, Radhakrishna and Subbarao (1997) conclude that universal PDS contributed only 0.3 per cent in the reduction of rural poverty and increased the daily caloric intake by only 41 kcal. Hence, in order to plug 'leakages' to the non-poor, supporters of TPDS advocate against a universal PDS (Anon 2009).

Lack of Affirmative Action in Universal PDS

Affirmative action refers to policies that take factors including 'race, colour, religion, gender, sexual orientation or national origin'[3] into consideration in order to benefit an under-represented group, usually justified as countering the effects of a history of discrimination.[4] Affirmative action programmes in the Indian context should be viewed as a compelling liberal response to the deprivation of adequate opportunities, and the consequent failure of achieving a decent human existence, for large sections of the underprivileged population, mainly due to their victimisation by a long history of repression, discrimination and marginalisation. Special privileges are essential to retrieving them from a deep sense of injustice, prejudice and existential marginality, and enabling them to live a more fulfilled human existence (Mohanty 2007).

People vary hugely in their ability to access sufficient food. There are enormous differences in food denials and vulnerabilities between households, based on divergent conditions of wealth, gender, caste, community, age, disability and a range of other factors. Many have questioned the justification for giving a rich rural landlord subsidised grain, and that too at the same levels and prices as a destitute rural landless widow, or a homeless disabled beggar (Mander 2011).

Arguments Supporting Universal PDS

Himanshu and Sen (2011) argue that universal access to the food distribution system is the only option that coheres with a rights-based approach to food security, citing other examples of socio-economic entitlements, such as MGNREGA and RTE, that are backed by legislation and guarantee universal access. Amartya Sen (1995) takes issue with 'targeting' as a method of poverty eradication, arguing that such policies render the poor as passive receivers rather than agents who have a claim to services and rights.

The natural question that may be put to these stances in a country as large as India is: does the government have the capacity to feed its people through a public food system? Himanshu and Sen (2011) make the case that it does. The requirement for 119 crore (1.19 billion) people as per 2010–11 data, they argue, is 100 million tons of grain if every person were to buy an average of 7 kilograms of rice and wheat per month from a ration shop. The total allocation of grains for the TPDS in 2010–11 was 48 million tons. The authors claim that it is possible to estimate the percentage of people who will actually buy PDS rations by looking at the rates of people who select out of the system and the average purchases per person in states with the strongest TPDS performance. Tamil Nadu, Andhra Pradesh, Himachal Pradesh and Kerala are all food-deficit states that have nearly 'universalised' their food distribution programmes through their own budgets, and have the highest percentage of purchases: Tamil Nadu (76 per cent), Andhra Pradesh (70 per cent), Himachal Pradesh (65 per cent) and Kerala (54 per cent). These percentages cohere with the national percentage of people with BPL and APL cards who buy any rations from the system (ibid.).

Second, the average purchases of current cardholders is 4.5 kilograms per person. If 70 per cent of the population, therefore, purchases the full quota of 7 kilograms of grains, the total allotment required would be 70 million tons. If, however, 70 per cent of the population purchases an amount closer to the national average of 5 kilograms per person per month, 50 million tons would be required. Hence, a universal PDS is well within the limits of feasibility in terms of the quantity of grain required.

High Rates of Exclusion and Inclusion Errors

Targeting has often been unsuccessful in identifying those who are most in need (ILO 2001; Krishna 2007), thus failing to ensure them social assistance. Krishna (2007) argues that targeting is more often unsuccessful than not, since it is based on the assumption that a fixed pool of disadvantaged people exists that can be identified fairly accurately and affordably. Because large numbers of people move in and out of poverty, identifying the poor for social programmes requires frequent revision of beneficiary lists. This seldom happens in India, where the last BPL census was conducted in 2002. Programmes should target not only those who are in poverty but also those who are likely to fall into it; however, it is more difficult to identify the latter group.

Reasons for failed targeting include complicated and time-consuming procedures, costs and difficulties in obtaining accurate information (FAO 2009a), inadequate availability of information on people's living standards, incomes and expenditure patterns, weak administrative capacities (Subbarao et al. 1995), large-scale administrative discretion resulting in favouritism and discrimination, and poor implementation (FAO 2009a).

Hence, targeting inevitably leads to some errors of exclusion (that is, exclusion of poor households from the list of those eligible for a programme), and errors of inclusion (inclusion of non-poor households in such lists). Due to the various benefits of the poverty reduction programmes, inclusion in the beneficiary lists of these programmes is considered highly desirable; thus, many non-poor try to get selected in targeting by manipulation or deceit (Hirway 2003). These tricks include feigning certain characteristics of poverty, for example, by under-reporting variable and undocumented incomes and movable assets (Krishna 2007; A. Sen 1995). Thus, a wealthy farmer or rich businessman may declare his adult sons landless and assetless to include their families in the BPL lists. Many non-poor people use their clout to control the process of preparation of BPL lists to get their names included, and these practices lead to inclusion errors. Usually the poorest households find it difficult to get themselves selected in targeting as they are often unaware of the process, unable to obtain the required documents and forms, unable to convince or assert themselves with the officials undertaking the process of identification, or they do not have the money to bribe the concerned officials to get their names included in the beneficiary lists, leading to exclusion errors and skewed targeting with a non-poor bias (Hirway 2003).

Errors of exclusion and inclusion can be quite large and the process of reducing these errors can be very costly (Baker and Grosh 1994; Cornia and Stewart 1995; Gaiha et al. 2001; Ravallion and Datt 1995). The extent of correct identification of poor households depends not so much on the targeting methodology as on the efficiency and honesty of the local administration (Hirway 2003). Usually, attempts to decrease one kind of error increase the chances of making another type of error. For instance, introducing stricter rules in the identification of the poor to screen out the non-poor may also make it more difficult for some of the poor to provide the necessary information which proves their poverty. Thus, in the process of trying to decrease inclusion errors, exclusion errors might increase.

In India, in the context of the PDS, at least three national household surveys highlight the high rates of exclusion and inclusion errors in the BPL lists. These surveys are the NSS (61st Round, 2004–05), the National Family Health Survey-3 (NFHS-3) (2005–06) and the Indian Human Development Survey (2004–05). These surveys show that about half of all poor households in rural areas do not have a BPL card, and in states such as Bihar and Jharkhand, the proportion of poor households[5] that did not have a BPL card was as high as 80 per cent in 2004–05.[6]

A study by the Planning Commission's Programme Evaluation Office reveals that more than half of the poor either have no BPL card or have been given APL cards, and are thus excluded from BPL entitlements. At the same time, almost 60 per cent of the BPL or Antyodaya cards have been given to non-poor households (Planning

Commission 2007). Similar results are reported from rural Gujarat (Hirway 2003). A study by Madhura Swaminathan and Neeta Mishra in rural Maharashtra highlighted that while the universal PDS excluded 5.5 per cent of BPL families, the proportion of this exclusion increased almost 10-fold to 54 per cent after targeting (Bunsha 2002). Dubiously, almost 60 per cent of households that are not poor possess BPL or Antyodaya cards (Planning Commission 2008). These findings have been corroborated by the 55th (1999–2000), 61st (2004–05) and 63rd (2006–07) Rounds of the NSS. The poorest of the poor, often comprising tribal villagers, female-headed households and people living in remote hamlets with poor reach of public services, have a greater chance of getting excluded rather than included in the BPL lists (Mander 2011).

Food-Insecure Households Far Exceed the Poverty Line

In India, targeting for PDS has been linked to the country's poverty line, which is criticised for being abysmally low (discussed in detail in the next chapter). The national poverty line at 2004–05 prices was taken as Rs 356 per capita per month in rural areas and Rs 539 per capita per month in urban areas, meaning that only 28.3 per cent of the population in 2004–05 was BPL. At this level of income, people in rural and urban areas are able to consume only 1,820 kcal a day, whereas the minimum calorie requirement for a healthy life is 2,400 kcal for rural areas and 2,100 kcal for urban areas. According to the 61st Round of the NSS (2004–05) (NSSO 2007), 79.8 per cent of the rural population and 63.9 per cent of the urban population was unable to achieve their daily consumption of 2,400 and 2,100 kcal, respectively, and this deprivation has been rising over the years.

The per capita calorie intake at the poverty line is not only grossly insufficient for a healthy life, but has declined significantly almost all over the country between 1972–73 and 1999–2000 (GoI 2009). According to NFHS-3 conducted in 2005–06, more than 40 per cent of India's women and children were undernourished, with almost 20 per cent of them in the richest quintile (Himanshu and Sen 2011); however, the central government allocations for providing BPL entitlements in the PDS is only at 33 per cent of the country's population (Drèze 2011b).

Concern about placing the poverty line so low has been expressed by many academics, activists and economists (Drèze and Khera 2010; Mander 2011; P. Sen 2005; Swaminathan 2001). These critics argue that PDS targeting has been purposely linked with the poverty line because, in setting the latter at a low level, the government can simultaneously decrease the recipients of the TPDS, enabling the food subsidy to be lowered over time. Karat (2006) adds that TPDS beneficiaries are linked not to a realistic poverty line but to an extremely low poverty line, for the same reason of reducing the number of people eligible for TPDS and thereby reducing expenditure on the programme.

High Costs of Targeting

Selecting recipients for a targeted intervention is an extremely expensive exercise (Jha and Srinivasan 2001). Since targeting has to be repeated every few years, with people moving in and out of poverty, the costs incurred in targeting are also of a recurring nature (Khera 2011b). The Right to Food Campaign also highlighted the costly nature of the BPL censuses used in India to target people for the TPDS and other government programmes. Thus, apart from being imperfect, targeting also involves various costs, as discussed below.

Administrative Costs and Opportunity Costs of Investigators

In the context of targeting, administrative costs refer to the costs of collecting information to identify suitable recipients of a programme (Coady et al. 2004). The understanding of administrative costs incurred in targeting should not be restricted to direct costs, such as those pertaining to the training and travel of the investigators and other inputs, but also needs to include the opportunity cost of the investigators' time. Investigators are often teachers or civil servants who perform the task of identifying households at the cost of other important duties (Drèze and Khera 2010). As administrative costs form a part of the total budget of the programme, they are incurred at the expense of the budget available to be distributed to recipients. Administrative costs can often be substantial, although they vary according to the precision and method of targeting (Coady et al. 2004).

Delays

The time taken to complete targeting processes depends on the size of the population from which recipients are to be selected, the precision of targeting, and the method of targeting. However, targeting is usually a lengthy process with frequent delays (Sen 1995), which may lead to delays in the commencement of the programme services. This in turn can results in economic costs for the recipients.

Political Costs

Narrowly targeted programmes have fewer and politically weaker recipients, who lack the voice to ensure the sustainability of the programme and quality of the services provided. This shortcoming of narrow targeting has been used to advocate for universal programmes in wealthier countries such as the US (Sen 1995). Smaller user bases of targeted programmes make it easier for governments to tamper with such programmes (Gelbach and Pritchett 2002; Gutner et al. 1999). For example, during times of recession, narrowly targeted programmes suffer most from budget cuts (Ravallion 1999). When universal programmes get converted into targeted ones, it leads to the termination of many people's access to certain benefits, which can lead to popular opposition and demands for compensation among the groups that get excluded, even if they are not disadvantaged (Subbarao et al. 1995).

Economic Costs to the Potential Recipients

Sometimes, targeting requires the potential programme recipients to obtain certification, such as ration cards, identity cards, proof of residence, disability documentation and others. Obtaining these documents is often time consuming and usually requires the applicants to incur transportation costs to reach various offices, and to pay other costs, for example, for photographs for application forms (Coady et al. 2004). The poorest sections, such as urban homeless persons, forest dwellers, distress migrants and bonded workers, are unlikely to possess these documents, and as a result find themselves excluded from targeted programmes.

Social Costs to Potential Recipients

Targeting, which requires people to be labelled as 'poor', may carry the implication that the poor cannot entirely meet their needs themselves and have to depend on some charity. This can hurt the self-respect of those targeted and have an adverse impact on the respect they receive from others. The stigma may actually cause those selected for a programme to not use the services of that programme, thereby defeating the purpose of targeting and the programme itself (Sen 1995). Countering the argument that stigmatisation is an issue of marginal interest, Amartya Sen quotes John Rawls, who argues that self-respect is probably the most important primary good. Targeting, which requires detailed disclosures about people's standard of living, can lead to the loss of individual privacy and autonomy. The degree of invasion increases with the precision of targeting (ibid.).

The asymmetric power relations between investigators and respondents in the process of identification of the poor has significant social costs. Investigation processes often provide the investigators with opportunities of exercising official authoritarianism and indulging in corruption, since they have significant scope to distribute targeted benefits in the form of favours (Sen 1995).

Incentive Costs

Incentive costs, also called indirect costs, occur when the eligibility criteria of a programme result in households changing their behaviour in an attempt to become beneficiaries. For instance, a programme only open to those below a certain income may cause some households to earn less, or report their earnings as less than they are, to be able to be selected in the programme. Recourse to such adverse incentives can become more serious when non-income indicators of poverty are deployed, such as a person being classified as poor if the family sends no children to school, or as not poor if they possess a sanitary toilet.

Better Performance of Universal PDS Than TPDS

The division of the population into categories of BPL and APL is considered divisive, with the potential of manufacturing class conflict.

This division prevents the emergence of a collective public demand which is essential for a well-functioning PDS (Drèze 2010; Drèze and Khera 2011). Bunsha (2002) notes that since the introduction of TPDS, food entitlements have been cut by more than half, which has consequently had a negative impact on food security. In 2011, a PDS survey was conducted in nine states on the following aspects of the PDS: degree of inclusiveness, purchase–entitlement ratio, quality of PDS grain, access to FPSs (measured in terms of time and distance), reliability (whether days of opening of the FPS were adequate), record keeping (maintenance of ration cards and of the sales and stocks register) and institutional arrangements (doorstep delivery of grain, and whether the FPS had its own physical space). The states with universal PDS (Himachal Pradesh and Tamil Nadu) had the best performance, followed by states with near-universalisation (Andhra Pradesh and Chhattisgarh), in turn followed by states with expanded PDS (Odisha and Rajasthan). The performances of states with TPDS (Uttar Pradesh, Jharkhand and Bihar) was the worst (Khera 2011b).

Another reason for the better performance of universal PDS than TPDS is that the former is easier to administer. Implementing TPDS requires FPSs to sell the same commodities to two different groups of people at substantially different prices, which is significantly more difficult than selling the same quantities to everybody at the same prices (Radhakrishna and Subbarao 1997).

Fewer Leakages in Universal PDS Than TPDS

The High Level Committee on Long-Term Grain Policy, appointed by the government in 2000 at a time of increasing food inflation, grain stocks and levels of hunger, found that targeting had reduced the efficacy of the PDS in meeting its original goal of price stabilisation, while not delivering fully in terms of focusing subsidies on the poor. Hence, in 2002, the committee recommended a return to a universal PDS (GoI 2002). This finding has been corroborated by subsequent data. For instance, the coefficient of variation of real cereal prices increased substantially after targeting, primarily because public grain management after 1997 has been much less successful in reducing the variability of per capita availability below variability in per capita production. The Planning Commission (2008), despite advocating for TPDS, reported that leakages in the PDS doubled after targeting.

A strong reason for moving towards a targeted PDS in 1997 was that the poorer states of the north received less subsidy than the richer states of the south. Yet targeting, instead of improving the geographical utilisation of the food subsidy, has led to much higher leakages in states which were already performing badly before targeting, particularly the poor northern states of Assam, Bihar, Jharkhand and Uttar Pradesh (Himanshu and Sen 2011).

Tamil Nadu, which is the only state that still has a universal PDS, has the largest proportion of population accessing the PDS (76.4 per cent in 2007–08, according to the 64th Round of NSS), and has almost no leakages. While this is predominantly due to better PDS governance in the state, universalisation has also contributed to this performance. Chhattisgarh, which had made its PDS near-universal by April 2007, almost doubled access to the PDS from 22.1 per cent in 2004–05 to 40.2 per cent in 2007–08. During the same time, Chhattisgarh succeeded in bringing down leakages in the PDS from the level of the national average to almost nil. Andhra Pradesh, which has a near-universal PDS with 80 per cent of households having BPL cards, also has low leakages.

Further analysis may be needed to understand exactly how the move towards universalisation reduces leakages, although it is clear that higher participation makes it more difficult to siphon off rations from FPSs. It also puts pressure on the government to carry out governance reforms.

Self-Targeting in Universal PDS Decreases Foodgrain and Subsidy Requirements

A very high proportion of the wealthy do not utilise their PDS entitlements because of the somewhat inferior quality of food rations, as well as other difficulties such as the large distances from FPSs and their infrequent opening. Such opting out reduces the actual offtake as well as expenditure in the PDS (Himanshu and Sen 2011). The extent of voluntary withdrawal from the PDS depends on the price of PDS rations compared to market prices. As the price of PDS rations decreases in comparison with market prices, it becomes more attractive to purchase these rations, and the rate of voluntary exclusion decreases. For instance, in 2007–08, only 76.4 per cent of the population purchased PDS rice at Rs 2 per kilogram in Tamil Nadu. While there was almost a 100 per cent offtake among the bottom deciles, this percentage gradually declined in the upper deciles and became 40 per cent in the top rural decile and almost nil in the top urban decile (ibid.).

Expanded PDS Encourages Revival of Agriculture

The Right to Food Campaign also argued that an expanded PDS had the potential to generate demand for foodgrains, resulting in higher production. This could contribute significantly to the revival of agriculture.

Process and Debates Leading to the Finalisation of PDS Entitlements in the NAC Draft

Draft Prepared by the Department of Food and Public Distribution

An initial draft of the NFSB was prepared by the Department of Food, Consumer Affairs and Public Distribution and cleared by the Empowered Group of Ministers for discussion in the cabinet in 2009. Provisions in this bill were restricted to the PDS, the scope of which was itself restricted to BPL and Antyodaya households. Recipients were to be entitled to 25 kilograms of foodgrains per month at Rs 3 per kilogram. The BPL entitlement proposed in the bill was even lower than that already provided by Supreme Court orders. As a result, following widespread criticism of the minimalistic provisions of the bill, the drafting of the NFSB by the Empowered Group of Ministers was deferred, and the task was taken up by the newly formed NAC.

Universalisation of PDS in One-Fourth of the Most Disadvantaged Districts

The Right to Food Campaign argues that an expanded PDS has the potential to boost the demand for foodgrains. This, in turn, would require higher production of foodgrains, which can significantly contribute to the revival of agriculture.

The idea of universalising the PDS in one-fourth of the poorest districts was subsequently abandoned. This move was condemned by supporters of universal PDS. Drèze (2011b; see also Khera 2011a, among others) argued that universalisation in the poorest 200 districts was an important initiative due to the especially poor administration of these districts. The process of targeting in these areas was likely to take a long time and would probably have large exclusion errors. He went on to argue that it would be redundant to utilise targeting in these areas which had very high proportions of food-insecure populations. Universalising the PDS in these areas would tackle the problem of India's food insecurity to a large extent, at a small extra cost. Drèze suggested that the government should consider universalising the PDS in the 200 poorest districts at least for the first 20 years after the enactment of the NFSA.

NAC's Working Group on Food Security: Proposal of Two Options

In August 2011, the NAC's Working Group on Food Security proposed two options. The first option was near-universalisation, in which 80 per cent of rural households would get Antyodaya entitlements (35 kilograms of foodgrain per household every month at Rs 3 per kilogram of rice, Rs 2 per kilogram of wheat, and Re 1 per kilogram of millet). The second option was universalisation with differential entitlements, in which 42 per cent of rural households were to get Antyodaya entitlements, and all the other households would get 25 kilograms of grain at Rs 5 or Rs 7.5 per kilogram. In both options, 33 per cent of urban households, which would include persons who were homeless, in vulnerable occupations, slum residents, or socially vulnerable, were to get 35 kilograms of foodgrains at Rs 3 per kilogram. These proposals were to be implemented in the first year of enactment of the NFSA in one-fourth of the poorest districts, and were to extend to the rest of the districts within four years.

While the first option did away with the requirement of a BPL census and was easier to administer, it had a higher subsidy bill, less regard for affirmative action, and carried the possible risk of misuse of the exclusion criteria. The second option cost less, had greater affirmative action, but relied on the problematic and discredited BPL census. It was also harder to administer because of the differential entitlements.

In September 2010, the Working Group proposed slight modifications to its second option. It now proposed providing Antyodaya entitlements to 50 per cent of the rural population, non-Antyodaya entitlements (25 kilograms of foodgrain at 50 per cent of the MSP, or Rs 7.5 per kilogram of rice, Rs 5.5 per kilogram of wheat and Rs 2 per kilogram of millet) to the remaining 50 per cent of the rural population, and Antyodaya entitlements to 33 per cent of the urban population. The proposals would have also allowed state governments to expand their Antyodaya coverage and increase the baseline entitlements by adding their own resources. The proposals were to be implemented in the 150 poorest districts in the first year and extended to all other districts by the fourth year of the implementation of the act. Antyodaya coverage was not to be reduced for at least 10 years, and Antyodaya price increases were to be linked to the consumer price index. By the final year, this proposal would have covered about 190 million households, required approximately 71 million tons of foodgrains (including non-PDS entitlements), and cost Rs 84,329 million. The estimate of 71 million tons of foodgrain was on the assumption of 90 per cent offtake (based on actual offtake ratios).

Final Drafts of the NFSB and Their Critique

NAC Draft

In October 2010, the NAC changed the terminology of the 'Antyodaya' category to 'priority' and the 'non-Antyodaya' category to 'general'. In the final NAC draft, which was submitted in July 2011, the coverage of PDS in rural areas was decreased from universal to 90 per cent, in such a manner that not less than 46 per cent of households would come under the 'priority' category and the rest under the 'general' category. In urban areas, coverage of the TPDS was increased from 33 per cent to 50 per cent, but 'priority' entitlements were restricted to a minimum of 28 per cent of total urban households. The remaining urban households under the TPDS were to come under the 'general' category of entitlements.

The entitlements of the 'general' households were decreased from 25 kilograms a month to 20 kilograms a month. Following the suggestion of the government that entitlements should be per capita instead of per household, the NAC equated the monthly entitlement of at least 35 kilograms of foodgrain for every 'priority' household to at least 7 kilograms for every person at prices not exceeding Rs 3 per kilogram for rice, Rs 2 per kilogram of wheat and Re 1 per kilogram for millets. Single-member 'priority' households were to receive a minimum of 14 kilograms of grain per month at the same prices. For the 'general' category, the monthly entitlement of least 20 kilograms of foodgrain for each household now became a minimum of 4 kilograms for every person at prices not exceeding 50 per cent of the MSP for rice, wheat and millets. Single-member general households were to receive a minimum of 8 kilograms of grain per month at the same prices.

Many have commented (see Drèze 2011a; see also Himanshu and Sen 2011; Khera 2011a) that the final NAC draft goes against the initial resolve of delinking the NFSA from the dysfunctional poverty lines and putting an end to targeting. The draft is seen to have been heavily diluted by the government's own concerns and priorities.

Draft Introduced in the Lok Sabha in December 2011

The NFSB draft introduced in the Lok Sabha gave monthly entitlements of *exactly* 7 kilograms to every person in the 'priority' category but *not less than* 3 kilograms to every person in the 'general' category. This draft also specified that TPDS entitlements would extend to *up to* 75 per cent of the rural population and *up to* 50 per cent of the urban population, of which *not less than* 46 per cent of rural households and *not less than* 28 per cent of urban households would get priority entitlements. It also mentioned that the government could provide wheat flour in lieu of foodgrains.

Critique of the Targeted Approach in the Drafts of the NFSB and NFSA

Supporters of universal PDS opposed the targeted approach of all the drafts of the NFSB. Apart from reiterating the general arguments in support of universal PDS (discussed earlier), they also argued that providing subsidised foodgrains to only some sections of the population went against the mandate of the NFSA, which was to provide a *universal* guarantee of adequate nutrition. The right to food, which is implicit in the fundamental right to life under Article 21 (interpreted by the Supreme Court as a right to live with dignity), should be realised for all. This understanding was also reflected in the president's speech to the Indian Parliament in 2009: 'My Government proposes to enact a new law—the National Food Security Act—that will provide a statutory basis for a framework which assures food security for all.'[7]

The Right to Food Campaign argued that the directive principles of state policy in the Constitution should never be compromised or undermined; instead, they need to be realised, strengthened and taken forward for all. The campaign went on to argue that food for all, health for all, education for all and work for all should be taken as the bottom line, and basic rights like those to food, education, social security and healthcare must be guaranteed by governments to all citizens as universal state-supplied and common entitlements. The campaign insisted that the NFSA should under no circumstances dilute these principles, and in no way should it be used to exclude people from their basic rights and needs. It should be the choice of citizens whether or not they wished to use these public services, but none should be excluded by law (Mander 2011).

The campaign also argued that by imposing the rigid 'priority' and 'general' categories on the state governments, the NFSB would thwart the efforts of various state governments to improve their PDS by moving away from BPL targeting and adopting a more inclusive approach with wider coverage. This restriction imposed in the bill also received opposition from state governments, especially those that had expanded or universalised the PDS (Sethi 2011). Strong arguments for universal or near-universal PDS were voiced by members of Parliament

(MPs) from Tamil Nadu and Chhattisgarh as well as by parties from the Third Front, the far left alliance.

The 2011 draft of the NFSB stated that not less than 46 per cent of rural households and not less than 28 per cent of urban households were to be provided 'priority' entitlements. However, as these proportions correspond to the poverty estimates of the Tendulkar Committee, with 10 per cent extra to account for targeting errors, the coverage of 'priority' households seems closely linked with the poverty line. As the government reduces the poverty estimates over time, linking the coverage of 'priority' households with the poverty line will lead to a reduction in the coverage of 'priority' entitlements over time.

In March 2012, a group of 36 development economists wrote an open letter to the prime minister suggesting certain modifications to the government draft of the NFSB (*The Hindu* 2012a). The economists argued that the division of the population into 'priority', 'general' and 'excluded' categories was not only artificial, but without any clarity. They pointed to the failures of the earlier BPL censuses in correctly identifying poor households, and the inability of the Socio-economic Caste Census (SECC) to perform any better, ultimately criticising the government draft (and by implication the NAC draft) for lacking simplicity and transparency.

The group instead suggested abolishing the distinction between general and priority households and providing the same PDS entitlements to all households, except those falling under the excluded category. The group argued that this proposal would do away with the need for a complicated identification process, except for employing the exclusion criteria which were relatively straightforward. They suggested that this framework would be simpler, more transparent, easier to understand and implement, and would better protect poor households from exclusion errors. The proposed common entitlement was a minimum of 25 kilograms of foodgrains at Rs 3 per kilogram of rice, Rs 2 per kilogram of wheat and Re 1 per kilogram of millets.[8] The group also recommended that the poorest households should continue to receive greater support under the Antyodaya scheme. The total foodgrain requirement for this proposal, 51.5 million tons, was marginally lower than that of the government draft at 52.4 million tons. However, as this proposal entitled everybody to foodgrains at 'priority' prices, its food subsidy, at Rs 81,524 million, was 4.6 per cent higher than the government draft at Rs 77,927 million (*The Hindu* 2012b).

Following this letter, the Standing Committee for the NFSB draft, taking the advice of the development economists, suggested folding together the 'priority' and 'general' categories, as well as vulnerable groups, into an 'included' category, besides making clear the criteria for exclusion. This was not implemented, however, and instead the Antyodaya classification replaced the 'priority' category, which in itself replaced the 'general' category. Thus, Antyodaya households were to receive 35 kilograms per household per month at Rs 3 per kilogram of rice, Rs 2 per kilogram of wheat and Re 1 per kilogram of millets; and 'priority' households would receive 5 kilograms per person per month at Rs 3 per kilogram of rice, Rs 2 per kilogram of wheat and Re 1 per kilogram of millets.

The final act rejected a universal PDS, keeping in place the coverage of 75 per cent of the rural population and 50 per cent of the urban population. Furthermore, it removed certain recommendations for specific targeting by omitting the provisions for vulnerable groups such as the homeless and destitute, leaving the NFSA without a universal PDS as well as with reduced affirmative action.

Targeting in Other Food Schemes

Integrated Child Development Services (ICDS)

The ICDS addresses the needs of children under the age of 6 along with the nutritional needs of adolescent girls, pregnant women and lactating mothers. It seeks to provide young children with an integrated package of services comprising supplementary nutrition, healthcare and pre-school education. As the health and nutrition needs of a child cannot be addressed in isolation from those of the mother, the programme also extends to adolescent girls, pregnant women and nursing mothers (Right to Food Campaign 2007).

The Government of India started the ICDS as a project in 33 blocks in 1975, and expanded it to almost 6,500 blocks by 2004 (Sinha 2006). On 28 November 2001, the Supreme Court passed an order which gave ICDS services the status of 'legal entitlements'. In the same order, the court also directed the government to 'universalise' the programme. This meant that every hamlet should have a functional anganwadi, and that the coverage of ICDS should be extended to all children under 6, pregnant women, lactating mothers and adolescent girls (Right to Food Campaign 2007).

Although the complete universalisation of ICDS is yet to happen, there is little dispute over the desirability of a universal ICDS, with an expectation that those who can afford to will not utilise the ICDS services, but instead use private services.[9] Despite the Supreme Court order on the universalisation of ICDS, not all of the services of the programme have been made legal entitlements in the various drafts of the NFSB, nor in the final act. The NAC draft accords the status of legal entitlements to all ICDS services for children and supplementary nutrition services for pregnant women and lactating mothers, but keeps services for adolescent girls in the section on 'progressive realisation'. The draft introduced in Parliament in December 2011 also only gave legal entitlements to supplementary nutrition for children, pregnant women and lactating mothers, and not for other ICDS services. The final act similarly gives legal entitlement only to supplementary nutrition for children, pregnant women and lactating mothers, with other ICDS services categorised under the non-binding section on 'Provisions for Advancing Food Security'.

Midday Meals Scheme

The MDMS entitles all children in the primary and upper primary classes of government and government-aided schools to free midday meals.[10] As in the case of the ICDS, the Supreme Court order of 28 November 2001 not only universalised MDMS in all government and government-aided schools, but also made it a legal entitlement. This entitlement has been retained in both drafts of the NFSB as well as in the final act. However, some are of the view that the MDMS should extend to private schools as well. A public interest litigation was filed by the Uttar Pradesh Basic Shikshak Sangh (an educational body) in April 2012 seeking the Supreme Court's direction to the central and state governments to provide midday meals (along with free textbooks, school uniforms and scholarships) to children in the age group of 6 to 14 studying in non-government schools. The petitioners argued that not providing children in private schools these facilities was a violation of Article 14 (equality) and of Article 15 (prohibition of discrimination) (*Deccan Herald* 2012).

Maternity Benefits

Maternity benefits are a form of cash assistance provided to women during pregnancy and/or lactation to assist them in meeting their nutrition needs. Currently in India there are two maternity benefits schemes—the National Maternity Benefit Scheme (NMBS) for home deliveries and the Janani Suraksha Yojana (JSY) for institutional deliveries. However, both these schemes are largely confined to women belonging to BPL households.[11] Both drafts of the NFSB entitled women to monthly maternity benefits for six months after delivery, finalised in the act as Rs 6,000 to be paid in instalments, as decided by the central government. However, women who are regularly employed with the government, public sector undertakings, state public sector undertakings, or those receiving similar benefits under any other law are to be excluded from the maternity benefits entitled under the NFSB draft and in the act. The Right to Food Campaign is of the opinion that these benefits should be universalised and provided to all women in maternity (Right to Food Campaign 2012). Like the PDS, other food schemes (ICDS, MDMS and maternity benefits) also need to be universal to ensure that people are able to access their food entitlements as a matter of right.

Notes

1. The term *antyodaya* is based on Gandhian idealism, and means 'development of the last person'.

2. Benefit is measured as change in consumer and producer surplus, and costs refer to government costs.

3. (US) Executive Order 11246: Equal Employment Opportunity, Federal Register, https://www.archives.gov/federal-register/codification/executive-order/11246.html (accessed 31 August 2017).

4. See 'Affirmative Action', http://en.wikipedia.org/wiki/Affirmative_action#cite_note-The_Federal_Register-0 (accessed 31 August 2017).

5. Those below the official poverty line as per the Tendulkar Committee constituted by the Planning Commission in 2005. This committee broadened

the aspects of well-being considered while determining the poverty line to include not just nutrition but also education, health, electricity, transport, etc.

6. NSS 61st Round, 2004–05, excerpted from the NAC's note on 'Framework for the National Food Security Bill'. The National Sample Surveys are large-scale multi-round sample surveys conducted across India by the National Sample Survey Office. They comprise primary data collection on various socio-economic indicators through households surveys. The NFHS is also a large-scale, multi-round survey conducted across the country to gather data on health indicators. The India Human Development Survey is a nationally representative multi-topic survey of 41,554 households in 1,503 villages and 971 urban neighbourhoods across India.

7. Address by President of India Smt. Pratibha Devisingh Patil to the Joint Session of the Fifteenth Lok Sabha, New Delhi, 4 June 2009.

8. In per capita terms, each person would be entitled to a minimum of 5 kilograms of foodgrains every month.

9. Harsh Mander, New Delhi, private communication.

10. Ministry of Human Resource Development, Department of School Education and Literacy, Government of India, 'Mid Day Meals Scheme', http://mdm.nic.in/ (accessed 1 September 2017).

11. Only in low-performing states (Assam, Bihar, Chhattisgarh, Jammu and Kashmir, Jharkhand, Madhya Pradesh, Odisha, Rajasthan, Uttar Pradesh and Uttaranchal), which have low in-facility birth coverage, are all women, irrespective of BPL status, entitled to the cash benefit under JSY.

References

Anon. 2009. 'Identifying the Poor'. *Economic and Political Weekly*, 44(29): 5–6.

Baker, J., and M. Grosh. 1994. 'Poverty Reduction through Geographic Targeting: How Well Does It Work?' *World Development*, 22(7): 983–95.

Bunsha, Dionne. 2002. 'The Human Face of Adjustment'. *Frontline*, 19(1): 5–18.

Coady, D., M. Grosh and J. Hoddinott. 2004. *Targeting of Transfers in Developing Countries: Review of Lessons and Experience*. Washington, D.C.: World Bank.

Cornia, G., and F. Stewart. 1995. 'Two Errors of Targeting', in D. van de Walle and K. Nead (eds), *Public Spending and the Poor*. Baltimore: Johns Hopkins University Press, pp. 350–86.

Datt, G., and D. Gunewardena. 1997. 'Some Aspects of Poverty in Sri Lanka: 1985–90'. Policy Research Working Paper no. WPS 1738, World Bank, Washington, D.C. http://documents.worldbank.org/curated/

en/826881468777305070/Some-aspects-of-poverty-in-Sri-Lanka-1985-90 (accessed 9 March 2018).

Deccan Herald. 2012. 'PIL seeks free meals, books for pvt students', 16 April. http://www.deccanherald.com/content/242625/pil-seeks-free-meals-books.html (accessed 1 September 2017).

Dev, S. Mahendra. 1995. 'Economic Reforms and the Rural Poor'. *Economic and Political Weekly*, 30(33): 2085–88.

Drèze, Jean. 2010. 'The task of making the PDS work'. *The Hindu*, 8 July.

———. 2011a. 'Mending the food security act'. *The Hindu*, 24 May.

———. 2011b. 'The poverty trap'. *Hindustan Times*, 23 September.

Drèze, Jean, and Reetika Khera. 2010. 'The BPL Census and a Possible Alternative'. *Economic and Political Weekly*, 45(9): 54–63.

———. 2011. 'Understanding the PDS'. *Frontline*, 28(27).

FAO. 2009a. *Guide on Legislating for the Right to Food*. Rome: FAO.

———. 2009b. *Methods to Monitor the Human Right to Adequate Food*, vol. I. Rome: FAO.

———. 2009c. *Right to Food Curriculum Outline*. Rome: FAO. http://www.fao.org/fileadmin/templates/righttofood/documents/RTF_publications/EN/4_toolbox_Curriculum_outline.pdf (accessed 31 August 2017).

Gaiha, R., K. Imai and P. Kaushik. 2001. 'On the Targeting and Cost-Effectiveness of Anti-poverty Programmes in Rural India'. *Development and Change*, 32(2): 309–42.

Gelbach, J., and L. Pritchett. 2002. 'Is More for the Poor Less for the Poor? The Politics of Means-Tested Targeting'. *Topics in Economic Analysis and Policy*, 2(1): 1–28.

George, P. S. 1994. 'Food Security in South Asia: Performance and Prospects'. *Economic and Political Weekly*, 29(18): 1092–94.

GoI (Government of India). 2002. 'Report of the High Level Committee on Long-Term Grain Policy'. Department of Food and Public Distribution, Ministry of Consumer Affairs, Food and Public Distribution, Government of India, New Delhi.

———. 2009. *Report of the Expert Group to Advise the Ministry of Rural Development on the Methodology for Conducting the Below Poverty Line (BPL) Census for Eleventh Five-Year Plan*. New Delhi: Ministry of Rural Development.

Gutner, T., S. Gomaa and S. Nasser. 1999. *The Political Economy of Food Subsidy Reform in Egypt*. Washington, D.C.: International Food Policy Research Institute.

Himanshu and Abhijit Sen. 2011. 'Why Not a Universal Food Security Legislation?' *Economic and Political Weekly*, 46(12): 38–47.

The Hindu. 2012a. 'A simple proposal on food security', 12 March. http://www.thehindu.com/opinion/op-ed/A-simple-proposal-on-food-security/article13072478.ece (accessed 9 March 2018).

———. 2012b. 'Simplifying the National Food Security Bill', 12 March. http://www.thehindu.com/migration_catalog/article13056807.ece/BINARY/Simplifying%20the%20NFSB (accessed 9 March 2018).

Hirway, Indira. 2003. 'Identification of BPL Households for Poverty Alleviation Programmes'. *Economic and Political Weekly*, 38(45): 4803–8.

ILO (International Labour Organization). 2001. *Social Security: A New Consensus*. Geneva: ILO. http://www.ilo.org/wcmsp5/groups/public/---ed_protect/---soc_sec/documents/publication/wcms_209311.pdf (accessed 9 March 2018).

Jha, Shikha, and P. V. Srinivasan. 2001. 'Taking the PDS to the Poor: Directions for Further Reform'. *Economic and Political Weekly*, 36(39): 3779–86.

Karat, Brinda. 2006. 'The PDS and eroding food security'. *The Hindu*, 6 June.

Khera, Reetika. 2011a. 'Going against the grain'. *Hindustan Times*, 26 January.

———. 2011b. 'Revival of the Public Distribution System: Evidence and Explanations'. *Economic and Political Weekly*, 46(44): 36–50.

———. 2011c. 'India's Public Distribution System: Utilisation and Impact'. *Journal of Development Studies*, 47(7): 1038–60.

Kochar, Anil. 2005. 'Can Targeted Food Programs Improve Nutrition? An Empirical Analysis of India's Public Distribution System'. *Economic Development and Cultural Change*, 54(1): 203–35.

Krishna, A. 2007. 'For Reducing Poverty Faster: Target Reasons before People'. *World Development*, 35(11): 1947–60.

Mander, Harsh. 2011. 'Ending Indifference: A Law to Exile Hunger?' *Economic and Political Weekly*, 46(25): 45–50.

Mohanty, Amarnath. 2007. 'Affirmative Action in India: An Alternative Perspective'. *Economic and Political Weekly*, 42(30): 3151–57.

Nawani, N. P. 1994. 'Indian Experience on Household Food and Nutrition Security'. Regional Expert Consultation, FAO-UN, Bangkok.

NSSO (National Sample Survey Organisation). 2007. 'Nutritional Intake in India 2004–2005'. Ministry of Statistics and Programme Implementation, Government of India. http://mospi.nic.in/sites/default/files/publication_reports/513_final.pdf (accessed 9 March 2018).

Parikh, Kirit. 1994. 'Who Gets How Much from PDS: How Effectively Does It Reach the Poor?' *Sarvekshana*, 17(3).

Planning Commission. 2007. *Eleventh Five Year Plan 2007–2012*, vol. 2, chapter 4. http://planningcommission.nic.in/plans/planrel/fiveyr/11th/11_v2/11th_vol2.pdf (accessed 2 September 2017).

Planning Commission. 2008. *Eleventh Five Year Plan (2007–12)*, vol. 2: *Social Sector*. New Delhi: Oxford University Press.

Radhakrishna. R., and K. Subbarao. 1997. 'India's Public Distribution System: A National and International Perspective'. World Bank Discussion Paper no. 380, Washington, D.C.

Ravallion, M. 1999. 'Monitoring Targeting Performance When Decentralized Allocations to the Poor are Unobserved'. World Bank Policy Research Working Paper no. 2080, Washington, D.C.

Ravallion, M., and G. Datt. 1995. 'Is Targeting through a Work Requirement Efficient?', in D. van de Walle, and K. Nead (eds), *Public Spending and the Poor*. Baltimore: Johns Hopkins University Press, pp. 413–44.

Right to Food Campaign. 2007. 'Anganwadis for All: A Primer'. Right to Food Campaign Secretariat, New Delhi.

———. 2012. 'Right to Food Campaign's Critique of the National Food Security Bill, Part II: Clause by Clause Critique of the National Food Security Bill, 2011'. http://www.righttofoodindia.org/data/right_to_food_act_data/events/March_2012_clause_by_clause_critique_final_18_february_2012.pdf (accessed 28 August 2017).

Sen, Amartya. 1995. 'The Political Economy of Targeting', in D. van de Walle and K. Nead (eds), *Public Spending and the Poor*. Baltimore: Johns Hopkins University Press, pp. 11–24.

Sen, Pronab. 2005. 'Of Calories and Things: Reflections on Nutritional Norms, Poverty Lines and Consumption Behaviour in India'. *Economic and Political Weekly*, 40(43): 4611–18.

Sethi, Nitin. 2011. 'Food security act to curb states' largesse'. *The Times of India*, 23 December.

Sharma, Alakh Narain, K. P. Kannan and S. Mahendra Dev. 2000. 'Concerns on Food Security'. *Economic and Political Weekly*, 35(45): 3919–22.

Sinha, Dipa. 2006. 'Rethinking ICDS: A Rights Based Perspective'. *Economic and Political Weekly*, 41(34): 3689–94.

Subbarao, K., J. Braithwaite and J. Jalan. 1995. 'Protecting the Poor during Adjustment and Transitions'. Human Capital Development and Operations Policy Working Paper no. 58. Background paper for the World Bank contribution to the Social Summit 'Advancing Social Development', World Bank, Washington, D.C.

Swaminathan, M. 2001. 'A Further Attack on the PDS'. *Frontline*, 18(2).

5

Selection for Targeted Entitlements

Ankita Aggarwal

The last chapter discussed the debate on whether the PDS should be universal or targeted. Although there are compelling reasons to universalise the PDS, the NFSA restricts the PDS entitlements to 75 per cent of the rural population and 50 per cent of the urban population. This leads to the necessity of identifying those households that are the most food insecure. Furthermore, as there are differential entitlements for the Antyodaya and 'priority' categories, there is a need to identify the more food insecure who would be eligible for the greater entitlements of the Antyodaya group, and the less food insecure for the 'priority' category entitlements.

In India, apart from the TPDS, various other central government programmes, such as the Indira Awaas Yojana,[1] pensions for old people, widows and persons with disabilities, health insurance, free medicines in public hospitals, and state government schemes have targeted beneficiaries. As it is practically impossible to separately select eligible households for each of these schemes, a common pool of poor households, known as BPL households, are identified for the purpose of providing social assistance.

Since 1992, poor people in rural areas have been periodically identified at intervals of at least five years through BPL censuses conducted by state governments with financial and technical assistance from the Ministry of Rural Development. Interestingly, there had been no similar censuses in urban areas until the 2011–12 BPL census, renamed the Socio-economic Caste Census. This was the first BPL census to identify poor households in urban areas. Due to this and other issues, the ability of BPL censuses to accurately identify all the poor households in the country has been a subject of much debate.

The Government of India has linked the exercise of identifying the poor with the task of estimating the number of poor. First a poverty line is decided, which is usually expressed in terms of per capita expenditure required to meet pre-specified calorie norms. Himanshu (2010) defines the poverty line as 'the monetary value of some normative expenditure that is essential for satisfying the basic needs'. However, as is discussed in a subsequent section, there is little consensus on where this poverty line should be drawn. The poverty line is fixed and the number of people living below it is estimated by the Planning Commission based on data on consumer expenditure collected in the thick rounds of the NSS, conducted by the NSSO at approximate intervals of five years. The state governments then ultimately identify poor households based on socio-economic indicators. This leads to a discrepancy in the number of poor households estimated by the Planning Commission on the one hand, and the number of poor households identified by state governments on the other. To deal with these discrepancies, the central government instructs the state governments to limit the number of poor identified by them to the number of poor estimated by the Planning Commission. This policy has received severe criticism from not only the state governments, as the numbers of poor identified by the states are invariably higher than the number estimated by the Planning Commission, but also from many policymakers, academics and members of civil society. Hence, the Planning Commission estimates are used as 'caps' on the number of BPL beneficiaries, and therefore on the central government's liability for meeting commitments, rather than as an actual validation of the BPL census estimates.

Targeting for Social Assistance Programmes

Targeting poor people for social assistance programmes is usually done by one of the following four methods—means testing, geographical targeting, community-based targeting and self-targeting. The choice of targeting method depends on the quality of information available about the poor, the level of geographical heterogeneity, available budget, and political viability.

Means Testing

Means testing, which includes methods such as verified means testing, simple means testing and proxy means testing, is a commonly used method of targeting. The verified means test is internationally the most common method of income-based targeting in which details of a person's income and/or wealth are collected and verified by sources such as pay stubs or income and property tax records. The ability to verify income is dependent on the availability of verifiable records and finances to collect and process the information; as both these requirements are usually not available in developing countries, verified means tests are rarely used in these countries. Simple means testing is utilised as an alternative as it does not involve independent verification (Coady et al. 2004). The revelation of personal circumstances in means testing can be a significantly humiliating experience for the respondents (Sen 1995), with another disadvantage being the high costs of collecting the required information (Krishna 2007).

Sen (1995) challenges the parameters of measurability and relevance, the alleged merits of income upheld by proponents of income-based targeting. He argues that accurate information about income is often hard to obtain and can be easily manipulated. Income may often be more difficult to observe than other characteristics of deprivation, such as morbidity, disability, undernourishment and illiteracy. With regard to relevance, as income is a means rather than an end, it may not have a direct relationship with people's capabilities. The shortcomings of income as an indicator of poverty also arise from the fact that poverty is not only low income but also 'the lack of real opportunities to have minimally adequate lives' (ibid., 17).

In order to avoid the shortcomings of income-based targeting, proxy means testing generates a score for the applicant individuals or households based on easy-to-observe household characteristics, such as the location and quality of the house, ownership of durable assets, demographic structure of the household, literacy levels of the household members, occupation of household members, etc. (Coady et al. 2004).

Geographical Targeting

Geographical targeting is the process of identifying the regions that have a higher incidence of poverty due to their inherent constraints and problems, and subsequently targeting everybody in those regions for poverty reduction programmes. Such regions usually suffer from depletion and degradation of natural resources, poor physical and economic infrastructure and/or poor human development (Hirway 2003). Geographical targeting is more suitable when poverty in the country is concentrated in certain areas, but is less useful in many parts of the world where high levels of income diversity exist within regions and even within communities (Bardhan and Mookherjee 1999; Coady and Harris 2004; Elbers et al. 2004; Nhate and Simler 2003). Geographical targeting is thus more appropriate when geography itself contributes to poverty, and migration is not a viable option (Ravallion and Wodon 1999), as well as when ethnic, historical and locational disadvantages overlap, as in some places in Latin America (Schady 2002). Geographical targeting is much more economical than means testing (Hirway 2003).

Poverty mapping based on small area estimations can help improve coverage and reduce leakages by lowering the population of targeted units (Bigman and Srinivasan 2002; Elbers et al. 2003). In the Indian context, geographical targeting has been resorted to for programmes intended to benefit tribal communities and minorities by targeting the areas in which they constitute significant numbers to provide additional funds.

Community-based Targeting

Community-based targeting makes use of the more accurate knowledge of poverty in a region among the community members (Esman and Uphoff 1984; Uphoff et al. 1998), although this method suffers from the danger of community knowledge not being utilised appropriately. For example, inequality within a community may actually worsen if local elites capture the processes of decision making and benefit allocation (Abraham and Platteau 2002; Conning and Kevane 2002; Galasso and Ravallion 2002; Pender and Ruben 2004). Inequalities in the exercise of power would always remain, but they can be decreased if the investment choices are publically justified on the basis of transparent analysis and do not remain only an exercise of arbitrary power (Krishna 2007). Radhakrishna and Subbarao (1997) have suggested community-based targeting for the identification of the poor in India, recommending that the panchayati raj institutions be made responsible for identifying the poor.

Self-targeting

Self-targeting is a method of restricting the participation of the non-poor in universal programmes. This is done by designing the programme in such a manner that it is more attractive to the poor than to those who are not poor (Krishna 2007). For instance, food provisioning programmes often provide commodities which are not preferred by the wealthy, such as coarser varieties of cereals. Self-selection not only erodes the self-esteem of the programme recipients less, it also does not involve administrative costs and delays (Sen 1995). Subbarao et al. (1995) suggest that countries which lack the resources required for identifying the poor should try and achieve self-targeting in ration schemes by selling only those commodities that are consumed more by the poor at subsidised prices. If even this is not possible, nutrition and health programmes in areas of high mortality should be restricted to children and women.

In many circumstances, a combination of targeting methods has resulted in more accurate identification of the poor. For instance, a combination of geographical and community-based targeting has been suggested, especially for communities where poorer sections are better organised (UNDP 2000). A wide range of studies conclude that no method of targeting is perfect and devoid of errors of exclusion and inclusion. However, the success of targeting and the rate of errors also depend to a large extent on the implementation of the targeting techniques. A strong and transparent implementation framework can contribute to reducing the possibility of tampering with and wilfully fudging the targeting process, which goes a long way in reducing the errors. An honest implementation of the targeting process requires committed people in the local bureaucracy, such as panchayats in the Indian context (Himanshu 2011a).

Brief Overview of Poverty Estimation in India

The Planning Commission of India was the official agency of the Government of India which made estimates of the percentage of the country's population below the poverty line in individual states and the country as a whole.[2] Such estimation is done on the basis of the large sample surveys of consumer expenditure carried out by the NSSO at approximate intervals of five years.

For the purpose of fixing poverty lines, the Planning Commission constituted various committees in the past. The first such committee was the Alagh Committee, which submitted its report in 1979, followed by the Lakdawala Committee which submitted its report in 1993. The Alagh Committee fixed the poverty line in terms of the value of a consumption basket that provided the desired number of calories per capita per day for rural and urban areas respectively at 1973–74 prices. In subsequent years, this poverty line has been indexed upwards to incorporate the impact of inflation in accordance with the recommendations of the Lakdawala Committee. Based on 1993–94 data, the poverty ratio was 37.3 per cent for rural areas and 32.4 per cent for urban areas, yielding a combined estimate of 36 per cent for the country as a whole.

When the Lakdawala method of poverty estimation was applied to the NSSO data of 2004–05, a national poverty ratio of 27.5 per cent was obtained. However, by the time these estimates became available, the methodology for estimating poverty was already under review, and the Planning Commission appointed another committee in 2005, this time under the chairpersonship of Professor Suresh Tendulkar, to re-examine the appropriateness of the poverty line. The Tendulkar Committee submitted its recommendations in 2009, concluding that while the urban poverty line was reasonable, the rural poverty line needed to be indexed upwards to take inflation into account; it also delinked the poverty line from any specific calorie norm (Himanshu 2011b). Hence, the Tendulkar Committee revised the poverty line derived from the NSSO data of 2004–05 from 27.5 per cent (28.3 per cent in rural areas) to 37.2 per cent (41.8 per cent for rural areas and 25.7 per cent for urban areas). The poverty line emerging from the Tendulkar methodology is a consumption of Rs 447 per capita per month (Rs 18 per day) for rural areas and Rs 579 per capita per month (Rs 20 per day) for urban areas at 2004–05 prices. These translate into 1,999 kcal per capita per day for the rural BPL (as opposed to the norm of 2,400 kcal suggested by the Indian Council of Medical Research [ICMR]), and 1,776 kcal per capita per day for the urban BPL (as opposed to the ICMR norm of 2,100 kcal). The poverty estimates of the Tendulkar Committee have come under heavy criticism from many quarters for being abysmally low (see, for example, Desai 2012; Drèze 2011a; Manna 2012).

Apart from the Tendulkar Committee, committees under the chairmanships of Arjun Sengupta and N. C. Saxena were set up to determine the proportion of the poor in India, as per the 2004–05 NSSO data. The Sengupta Committee estimated the country's poverty at 77 per cent. This estimate is based on a poverty line of Rs 20 per capita per day. Himanshu (2010) argues that the rationale for choosing this figure has not been explained, and that the reason why the poverty estimate by the Sengupta Committee is so high is because, unlike the Tendulkar Committee, it used the same figure of Rs 20 across all states and for rural as well as urban areas, without making adjustments for varying prices across states and across rural and urban areas. Not making this adjustment results in a person spending the same amount of money in real terms being treated as poor in urban areas, but not in rural areas.

The Saxena Committee concluded that the proportion of India's population considered BPL should be at least 50 per cent, also noting that if the poverty line is fixed at the calorie norm of 2,400 kcal, about 80 per cent of the country's population would be considered BPL (GoI 2009). Himanshu (2010) criticises the Saxena Committee's proposal of 50 per cent for not corresponding to a poverty line and for lacking any justification.

As per the World Bank's international poverty line of an income of $1.25 per day (adjusted for each country's purchasing power parity), 41.6 per cent of India's population is poor, based on 2004–05 NSSO data (GoI 2009).

On 14 May 2011, the Supreme Court ordered the Planning Commission to revise its poverty estimates based on the Tendulkar Committee. This order came as the result of a petition filed by the PUCL in the right to food case (Mahapatra 2011a). In an affidavit to the Supreme Court on 20 September 2011, the Planning Commission provisionally placed the poverty line at Rs 965 per capita per month (Rs 32 per capita per day) for urban areas and Rs 781 per capita per month (Rs 26 per capita per day) for rural areas. These figures were arrived at by applying the price level in June 2011. In the affidavit, the Planning Commission made the claim that the revised poverty line was adequate for meeting the basic needs of food, education, health, etc. (Mahapatra 2011c), also adding that the BPL list was expected to be reduced over time according to the official poverty estimates based on the government's poverty line (Drèze 2011a).

The Planning Commission's revised poverty line received large-scale criticism, which forced the commission and the Rural Development Ministry to take the decision of not using the current poverty line to decide the recipients of food entitlements (and other government entitlements), but instead to wait until the completion of SECC 2011 to decide on the cut-off limit (Mahapatra 2011b). In 2011, the Abhijit Sen Committee was set up to consider options on how the deprivation criteria in SECC 2011 would be used to determine the cap on recipients of government schemes (Balchand 2012), although the Right to Food Campaign noted that this committee did not have representatives of vulnerable groups such as single women and persons with disabilities.

In March 2012, the Planning Commission released the new poverty line of Rs 28.65 for urban areas and Rs 22.43 for rural areas, based on 2009–10 data. It also declared that the all-India poverty ratio had declined from 37.2 per cent in 2004–05 to 29.8 per cent in 2009–10, with rural poverty declining from 41.8 per cent to 33.8 per cent and urban poverty declining from 25.7 per cent to 20.9 per cent (GoI 2012). The extent of poverty reduction declared by the Planning Commission was challenged by many, with Himanshu (2012) noting that the commission had manipulated the consumption data of 2009–10 used for estimating poverty, undercounting the poor by 18 million people.

While the Parliament's Standing Committee on Food, Consumer Affairs and Public Distribution recommended using SECC 2011 to determine the coverage in the NFSA, it could not agree upon any concrete census to be used for the eligibility criteria. The Standing Committee also recommended that the central government coordinate with the state governments in identifying households to ensure that deserving persons were not excluded, as well as to include clearly defined exclusion criteria. Despite these recommendations, the NFSA merely states that population estimates as per the census shall determine the total number of persons to be covered, and that state governments will specify guidelines for identification. Moreover, state governments had to identify eligible households within 180 days of the passing of the act. The lack of clear census data at the state level, as well as the stringent deadline for identification, spurred criticism from members of Parliament cutting across state and party lines, who wanted significant changes in eligibility criteria. Despite the vocal criticism, no amendments were passed.

Methodologies and Critiques of Previous BPL Censuses

BPL Survey 1992

Methodology

In 1992, a detailed procedure was prescribed by the Ministry of Rural Development to identify BPL families in rural areas for the Eighth Five Year Plan. The 1992 survey used income as the criterion, and guidelines were issued to assess annual household incomes. All households with an annual income of less than Rs 11,000 were considered poor.

Critique

One problem with using household income as the indicator was that it did not take into account household size. It is very difficult to accurately estimate the income of households which are in the informal sector, more so when they are casually employed. The tendency to under-report incomes also contributed to inclusion errors (GoI 2009). The number of households identified as poor in this survey (52.49 per cent) was almost double the poverty estimate of the Planning Commission and was considered highly exaggerated by the government (ibid.). Hence, the Ministry of Rural Development set up an expert group to guide the process of another BPL survey.

BPL Census 1997

Methodology

To avoid the pitfalls of determining poverty on the basis of income, as was done in the first BPL survey of 1992, the expert group constituted for the BPL census of 1997 used an 'expenditure' approach involving 'exclusion criteria'. The schedule for the BPL census had two parts.

Part A covered all rural households and aimed to assess the status of households quickly with the following information:

1. Size of landholding (whether it was more than 2 hectares)
2. Ownership of a *pucca* house[3] as defined in the population census
3. Receipt of annual income of Rs 20,000 or more
4. Ownership of consumer durables like television, refrigerator, ceiling fan, motorcycle/scooter and three-wheelers
5. Ownership of farm equipment such as tractor, power tiller or combined thresher/harvesters

Any household which gave an affirmative answer to any of the categories listed above was classified as 'visibly non-poor' and excluded from the BPL list. For the remaining households, data was collected on the total consumption of the household, both purchased and home grown. All households whose per capita consumption was equal to or less than the Planning Commission's state-level poverty estimates were classified as poor (GoI 2009). The census also collected information on age, gender, the educational status of households, social group affiliation, housing conditions, skills training and so on, to identify BPL households (Mehrotra and Mander 2009).

Critique

Although this was an improvement on the income-based methodology, the adoption of uniform criteria without allowing for inter-state variations, particularly in respect of hilly and remote areas, contributed to several exclusion errors. The exclusion criteria of possessing a ceiling fan and 2 hectares of land (which is quite unproductive in a few regions of the country) were considered overly stringent (GoI 2009).

BPL Census 2002

Methodology

Another expert group[4] was set up in 2001 for the identification of BPL households (Hirway 2003), which recommended a 'score-based ranking' based on proxy means testing which indicated the quality of life of each household. For every household, a score of 1, 2, 3 or 4 was assigned to each of the 13 scorable economic or social indicators (Table 5.1). By simple aggregation, the total score for each household

Table 5.1. Indicators and scoring scheme for identification and subcategorisation of the poor in the 2002 BPL census

S. no.	Characteristics	Scores 0	1	2	3	4
1	Size group of operational holding of land	Nil	Less than 1 ha of unirrigated land (or less than 0.5 ha of irrigated land)	1–2 ha of unirrigated land (or 0.5–1.0 ha of irrigated land)	2–5 ha of unirrigated land (or 1.0–2.5 ha of irrigated land)	More than 5 ha of unirrigated land (or 2.5 ha of irrigated land)
2	Type of house	Houseless	Kutcha (temporary)	Semi-pucca	Pucca	Urban type
3	Average availability of normal clothing (per person in pieces)	Less than 2	2 or more, but less than 4	4 or more, but less than 6	6 or more, but less than 10	10 or more
4	Food security	Less than one square meal per day for major part of the year	Normally, one square meal per day, but less than one square meal occasionally	One square meal per day throughout the year	Two square meals per day, with occasional shortage	Enough food throughout the year
5	Sanitation	Open defecation	Group latrine with irregular supply of water	Group latrine with regular water supply	Clean group latrine with regular water supply and regular sweeper	Private latrine

(contd.)

(Table 5.1 continued)

S. no.	Characteristics	Scores 0	1	2	3	4
6	Ownership of consumer durables: Do you own (tick (√) • TV • Electric fan • Kitchen appliances like pressure cooker • Radio	Nil	Any one	Two items only	Any three or all items	All items and/or ownership of any one of the following • Computer • Telephone • Refrigerator • Colour TV • Electric kitchen appliances • Expensive furniture • LMV/LCV[a] • Tractor • Mechanised two-wheeler/three-wheeler • Power tiller • Combined thresher/ harvester
7	Literacy status of the highest literate adult	Illiterate	Up to primary (Class V)	Completed secondary (passed Class X)	Graduate/ professional diploma	Postgraduate

(contd.)

(Table 5.1 continued)

S. no.	Characteristics	Scores 0	1	2	3	4
8	Status of the household labour force	Bonded labour	Female and child labour	Only adult females and no child labour	Adult males only	Others
9	Means of livelihood	Casual labour	Subsistence cultivation	Artisan	Salary	Others
10	Status of children (5–14 years) [any child]	Not going to school[b] and not working	Going to school[b] and working			Going to school[b] and not working
11	Type of indebtedness	For daily consumption purposes from informal sources	For productive purposes from informal sources	For other purposes from informal sources	Borrowing only from institutional agencies	No indebtedness and possesses assets
12	Reason for migration from household	Casual work	Seasonal employment	Other forms of livelihood	Non-migrant	Other purposes
13	Preference of assistance	Wage employment/ TPDS	Self-employment	Training and skill upgradation	Housing	Loan/subsidy more than Rs 1 lakh or no assistance needed

Notes: The total score for a house will vary between 0 and 52.

[a] Four-wheeled light motor vehicle and light commercial vehicle, respectively.

[b] Including non-formal education.

Source: GoI (2002).

was calculated, indicating the relative position of that household in the village (ibid.).

Critique

The 2002 BPL census methodology has been criticised for not only having a uniform list of 13 indicators, but also a uniform categorisation for all these indicators for assigning scores to households. The removal of exclusion criteria expanded the required coverage of the elaborate enquiry by twofold to 16-fold (Sundaram 2003). Because of the unwieldy, elaborate and confusing methodology of the 2002 BPL census, it was implemented in a haphazard manner (Drèze and Khera 2010). The Ministry of Rural Development was unable to correctly assess the manpower requirement for the 2002 BPL census, which led to chaos, bungling and dissatisfaction on an unprecedented level (GoI 2009).

Some of the indicators have been considered non-transparent and non-verifiable, which increases the possibility of cheating by misreporting (Drèze and Khera 2010). Hirway (2003) argued that the power politics of villages would prevent objective scoring of the indicators. The scoring method of the BPL census of 2002 also suffered from a technical error, with two of the 13 indicators, those relating to the status of children and the reasons for migration from households, not applicable to all households. This raised the question of what scores should be assigned to households which did not have any children or migrants. Assigning no score to these indicators translated into giving these indicators the score of 0, which would artificially increase the extent of poverty assigned to these households.

Problems have also been identified in the suggested scoring pattern for the individual indicators. For instance, the assumption that artisan households are less deprived than those practising subsistence cultivation is not always correct. There was also ambiguity over various terms as they were not defined. Thus, 'self-employed service providers' were given the score of least deprivation, irrespective of whether the services that were provided were skilled or unskilled. In the case of the indicator relating to the size of the operational holding, the size classes for scoring were fixed and had to be adhered to uniformly, without taking into consideration local conditions. Hence, two pieces of land of the same size, one in a high-rainfall area and the other in an arid region, were both assigned the same score, even though the productivity and thus the incomes generated from these two pieces of land were vastly different.

The aggregation of the individual scores of each indicator to arrive at the final score for each household has the absurd implication that all the disparate conditions at a particular score, for example, being a bonded labourer and being illiterate, can be equated. This aggregation also does not give any consideration to the concept of basic needs (Sundaram 2003).

Due to the various conceptual and technical flaws in the 2002 BPL census, it came under considerable criticism and became a case of 'hit or miss' (Drèze and Khera 2010; Kumar 2011). The petitioners of the right to food case highlighted the serious flaws of the 2002 BPL census, which led to the Supreme Court putting a stay on the use of the census in May 2003. The stay was lifted almost three years later in February 2006 (Himanshu 2008).

Proposals Regarding BPL Census for the Eleventh Five Year Plan

Saxena Committee (2008–09)

Methodology

In light of the various shortcomings of the methodology of the 2002 BPL census, the Ministry of Rural Development constituted an expert group, chaired by N. C. Saxena, in August 2008 to advise it on a more suitable methodology for conducting a BPL census.[5] The committee proposed a three-step methodology for the identification of poor households:

1. automatic exclusion of the visibly non-poor;
2. automatic inclusion of the most disadvantaged; and
3. grading the rest of the households and identifying the poorest amongst them based on proxy means testing so that the total number of poor corresponds with the *gram* (village) panchayat-level poverty caps.

Automatic exclusion: Automatic exclusion of the visibly non-poor aimed at minimising inclusion errors and saving time by excluding the most visibly non-poor rural households. By not surveying these non-poor, they are unable to be wrongly included in the BPL lists. The exclusion criteria include:

1. Households having at least one member working in the formal economy, either public or private. Such households constitute only 7 per cent of the national population and by definition do not experience the kinds of vulnerabilities and exogenous shocks that are faced by many of the poor.
2. Households (not including STs because of the problem of land alienation) that own double (triple in the case of completely unirrigated land) the land of the district (or the block, if a state desires) average of agricultural land per agricultural household[6] if partially or wholly irrigated.
3. Households owning three- or four-wheeled motorised vehicles.
4. Households owning at least one mechanised farm equipment, such as a tractor, power tiller, thresher or harvester.
5. Households having any person drawing a monthly salary of more than Rs 10,000 in non-government or private organisations or employed in government or parastatal institutions on a regular basis with pensionary or equivalent benefits.
6. Income tax payers.

States are permitted to add to the list of exclusion criteria, with examples of such criteria including ownership of motorised two-wheeled vehicles or ownership of a running borehole. Based on NSS 2004–05 data, only 8.8 per cent of all rural households would be automatically excluded by the method described above.

Automatic inclusion: Based on the experience that often the most disadvantaged get excluded from BPL lists because of their powerlessness, the committee recommended automatic inclusion of the following categories:

1. Destitute people who are forced to live by alms: Such people are often old, without caregivers, disabled or infirm. Destitute people, who are the poorest of the poor, lack access to property, possessions, or the opportunity to earn income by applying their own labour (Harriss-White 2002).
2. Particularly Vulnerable Tribal Groups (PVTGs): Extremely disadvantaged STs who practise a pre-agricultural level of technology (shifting agriculture), have extremely low levels of

literacy, experience stagnant or declining populations (Standing Committee on Labour and Welfare 2002), and suffer increasing dispossession from forests and alienation from their traditional systems of livelihood, coupled with exclusion from mainstream governance systems. Their lives are characterised by chronic hunger and reliance on private moneylenders.

3. Most discriminated Dalit groups or Mahadalits: State-identified categories comprising some of the most dispossessed communities in the country, like the Musahars of Bihar and Uttar Pradesh.
4. Single women (widows, women who are deserted, separated, divorced or unmarried and above 35 years of age) and their dependents: A large proportion of single women face exclusion and discrimination, with widows and older women particularly neglected and having to bear hardship. The powerlessness and marginalisation of single women are evident from the fact that they consume less than 90 per cent of the minimum calorie norm for married women.
5. Households with disabled persons as bread earners: People with disabilities face invisibility in everyday life as well as in policy space. The question of poverty is intricately tied up with the question of disability, being both a cause and consequence of disability. There is high chance of impoverishment in families in which either the breadwinner or a dependent is disabled. A large proportion of households with self-reported disabled members are below the poverty line, having fewer assets, smaller landholdings and greater debt than other households (Elwan 1999). People living with disability have greater needs than others for healthcare, basic needs, accessibility and transport, but have a lower earning capacity due to their impairment and lack of access to opportunities, in spite of any desire to work. The cost to caregivers is also high in such situations as often an adult member is unable to work due to caring for the disabled members.
6. Households headed by a minor: These are households wherein every member is below 18 years of age. Such households also include unrelated children living together. A situation where children have to take on the responsibility of looking after themselves and others incurs a definite loss of rights for at least one of the members of the household—the one who heads it, the one who earns money, who

makes decisions, and who, at the same time, should be in the care of adults (MacLellan 2005). Given the extreme vulnerability of this category of persons, in terms of financial, social and psychological requirements, there is a need for support systems to be put in place to alleviate their distress.

7. The homeless: In situations where this already marginalised group of people is denied access to legal documents such as ration cards, they face further hardship resulting from their inability to access government schemes and services. In the absence of access to nutritious food, they are forced to beg, depend on charity, or forage for whatever food they can get.
8. Bonded labourers: Although bonded labour has been outlawed in the country, and its existence is denied in official spheres, it continues to exist. The bonded labour system is based on feudal social relations and the caste system. Heavy debts, compounded rates of interest, and the exploitation of the illiterate bonded labourer by the landlord ensure that families whose members are engaged as agricultural bonded labourers remain trapped in the system.

Survey of the remaining rural households by scoring: As mentioned earlier, districts with more than 80 per cent poverty should only exclude the non-poor by the automatic exclusion criteria and include all the rest of the households in the BPL lists. For the rest of the districts, the process of scoring is explained below. The survey would include all households in a village except those which are automatically included or excluded. The scores for different groups are as follows:

- SCs and STs: 3 points
- Muslim/OBCs: 1 point
- Denotified tribes and designated 'Most Backward Castes': 2 points
- Landless agricultural workers: 4 points
- Agricultural labourers (with some land): 3 points
- Casual workers (households doing both agricultural labour and casual work would be classified as agricultural workers): 2 points
- Self-employed artisans or self-employed fisherfolk (including those employed by others in such professions): 2 points
- Households headed by an old person of age 60 years or above: 1 point
- Where any member of the household has tuberculosis (TB), leprosy,[7] disability, mental illness or HIV/AIDS: 1 point

- Where there is no adult above 30 years of age who has studied up to Class V in the household: 1 point

Critique of the Recommendations of the Saxena Committee

The recommendations of the Saxena Committee have been critiqued, first and foremost, by some of its own members, who expressed the concern that some of the exclusion criteria would exclude the poor. They also suggested some additional indicators that should be included in the list of inclusion criteria. On the whole, while the methodology recommended by the Saxena Committee is widely considered a major improvement over the confused methodology of the 2002 BPL census (Drèze and Khera 2010; Mishra 2011), even the simplified scoring method prescribed by the committee is complicated to implement (Drèze and Khera 2010). It divides households in 11 categories, each with a different score that does not have a simple interpretation, as a particular score can be obtained in different ways from the component indicators (Roy 2011).

Drèze and Khera (2010) are of the opinion that the inclusion criteria are very narrow, arguing that the recommendations do not address the issue that many households would meet both the inclusion as well as the exclusion criteria. This issue was also raised by Raghav Gaiha, a member of the Saxena Committee. Some of the indicators proposed by the committee are non-transparent and non-verifiable, such as 'any member of the household having TB, leprosy, disability, mental illness or HIV/AIDS'. Such indicators increase the scope of cheating by misreporting (ibid.).

The NAC endorsed all the categories of automatic inclusion proposed by the Saxena Committee, and also agreed with the committee's recommendation of giving highest priority (after including the categories of automatic inclusion) to SCs and STs in the list of BPL households.

Proposal by Jean Drèze and Reetika Khera

Drèze and Khera (2010) give alternative proposals for identifying households for social support. They suggest terming the selected households as the 'social assistance base' and avoiding the terminology of 'below poverty line', as they argue that there is no necessity of selecting households based on the poverty line. They propose four approaches for the identification of social assistance base households based only on exclusion and inclusion criteria, rejecting the imposition of caps. The four approaches are:

1. Exclusion approach: Households except those that meet any of the exclusion criteria are selected.
2. Inclusion approach: Only those households are selected which meet any of the inclusion criteria.
3. Play safe approach: Households except those that meet any of the exclusion criteria, but not any of the inclusion criteria, are selected.
4. Restrictive approach: Only those households are selected that meet any of the inclusion criteria, but not any of the exclusion criteria.

Ownership of any of the following assets constitutes an exclusion criterion: car, refrigerator, landline telephone, scooter, colour television and all of the following three amenities (together)—electricity, piped water and flush toilet. The following are the inclusion criteria: households belonging to SCs or STs, households with no agricultural land, households with no adult member educated beyond Class V, households headed by single women, and households with at least one adult member working as an agricultural labourer.

Socio-economic Caste Census 2011–12

Methodology

The Ministry of Rural Development adopted a methodology for identifying BPL households as part of SECC 2011. It followed the Saxena Committee recommendations—automatic exclusion, automatic inclusion, and ranking of the remaining households on the basis of deprivation indicators.

Any household meeting any one of the following exclusion criteria would be automatically excluded:

1. Owning motorised vehicles: two-/three-/four-wheelers/fishing boats (which require registration)

2. Owning mechanised three-/four-wheeler agricultural equipments such as tractors and harvesters
3. Having a *kisan*[8] credit card with credit limit of Rs 50,000 and above
4. Households with any member who is a government employee: gazetted and non-gazetted employees of the central government, state government, public sector undertakings, government-aided autonomous bodies and local bodies. This excludes incentive and other honorarium-based workers like ASHAs (accredited social health activists), anganwadi workers, etc.
5. Households with enterprises registered with the government for any purpose; any non-agricultural enterprise registered with the central or state governments
6. Households with any member in the family earning more than Rs 10,000 per month
7. Households paying income tax or professional tax
8. Households with three or more rooms with pucca walls and pucca roof
9. Households owning a refrigerator
10. Households owning landline phones
11. Households owning[9] 2.5 acres or more of irrigated land with at least one irrigation equipment such as diesel/electric-operated borehole/tube well
12. Households owning 5 acres or more land irrigated for two or more crop seasons
13. Households owning 7.5 acres or more land with at least one irrigation equipment such as diesel/electric-operated borehole/tube well

A household which did not meet any of the exclusion criteria mentioned above but fulfilled any of the inclusion criteria listed below would be automatically included:

1. Households without shelter
2. Destitute/living on alms
3. Manual scavengers
4. PVTGs
5. Legally released bonded labourers

Households which met neither the automatic exclusion nor the automatic inclusion criteria would be assigned deprivation scores depending on the number of the following criteria they met:

1. Households with only one room with kutcha[10] walls and kutcha roof
2. Households with no adult member between 16 and 59 years of age
3. Female-headed households with no adult male member between 16 and 59 years of age
4. Households with any disabled member and no able-bodied adult member
5. SC/ST households
6. Households with no literate adult above 25 years
7. Landless households deriving the major part of their income from manual casual labour

Households meeting any of the automatic inclusion criteria would get the highest priority. For households eligible for ranking under the deprivation indicators, a deprivation score would be derived for each household by adding up the number of deprivation criteria satisfied by the household. As there are seven indicators, the score would vary from a minimum of 0 to a maximum of 7. For inclusion in the priority group mentioned in the NFSB, a cut-off would be chosen such that 46 per cent of the households in rural areas (Tendulkar Committee poverty line + 10 per cent to account for targeting errors) having the highest scores would be selected (Drèze 2011b).

Critique

All aspects of SECC 2011—the criteria for automatic inclusion and exclusion, the deprivation criteria for ranking households, definitions of several terms and the process of implementation of the census—have come under heavy criticism. Although there are views that the discrepancies can be sorted out, most economists are of the view that faulty questionnaires and rubrics result in bad-quality fieldwork (Raman 2011). The Right to Food Campaign advocated that an 'exclusion criteria only' approach should have been used that treated all the non-excluded households as priority groups. In August 2011, the campaign sent a letter to the Ministry of Rural Development containing its critique of the SECC along with the following demands:

- Re-examination of the problematic criteria of exclusion, inclusion, ranking and definitions, and removal of criteria such as ownership of a two-wheeler, motorised boat, three-roomed pucca house, etc.

- Provision of opportunities (including appeal procedures) for excluded households to be reinstated in the BPL list.
- Mandating of procedures to ensure participatory verification of SECC 2011 data by *gram sabhas* (village councils), including pre-survey gram sabha meetings where the participation of villagers in the survey process is ensured
- Acceptance of any list of names for inclusion suggested by a gram sabha
- Guaranteeing of high standards of transparency in the entire process

Critique of the Exclusion Criteria

Many critics have found some of the criteria for automatic exclusion flawed as they excluded poor households as well. Brinda Karat (2011) argues that with widely fluctuating agricultural incomes, large debt burdens on poor peasant households, unpredictable weather, and natural calamities such as floods and droughts, even 2.5 acres of irrigated land with at least one irrigation equipment such as a diesel/electric-operated borehole/tube well would not be adequate to pull agricultural households out of poverty. She also questions the appropriateness of excluding households in which a non-agricultural enterprise is registered with the government. For example, micro-enterprises run by women's self-help groups are registered with the government, although many of the women in such groups are not non-poor.

Mishra (2011) highlights the objection of many commentators to the exclusion criteria of ownership or part-ownership of mechanised boats. For example, there are groups of extremely poor fishermen who pool money to buy a ragtag boat for them to use together. The Right to Food Campaign argues that many households that own motorised boats are landless and extremely vulnerable due to the decreasing fish catch as a result of competition from trawlers and climate change induced calamities. The criterion of owning two-wheelers would exclude many small vendors who carry vegetables, milk and suchlike from villages to nearby markets. Excluding households owning water pumps would unfairly exclude many tribal households who have been pushed into buying them through subsidies. Also, PVTG households that have walled the verandahs of their Indira Awaas Yojana houses to create a third room would also get excluded.

Critique of the Inclusion Criteria

The Right to Food Campaign considers the inclusion criteria to be too narrow and virtually irrelevant, a criticism echoed by politicians such as Karat (2011), who argues that they barely cover even 5 per cent of the rural poor. As the destitute are defined only as those living on alms, old people who are forced to work for their survival will not get automatically included. Karat also questions the logic of automatically including only those who have been legally released from bonded labour, but not those who have escaped from bondage. Vulnerable categories such as STs, SCs, the disabled, widows and casual manual workers are not automatically included.

Mishra (2011) questions the removal of Mahadalits, single women, persons with disabilities and households headed by minors—groups recommended for automatic inclusion by the Saxena Committee. Patnaik (2011) adds that as per the Supreme Court orders, persons with disabilities and widows along with PVTGs and destitute households are categorised as 'poorest of the poor'. Hence, the removal of persons with disabilities and widows from the list of those who are automatically included is actually a violation of this order.

Critique of the Deprivation Indicators in the Ranking System

The Right to Food Campaign considers the deprivation indicators to be patchy and arbitrary and states that even if they were applied correctly, they would exclude a large number of poor households. It adds that except for the indicator of SC/ST, all the other indicators are inordinately conditional. Karat (2011) too criticises the indicators for excluding many poor households. The criterion of only including households which have only one kutcha room house excludes many poor households, such as artisans like potters, weavers and washermen who create a small room as their workspace.

The indicators are so designed that they fail to give scores to various cases of deprivation. For example, a 16-year-old boy heading a household will not earn that household any deprivation indicator; a family that has a disabled husband and an able-bodied wife will not get any score either for the household being headed by a female, or for having a disabled person; a fisherman, weaver, blacksmith or any other artisan who does not own any land will not get any score for being landless; and a household with a one-room house with kutcha walls and a tin roof (extremely common in north-eastern states) will not get any score.

While there have certainly been improvements in the methodologies of identifying the poor over the successive poverty censuses, it has been repeatedly established that no such exercise can be without errors of inclusion or exclusion. Also, as the number of people in need of a particular kind of social assistance may surpass the number of people considered poor, poverty lines need to be delinked from the exercise of identification of beneficiaries for social programmes.

Notes

1. A free rural housing programme. The Indira Awaas Yojana has now been renamed the Pradhan Mantri Gramin Awaas Yojana.

2. In 2015, the NDA government at the centre dismantled the Planning Commission and replaced it with the NITI Aayog.

3. A house designed to be solid and permanent.

4. Chaired by P. L. Sanjeev Reddy, former secretary to the Government of India and director, Indian Institute of Public Administration, the Expert Group had among its members a number of leading economists (Professors Subashish Gangopadhyay, K. Krishnamurty, R. Radhakrishna, V. M. Rao, and Rohini Nayyar), besides senior officials from the Central Statistics Office, the NSSO and concerned government departments of the central and (five) state governments.

5. The 17-member expert group, chaired by N. C. Saxena, had the following members: Raghav Gaiha, N. J. Kurian, P. Sainath, Harsh Mander, S. C. Seddey, M. N. Roy, K. Raju, Manoj Kumar Singh, G. S. Sandhu, A. K. Misra, Amar Singh, Arun Kumar, Wilfred Lakra, Aruna Roy, S. C. Gautam and K. L. Datta.

6. To calculate the district average, the total gross cultivated land has to be divided by the total number of households that own some land.

7. Diseases which generally afflict the poor and carry stigma with them were chosen.

8. 'Kisan' is Hindi for 'farmer'. Kisan credit cards are special credit cards given to farmers.

9. Even land possessed as a matter of traditional rights, without legal title, is to be counted as 'owned' (Drèze 2011b).

10. 'Kutcha' refers to an abode built with non-durable, flimsy material.

References

Abraham, Anita, and Jean-Philippe Platteau. 2002. 'Participatory Development in the Presence of Endogenous Community Imperfections'. http://siteresources.worldbank.org/INTABCDEWASHINGTON2001/Resources/abraham.pdf (accessed 10 March 2018).

Balchand, K. 2012. 'Centre defers SECC household count in poll-bound U.P.' *The Hindu*, 3 January.

Bardhan, P., and D. Mookherjee. 1999. 'Relative Capture of Local and Central Governments: An Essay in the Political Economy of Decentralization'. CIDER Working Paper, Center for International and Development Economics Research, University of California, Berkeley.

Bigman, D., and P. Srinivasan. 2002. 'Geographical Targeting of Poverty Alleviation Programs: Methodology and Applications in Rural India'. *Journal of Policy Modeling*, 24(3): 237–55.

Coady, D., and R. Harris. 2004. 'Evaluating Targeted Cash Transfer Programs: A General Equilibrium Framework with an Application to Mexico'. Discussion Paper 137, Food Consumption and Nutrition Division, International Food Policy Research Institute, Washington, D.C.

Coady, David, Margaret E. Grosh and John Hoddinott. 2004. 'Targeting of Transfers in Developing Countries: Review of Lessons and Experiences'. World Bank, Washington, D.C.

Conning, J., and M. Kevane. 2002. 'Community-Based Targeting Mechanisms for Social Safety Nets: A Critical Review'. *World Development*, 30(3): 375–94.

Desai, Sonalde. 2012. 'The great Indian poverty game'. *Business Standard*, 20 January.

Drèze, Jean. 2011a. 'Struggling to enter the BPL club'. *The Hindu*, 2 October.

———. 2011b. 'Kaun banega scorepati?' *The Hindu*, 28 November. http://www.thehindu.com/opinion/op-ed/article2665893.ece (accessed 17 September 2017).

Drèze, Jean, and Reetika Khera. 2010. 'The BPL Census and a Possible Alternative'. *Economic and Political Weekly*, 45(9): 54–63.

Elbers, C., P. Lanjouw, J. Mistiaen, B. Ozler and K. Simler. 2004. 'On the Unequal Inequality of Poor Communities'. *World Bank Research Observer*, 18(3): 401–21.

Elbers, C., T. Fujii, P. Lanjouw, B. Ozler, and W. Yin. 2003. 'Poverty Alleviation through Geographic Targeting: How Much Does Disaggregation Help?' World Bank Policy Research Working Paper no. 3419, World Bank, Washington, D.C.

Elwan, Ann. 1999. 'Poverty and Disability: A Survey of the Literature'. Social Protection Unit, World Bank, Washington, D.C., 18 December.

Esman, M. J., and N. T. Uphoff. 1984. *Local Organizations: Intermediaries in Rural Development*. Ithaca: Cornell University Press.

Galasso, E., and M. Ravallion. 2002. 'Decentralized Targeting of an Anti-poverty Program'. World Bank Policy Research Working Paper, World Bank, Washington, D.C.

GoI (Government of India). 2002. 'Report of the Expert Group on Identification of Households below Poverty Line (BPL Census 2002)'. Ministry of Rural Development, Government of India, New Delhi, September.

———. 2009. 'Report of the Expert Group to Advise the Ministry of Rural Development on the Methodology for Conducting the Below Poverty Line (BPL) Census for Eleventh Five-Year Plan'. Ministry of Rural Development, Government of India, New Delhi.

———. 2012. 'Press Note on Poverty Estimates, 2009–10'. Planning Commission, New Delhi.

Harriss-White, Barbara. 2002. 'A Note on Destitution'. Working Paper no. 86, QEH Working Paper Series. http://www3.qeh.ox.ac.uk/pdf/qehwp/qehwps86.pdf (accessed 10 March 2018).

Himanshu. 2008. 'Counting the poor: Some methodological issues'. *Mint*, 24 September.

———. 2010. 'Demystifying poverty: Which estimate and why'. *Mint*, 19 January.

———. 2011a. 'Is the BPL census correctly structured?' *Business Standard*, 17 August. http://www.business-standard.com/india/news/isbpl-census-correctly-structured/445977/ (accessed 17 September 2017).

———. 2011b. 'Poverty line: Can we do without one?' *Business Standard*, 5 October. http://www.business-standard.com/india/news/poverty-line-can-we-do-without-one/451418/ accessed 17 September 2017).

———. 2012. 'India undercounts its poor'. *Mint*, 26 March.

Hirway, Indira. 2003. 'Identification of BPL Households for Poverty Alleviation Programmes'. *Economic and Political Weekly*, 38(45): 4803–8.

Karat, Brinda. 2011. 'An exercise in undercounting'. *The Hindu*, 9 July.

Krishna, A. 2007. 'For Reducing Poverty Faster: Target Reasons before People'. *World Development*, 35(11): 1947–60.

Kumar, Ashwini. 2011. 'Poverty line: Can we do without one?' *Business Standard*, 5 October. http://www.business-standard.com/india/news/poverty-line-can-we-do-without-one/451418/ (accessed 17 September 2017).

MacLellan, Marion. 2005. 'Child Headed Households: Dilemmas of Definition and Livelihood Rights', Better Care Network.

Mahapatra, Richard. 2011a. 'Poverty Begets Poverty'. *Down to Earth*, 31 May. http://www.downtoearth.org.in/content/poverty-begets-poverty (accessed 17 September 2017).

Mahapatra, Richard. 2011b. 'Poverty Line Delinked from Food and Other Entitlements Given to the Poor'. *Down to Earth*, 3 October. http://www.downtoearth.org.in/content/poverty-line-delinked-food-and-other-entitlements-given-poor (accessed 17 September 2017).

———. 2011c. 'Planning Commission to SC: Rs 26 a Person a Day the "Adequate" Poverty Line'. *Down to Earth*, 20 September. http://www.downtoearth.org.in/content/planning-commission-sc-rs-26-person-day-adequate-poverty-line (accessed 17 September 2017).

Manna, G. C. 2012. 'On Some Contentious Issues of the New Poverty Line'. *Economic and Political Weekly*, 47(15): 11–14.

Mehrotra, Santosh, and Harsh Mander. 2009. 'How to Identify the Poor? A Proposal'. *Economic and Political Weekly*, 44(19): 37–44.

Mishra, Neelabh. 2011. 'Trash that poverty register'. *Outlook*, 25 July. http://www.outlookindia.com/article.aspx?277648 (accessed 17 September 2017).

Nhate, V., and K. Simler. 2003. 'Poverty, Inequality, and Geographic Targeting: Evidence from Small-Area Estimates in Mozambique'. Discussion Paper no. 192, Food Consumption and Nutrition Division, International Food Policy Research Institute, Washington, D.C.

Patnaik, Biraj. 2011. 'Is the BPL census correctly structured?' *Business Standard*, 17 August. http://www.business-standard.com/india/news/isbpl-census-correctly-structured/445977/ (accessed 17 September 2017).

Pender, J., and R. Ruben. 2004. 'Rural Diversity and Heterogeneity in Less-Favored Areas: The Quest for Policy Targeting'. *Food Policy*, 29(4): 303–20.

Radhakrishna, R., and K. Subbarao. 1997. 'India's Public Distribution System: A National and International Perspective'. World Bank Discussion Paper no. 380, Washington, D.C.

Raman, Anuradha. 2011. 'Find me some destitutes'. *Outlook*, 17 October. http://www.outlookindia.com/article.aspx?278575 (accessed 17 September 2017).

Ravallion, M., and Q. Wodon. 1999. 'Poor Areas or Only Poor People?' *Journal of Regional Science*, 39(4): 689–711.

Roy, Indrajit. 2011. '"New" Lists for "Old": (Re-)constructing the Poor in the BPL Census'. *Economic and Political Weekly*, 46(22): 82–91.

Schady, N. 2002. 'Picking the Poor: Indicators for Geographic Targeting in Peru'. *Review of Income and Wealth*, 48(3): 417–33.

Sen, A. 1995. 'The Political Economy of Targeting', in D. van de Walle and K. Nead (eds), *Public Spending and the Poor*. Baltimore: Johns Hopkins University Press, pp. 11–24.

Standing Committee on Labour and Welfare. 2002. 28th Report. Ministry of Tribal Affairs, Development of Primitive Tribal Groups, 13th Lok Sabha, Lok Sabha Secretariat, New Delhi.

Subbarao, K., J. Braithwaite and J. Jalan. 1995. 'Protecting the Poor during Adjustment and Transitions'. Human Capital Development and Operations Policy Working Paper no. 58, Background Paper for the World Bank contribution to the Social Summit 'Advancing Social Development', World Bank, Washington, D.C.

Sundaram, K. 2003. 'On Identification of Households below Poverty Line in BPL Census 2002: Some Comments on the Proposed Methodology'. *Economic and Political Weekly*, 38(9): 896–901.

UNDP (United Nations Development Programme). 2000. *Overcoming Human Poverty: UNDP Poverty Report 2000*. New York: UNDP.

Uphoff, N. T., M. J. Esman and A. Krishna. 1998. *Reasons for Success: Learning from Instructive Experiences in Rural development*. West Hartford, CT: Kumarian Press.

6

Conditionalities

ANKITA AGGARWAL

CONDITIONAL CASH TRANSFERS

Conditional cash transfers (CCTs) are safety net programmes that have been implemented in many developing countries such as Bangladesh, India, Indonesia, Nepal and Malawi, and by countries in Latin America and the Caribbean (Attanasio et al. 2005; Government of Nepal 2005; Government of the People's Republic of Bangladesh 2007; Morris et al. 2004; Powell-Jackson et al. 2009). Under these programmes, cash is transferred to people on the condition that they fulfil certain requirements, which usually pertain to education or health.

Most CCTs have the following traits in common:

- They are targeted, with the known problems of both inclusion and exclusion errors.
- They demand that the beneficiaries adhere to certain conditions.
- The transfers are small and are subject to an income ceiling that cannot be crossed.
- Their behaviour is pro-cyclical, and the transfers are strongly determined by the evolution of economic growth.
- They lack a preventive dimension with regard to insufficient income, i.e., they are not designed to address and mitigate the cause of poverty but only provide some relief to the recipients.
- They favour children of schoolgoing ages and/or women in the reproductive cycle.
- They are not conceived or designed as rights.
- They are not permanent or stable, but transitory.

There are differences, however, among various CCT programmes in their emphasis on how their objectives are to be achieved. In some countries, the introduction of a targeted cash transfer was originally motivated by a principle of universalism (e.g., Bolsa Família in Brazil), aimed at guaranteeing a minimum income and common standardised public services nationally. In others, the prevailing objective has been that of promoting human capital accumulation among low-income families by improving education and health service utilisation, especially among children (e.g., Progresa/Oportunidades in Mexico). In still others, the rationale for the introduction of cash transfers is to reach out to the extremely poor excluded from existing safety nets (e.g., Chile Solidario), or to reach narrowly defined vulnerable groups affected by specific adverse events (e.g., Familias en Acción in Colombia). In such cases, cash transfers were originally envisaged as temporary, compensatory or emergency-based measures.

Variations in the rationale of CCTs shape the transfer amounts, the coverage and the duration of programme participation, targeting practices, the definition of conditionality and the response to non-compliance with the behavioural requirements. While basic behavioural definitions are common to most CCTs, the rationale varies in terms of whether the emphasis is placed on beneficiary behaviour or the behaviour and responsibilities of service providers. This, in turn, influences government decisions on conditionality monitoring and the regulation of beneficiary non-compliance.

Hypotheses Underlying Conditionalities

Conditionalities are based on the hypotheses that poverty is a consequence of a lack of human capital (see Yanes 2011), and that poor people do not know what is best for them (Fiszbein and Schady 2009).

Poverty as a Consequence of a Lack of Human Capital

Yanes (2011) argues that conditionalities are based on the hypothesis that poverty is a consequence of a lack of human capital and that poor people require a system of rewards and punishments in order to send their children to school and take care of their health. Boltvinik (2004) challenges this hypothesis in the context of a CCT programme in Mexico, Oportunidades, which makes cash transfers conditional on school attainment. Boltvinik argues that there is no direct relation between literacy and poverty: while educational poverty in Mexico decreased rapidly from 1969 to 2000, income poverty decreased only marginally—from 72.6 per cent in 1968 to 68.5 per cent in 2000. In fact, while educational poverty declined by as much as 60 per cent, from 43.7 per cent to 26.3 per cent between 1980 and 2000, income poverty actually increased by 41 per cent—from 48.5 per cent in 1981 to 68.5 per cent 2000.

Inability of Poor People to Make Good Decisions for Themselves

The second hypothesis on conditionalities is somewhat linked to the first, that poor people do not know what is best for them and hence need a system of rewards and punishments in order to send their children to school and take care of their heath. Fiszbein and Schady (2009) strongly defend the idea of conditionality, based on their idea of the 'microfoundations of paternalism'. According to them, a reason to justify the introduction of conditions is that poor people do not always behave exactly as could be expected from rational agents with perfect information:

> What imperfect information, myopia, and incomplete altruism have in common, for our purposes, is that they may cause a family's privately chosen level of investment in human capital to be too low, compared with its own 'true' private optimal. If they are pervasive, then these distortions in private decision-making provide some contemporary support to the time-honoured notions that governments may 'know better' what is privately good for poor people than do the poor themselves, at least in some realms. (Ibid., 50)

Contrary to the argument made by Fiszbein and Schady, Hanlon et al. (2010) find that households make good use of transfers, stating that transfers are followed by increased consumption and better nutrition among recipients. When money is given to mothers, it results in increases in expenditure on children. Yanes (2011) is of the opinion that the hypotheses underlying conditionality in cash transfers reflect a classist and racist mindset of policymakers towards poor people and towards poverty itself.

Conditionalities and Support for Programmes

In a World Bank–sponsored study in support of conditionalities, Fiszbein and Schady (2009) argue that making transfers conditional upon 'good behaviour' practices solicits the cooperation of taxpayers and the more affluent members of society in funding these programmes:

> The second main conceptual argument for conditioning a cash transfer is that governments typically do not behave like textbook benevolent dictators. Policy decisions generally result from decision-making processes that involve voting, lobbying, bureaucratic and inter-agency bargaining, and a variety of other forms of what one broadly might call political economy. Under some circumstances, conditioning cash transfers on 'good behaviour' may increase public support for them, making the programme either feasible or better-endowed. (Ibid., 50)

Yanes (2011) notes that after years of advocacy of rational expectations or public choice theory, of designing economic and social policies on the hypothesis of *homo economicus* and considering individual economic decisions as based on rational decisions for cost minimisation and benefit maximisation, the study by Fiszbein and Schady and the design of CCTs established an exception. The poor do not behave in a rational manner, cannot make correct decisions, do not know what they really need, and must thus be guided. Should they fail to meet the conditions set forth for their own benefit, they must be punished with exclusion from the cash transfer programme. Since the dominant social mentality is that the poor make bad decisions and tend towards vagrancy, irresponsibility, laziness and vice, it is necessary, for political and argumentative reasons, to include conditions in order to keep taxpayers from complaining about these programmes.

Yanes (2011) argues, however, that the concept of conditionalities is incongruent with a rights-based perspective, in which people have rights they can demand and the state has a legal obligation to fulfil them. Conditionalities, which reduce people to 'beneficiaries', violate their unconditional right to social security and make programme 'benefits' contingent upon the fulfilment of co-responsibilities.

Success of Conditionalities in Fulfilling their Intended Purposes

Yanes (2011) points out that the proponents of CCTs have failed to demonstrate that the improvements observed in some programmes in school attendance or attendance in health clinics are derived from the conditionalities rather than the improved income of the family due to the transfers. In studies about conditionalities, there is no evidence supporting a link between the benefits of the transfer and the associated conditions. The introduction of conditions seems linked to political reasons, to the punitive perspective, and to the logic that binds together aid and control (Lo Vuolo 2010). Hanlon et al. (2010, 131) note in their comparative study on conditional transfers in different parts of the world: 'Despite the importance of conditions, their effectiveness has not been studied separately from the programmes that include them, and it tends simply to be assumed that they work. In fact, there is almost no evidence that conditions make any major difference.'

Costs Associated with Conditionalities

Hanlon et al. (2010) highlight the worrying fact that Latin American governments assign millions of dollars to the administration of conditions without rigorously analysing how much these conditionalities contribute to programmes, and if the said contributions match the cost involved. Bastagli (2011) too notes that implementation and monitoring of conditionalities are associated with several administrative and social costs, some of which are discussed below.

Conditionalities and the Risk of Excluding the Most Disadvantaged

Poor households often face high opportunity costs of meeting the conditionalities of cash transfers, and are thus unable to fulfil them. Apart from economic constraints, the poor often face physical and social barriers in fulfilling the conditionalities. Conditional cash transfers may lead to penalising the poor by excluding them from the transfers for not fulfilling the conditionalities, through no fault of their own.

Alvarez et al. (2006) find that indigenous populations and the extreme poor in communities where there is likely to be greater inequality have higher odds of dropping out of programmes based on conditionalities. Other studies have drawn attention to the ways in which conditionalities impose burdens disproportionately among household members, with additional responsibilities falling mainly on women (Gonzalez de la Rocha 2005; Molyneux 2006).

Concerns that conditionalities may additionally penalise vulnerable households have led to the adoption of procedures aimed at minimising their sanctionary nature vis-à-vis beneficiaries. These procedures include processes of verification of the reasons for non-compliance and initiatives to assist beneficiaries in complying with education and health conditions.

If one takes the position that poor people make rational decisions within the circumstances in which they find themselves, then the failure of the poorest and most disadvantaged households and persons to fulfil conditionalities may be the direct outcome of the difficult circumstances in which they are placed. A street child cannot enter school, for instance, unless she is protected, and assured other rights such as security, shelter, food and healthcare; to penalise her or her homeless mother for not sending her to school would amount to punishing the victim, a highly unjust 'double whammy' of deprivation and injustice against the most marginalised.

Supply Constraints and Quality of Services in Conditional Schemes

If the services required to be utilised as conditionalities of transfers are in short supply and/or of poor quality, they may prevent or discourage potential CCT recipients from using them, which may lead to the intended educational and health outcomes not materialising. For instance, there are huge regional disparities in the quality of schools in Brazil, with private schools being considered better and people preferring to send their children to them if possible. Brazil's CCT programme, Bolsa Família, aims to bring children to school, but here they run into a supply constraint. The quality of public schools is poor and consequently the quality of education is poor. There is also a problem of violence in some of these schools, leading to a shortage of trained and competent teachers who also need to be paid adequately. Further, due to teachers' social status being considered low, it is necessary to raise their social standing. For the health services as well, there is a need to improve the quality of services by increasing the resources allocated to the sector (Vyasulu 2010).

Unintended Impacts of Conditionalities

Sometimes, people misunderstand the requirements and purpose of conditionalities, which leads to unintended behavioural effects. In Nicaragua, a CCT programme required children to gain weight. If the children fell below an established weight gain on two successive assessments of weight, parents could be sanctioned by the suspension of benefits. Adato (2000) reports how this requirement led to the overfeeding of children before they were weighed. In education, Nicaragua's CCT initially required beneficiary pupils to pass their grade successfully, leading some schools to advance children to the next grade even when such a move was not deserved (Maluccio and Flores 2005). In some countries, conditionality in cash transfers has been revised in response to such unintended effects. In Nicaragua, for example, the requirement for beneficiary children to pass their grades successfully was suspended once it was observed that this condition was associated with schoolteachers promoting beneficiary children in order to secure their benefit receipt (ibid.).

Cash Transfer Schemes in India: NMBS and JSY

This section discusses two maternity benefit schemes—the National Maternity Benefit Scheme, an unconditional cash transfer programme, and Janani Suraksha Yojana, a CCT programme. While the government wanted to subsume NMBS in JSY, citing the arguments in favour of conditionalities discussed above, the actual experience of JSY has validated the concerns with linking social supports with certain 'good behaviours'.

An unconditional cash transfer programme in the form of the NMBS has existed in India since 1995. It was introduced as a part of the wider National Social Assistance Programme under the Ministry of Rural Development and was entirely funded by the central government. The programme was transferred to the Ministry of Health and Family Welfare in 2001–02. Under the scheme, pregnant BPL women are entitled to Rs 500 8–12 weeks before delivery, for up to two live births, if they are above 19 years of age. This payment is meant to facilitate the purchase of food to meet specific nutritional requirements during pregnancy.[1]

The JSY, on the other hand, is a CCT scheme for institutional deliveries introduced in 2005 by the Government of India as a response to the slow and varied progress in improvement of maternal and neonatal[2] health in the country under NMBS. Conditionalities and the amount of money given differ by state, based on rates of institutional delivery. In low-performing states (Assam, Bihar, Chhattisgarh, Jammu and Kashmir, Jharkhand, Madhya Pradesh, Odisha, Rajasthan, Uttar Pradesh and Uttarakhand), all pregnant women delivering in government or accredited private institutions are entitled to Rs 1,400 in rural areas and Rs 1,000 in urban areas, after delivery for up to two live births. In the other states, only BPL and SC/ST women are eligible for cash assistance of Rs 700 and Rs 600 in rural and urban areas respectively. In low-performing states, ASHA workers are also paid a sum of Rs 600 in rural areas and Rs 200 in urban areas as an incentive to bring women in labour to institutions and compensate them for commuting costs. The JSY programme is the largest CCT programme in the world in terms of the number of beneficiaries (Lim et al. 2010).

Conditionalities in the Context of JSY and NMBS

The rationale for introducing a second scheme for pregnant women was that the NMBS did not address the health needs of women who were about to deliver. It offered no protection from avoidable maternal mortality and morbidity due to the lack of neonatal and postnatal care. Post-introduction, the JSY is credited with an increase in rural institutional deliveries from 29.8 per cent in 2002–04 to 37.8 per cent in

2007–08 (IIPS 2008). Coverage under the scheme is steadily improving; users of JSY multiplied more than 12 times, from 0.74 million in 2005–06 (Paul 2010) to 9.5 million in 2009–10 (this covers 36 per cent of the 26 million women who deliver in the country annually). A study by Lim et al. (2010) showed that two to three years after the introduction of JSY, the programme led to a reduction of about four perinatal deaths (death of the foetus or neonate) and two neonatal deaths per 1,000 live births.

The government envisaged the JSY as a package of schemes that would replace the NMBS, subsuming its cash entitlement. Its efforts to discontinue the latter led to a discussion of conditionalities in the context of reproductive rights and women's rights to health and nutrition. Due to a prior Supreme Court ruling, schemes could not be discontinued or restricted by the government without prior approval of the court.[3] On 9 May 2005, the Supreme Court rejected a petition filed by the Government of India to phase out NMBS and provide maternity benefits under JSY, as it was not clear whether JSY preserved all the benefits available under NMBS, especially for deliveries at home (Right to Food Campaign 2008). Further, on 20 November 2007, the Supreme Court passed an order allowing the NMBS to continue to provide Rs 500 per pregnancy to all pregnant BPL women, irrespective of their age and number of children.

In an interim application,[4] the Ministry of Health and Family Welfare sought to reverse the order of 20 November 2007 by arguing that the NMBS be subsumed under the JSY. The ministry cited two reasons: giving cash before delivery (as with the NMBS) would disincentivise institutional deliveries,[5] and there was no way to ensure that the money would be used to buy nutritious food for pregnant women.[6] Therefore, cash assistance formerly given 8–12 weeks before delivery under the NMBS should be given for home births, at the time of delivery or one week before, under the new JSY. A higher amount of JSY cash assistance would be provided to women who delivered in institutions, making more resources available for food and medical care for nursing women. The ministry also sought to restrict NMBS and JSY assistance to women who were 19 years and above, up to two live births. If these conditions were not imposed, it argued, the scheme would discourage family planning and contravene the legal age for marriage.

Issues Needing to Be Addressed to Improve Utilisation of JSY

In IA no. 98/2010 in WP(C) no. 196/2001,[7] the Government of India acknowledged the infrastructure gaps required for full implementation of the JSY. It reported that under the National Rural Health Mission (NRHM), several new initiatives were being taken up to make primary health centres operational for round-the-clock delivery of services, and to establish first referral units to provide emergency obstetric care for complicated pregnancies. Contractual appointments of doctors, paramedics and medical specialists were also allowed.

In April 2010, the Supreme Court Commissioners' response (see Birchfield and Corsi 2010) to the government's application[8] highlighted the implications of merging the NMBS and JSY schemes due to the nature of conditionalities. The NMBS was conceived to provide some form of social security to pregnant women to enable them to access better nutrition during pregnancy. Making the cash entitlement contingent on site of delivery would defeat its purpose. Institutional deliveries undoubtedly help improve maternal health, but women also need nutritional support during and after pregnancy. The commissioners noted that although there had been no large-scale study carried out by the Government of India, field evidence collected by the commissioners' office showed that this money, if provided in time, was used in many cases towards food and/or health expenses during pregnancy. Furthermore, with this benefit being given only to those women who were BPL, it is very likely that any additional cash would be used on essentials such as food.

Reasons for money not being spent on women's health and nutrition also need to be taken into account. Women often have no control over how family resources are spent, and specific needs during pregnancy may not be considered a high priority. Institutional difficulties may prevent families from spending the cash where required. For instance, a study (Tripathi 2013) on maternity benefit schemes conducted in 40 villages of Madhya Pradesh revealed that 41 per cent of women sampled spent a part of the entitlement on bribes for nurses, doctors, sweepers and drivers. In one district, they were forced to spend 80 per cent of the money to this end. A significant proportion—29.4 per cent—also spent the money on food for the whole family, and a little over 7 per cent reported spending the money on their own food.

The commissioners also argued against linking the cash assistance to the number of live births, as this is unethical and violates the rights of women. It is also modelled on schemes associated with 'population control policies' that are based on incentives and targets. The Government of India has accepted this position in policy documents, and committed itself to a target-free approach regarding population at the UN International Conference on Population and Development (Raman 2011). The National Population Policy of 2000 affirms the commitment of the government towards voluntary and informed choice and consent of citizens while availing of reproductive healthcare services, and continuation of the target-free approach in administering family planning services. It has been stated that having targets for sterilisation, etc., would be done away with, and that benefits under other schemes cannot be linked to the number of children.[9]

Excluding women who already have two children or more would deprive 60 per cent of pregnant BPL women from the scheme (Raman 2011). This would have been tantamount to putting their lives at risk and contributing to the high rate of maternal mortality. The problem of population growth could be addressed appropriately by streamlining the services of family planning and addressing the unmet needs of couples for spacing/terminal methods. On the other hand, JSY is meant for reducing maternal mortality, an objective that is different from containing population growth. It logically follows from this argument that there should be no restriction on the number of births for the NMBS, whose objective is providing nutritional support to pregnant women and not containing population growth. In fact, the NMBS could also contribute to the JSY objective of reducing maternal mortality by ensuring a healthier pregnancy.[10] Malnutrition through life is known to cause a range of health problems, including anaemia, which contributes to a significant number of maternal deaths. One study estimates that around 22,000 deaths in 2004 were attributable to anaemia (Gragnolati et al. 2005).

Many women do not have any control over the number of children they conceive. By restricting maternity benefits to two live births, such women are being unfairly penalised for having more than two children (Raman 2011). Further, the data from NFHS-3 shows that anaemia is prevalent among a higher percentage of women who have more than three children than among women who have fewer children. While 54.9 per cent of women with two to three children ever born are anaemic, 58 per cent of women with four to five children and 59 per cent of women with six or more children ever born are anaemic.

The commissioners also recommended removing the age restriction from the maternity benefits. The application of the Government of India itself stated: 'data from the NFHS 3 and Registrar General of India shows that nearly 26 per cent of women get married by 15 years of age and that 12 per cent of mortality is among women aged 15–19 years.' At the same time, the government tried to make a case against giving benefits to women under the age of 19. However, girls who are married and pregnant before the age of 19 are already victims of a social system that does not protect their rights to childhood, health and education. Punishing the victim for the crime cannot be accepted; younger women instead need additional support, given their vulnerability during pregnancy. There should therefore be no restriction in giving benefits under the NMBS to women based on their age.

Also, as highlighted earlier, various barriers, such as lack of physical access and poor quality of health services, prevent or discourage pregnant women from accessing health centres for institutional deliveries. The study by Lim et al. (2010) highlights the poor quality of maternal and neonatal care in health facilities. Despite substantial inputs under the NRHM, most public sector facilities continue to be understaffed and do not meet desired functional standards (Mohapatra et al. 2008). Deliveries might often be done by unskilled support staff rather than skilled nurses or doctors. Best practices, such as partogram, neonatal resuscitation and kangaroo care, are not followed. The system of referral to a higher level for emergencies is inadequate. Most mothers and babies are discharged within hours after delivery because the hospitals lack amenities, and families want to return home having received the cash incentive (UNFPA and CORT 2007a, 2007b). As a result, there is inadequate time for newborn care counselling, stabilisation of the postpartum mother, and detection of danger signs in the mother and the infant (Paul 2010). Studies have suggested that JSY has led to increased workloads and reduced quality of care in health facilities, such as early discharge after delivery (Devadasan et al. 2008; Lahariya 2009). There have been many instances of delays in payments to mothers and health workers and corruption in the JSY (Malini et al. 2008). This inadequacy in health services discourages their use by the intended demographic.

The JSY suffers from the same problems as state programmes across the world—it does not reach the most disadvantaged, i.e., the poorest and least educated women (Gakidou et al. 2006; Victora et al. 2000). There are several possible explanations for this trend. Due to poor physical access, many of the poorest women may be unable to access health facilities. Anecdotal evidence suggests that government health centres do not have adequate vehicles to transport pregnant women, which forces many women to give birth at home, even if they wish to deliver in institutions. Cultural barriers against institutional deliveries are also prevalent among many poor and uneducated families in India. The lower uptake among Muslims and Christians than among women of other religious communities might be a result of the poor reach of ASHAs in these communities, or poor access of these minorities to accredited health facilities (Lim et al. 2010).

Linking cash assistance to institutional deliveries unfairly penalises those women who are unable to have institutional deliveries through no fault of their own. Poor households often face high opportunity costs of meeting the conditionalities of cash transfers, and are thus unable to fulfil them. Hence, penalising the poor by excluding them from cash transfers for not fulfilling the conditionalities defeats the purpose of poverty alleviation—the intended objective of CCTs.

Conclusions

Although the JSY has substantially increased the rate of institutional deliveries, the conditional nature of the cash transfers provided in the scheme raises some of the concerns discussed in the earlier part of this chapter. Utilisation of the scheme and the realisation of its benefits would be more effectively achieved by improving the implementation of the scheme, rather than by imposing conditionalities. The quality of healthcare provided to women delivering in hospitals to avail of the cash benefit of the scheme is far from satisfactory, acting as a deterrent to women considering delivering in institutions. The poorest women, who need the maternity benefits the most, often live in villages located very far from institutions where children can be delivered. Hence, inter alia, because of lack of physical access, many such women are unable to deliver in institutions and to receive the maternity benefit cash transfer.

While on the one hand there are logistical reasons preventing conditionalities from attaining their objective of enhanced utilisation of education and health services, on the other hand, there is also strong ideological and ethical opposition to this policy as well. Conditioning cash transfers upon certain behaviours violates people's right to unconditional social security.

The NAC draft (NAC 2011) entitled every child under 14 years of age to approach a feeding facility, such as anganwadi centres, school midday meals, destitute feeding centres and suchlike, for a freshly cooked nutritious meal, and disallowed any such institution from denying the meal on any grounds. However, the government's 2011 draft and the final NFSA only give legal entitlements to midday meals in government and government-aided schools, up to the eighth standard. Thus, effectively, the government draft imposes the conditionality of school attendance for receiving the midday meals, which deprives children who are unable to attend school of nutritious hot cooked meals.

However, other entitlements under the NAC draft are not conditional upon the utilisation of education or health services or on 'good behaviour' regarding age of marriage and number of children conceived. While the Standing Committee on Food, Consumer Affairs and Public Distribution (2013) recommended that maternity benefits be 'admissable only up to the birth of the second child to stabilise the population', the NFSA entitles 'every pregnant and lactating mother' to one free meal through the local anganwadi. With the exception of conditions on the access to meals of children of schoolgoing age, there seems to be a broad consensus against conditionalities on entitlements under the right to food law. However, the NFSA does prohibit access to maternity benefits for pregnant women employed by central or state governments or public sector undertakings.

Notes

1. IA No. 98/2010 in *People's Union of Civil Liberties v. Union of India & Ors* (henceforth *PUCL v. UoI*), WP(C) No. 196/2001, order dated 11 November 2007.

2. Relating to infants below 28 days of age.
3. SC order dated 27 April 2004 in *PUCL v. UoI*, WP(C) No. 196 of 2001.
4. IA No. 98/2010 in *PUCL v. UoI*, WP(C) No. 196/2001.
5. Order dated 20 November 2007, *PUCL v. UOI*, WP(C) No. 196/2001.
6. IA No. 98/2010, *PUCL v. UOI*, WP(C) No. 196/2001.
7. Order dated 11 November 2007, *PUCL v. UOI*, WP(C) No. 196/2001.
8. IA No. 98/2010, *PUCL v. UOI*, WP(C) No. 196/2001.
9. Commissioner's response on 5 April 2010 (see Saxena and Mander 2010).
10. Commissioner's response on 5 April 2010.

References

Adato, M. 2000. 'The Impact of Progresa on Community Social Relationships'. Final Report, International Food Policy Research Institute, Washington, D.C.

Alvarez, Carola, Florencia Devoto and Paul Winters. 2006. 'Why Do the Poor Leave the Safety Net in Mexico? A Study of the Effects of Conditionality on Dropouts'. Working Paper no. 2006–10, Department of Economics, American University. http://w.american.edu/cas/economics/repec/amu/workingpapers/2006-10.pdf (accessed 10 March 2018).

Attanasio, O., E. Fitzsimons, A. Gomez, D. Lopez, C. Meghir and A. Mesnard. 2005. 'The Short-Term Impact of a Conditional Cash Subsidy on Child Health and Nutrition in Colombia'. Institute of Fiscal Studies, London. https://www.ifs.org.uk/edepo/rs_fam03.pdf (accessed 10 March 2018).

Bastagli, F. 2011. 'Conditional Cash Transfers as a Tool of Social Policy'. *Economic and Political Weekly*, 46(21): 61–66.

Birchfield, Lauren, and Jessica Corsi. 2010. 'Between Starvation and Globalization: Realizing the Right to Food in India'. *Michigan Journal of International Law*, 31(4): 691–764.

Boltvinik, J. 2004. 'Políticas Focalisadas de Combate a la Pobreza en México. El Progresa/Oportunidades' (Targeted Poverty Fighting Policies in Mexico), in J. Boltvinik and A. Damián (eds), *La Pobreza en México y el Mundo: Realidades y Desafíos* (Poverty in Mexico and the World: Realities and Challenges). México: Siglo XXI editores.

Devadasan, N., M. A. Elias, D. John, S. Grahacharya and L. A. Ralte. 2008. 'Conditional Cash Assistance Programme for Promoting Institutional Deliveries among the Poor in India: Process Evaluation Results', in F. Richard, S. Witter and V. de Brouwere (eds), *Reducing Financial Barriers to Obstetric Care in Low-Income Countries*. Antwerp: ITG Press, pp. 257–73.

Fiszbein, Ariel, and Norbert Schady (with Francisco H. G. Ferreira, Margaret Grosh, Nial Kelleher, Pedro Olinto and Emmanuel Skoufias). 2009.

'Conditional Cash Transfers: Reducing Present and Future Poverty'. World Bank Policy Research Report, Washington, D.C.

Gakidou, E., R. Lozano, E. Gonzalez-Pier et al. 2006. 'Assessing the Effect of the 2001–06 Mexican Health Reform: An Interim Report Card'. *Lancet*, 368: 1920–35.

Gonzalez de la Rocha, Mercedes. 2005. 'Familias y Politica Social en Mexico El Caso de Oportunidades'. Serie Seminarios y Conferencias 46, ECLAC, Santiago.

Government of Nepal. 2005. 'Operational Guidelines on Incentives for Safe Delivery Services'. Ministry of Health and Population, Kathmandu.

Government of the People's Republic of Bangladesh. 2007. 'Demand Side Financing Pilot Maternal Health Voucher Scheme Proposal'. Ministry of Health and Family Welfare, Government of the People's Republic of Bangladesh, Dhaka.

Gragnolati, Michele, Meera Shekar, Monica Das Gupta, Caryn Bredenkamp and Yi-Kyoung Lee. 2005. 'India's Undernourished Children: A Call for Action'. HNP Discussion Paper no. 5. http://siteresources.worldbank.org/SOUTHASIAEXT/Resources/223546-1147272668285/IndiaUndernourishedChildrenFinal.pdf (accessed 18 September 2017).

Hanlon, Joseph, Barrientos Armando and David Hulme. 2010. *Just Give Money to the Poor: The Development Revolution from the Global South*. Sterling, VA: Kumarian Press.

IIPS (International Institute for Population Sciences). 2008. District Level Household Survey (DLHS-3), India 2007–08. Mumbai IIPS.

Lahariya, C. 2009. 'Cash Incentives for Institutional Delivery: Linking with Antenatal and Postnatal Care May Ensure "Continuum of Care" in India'. *Indian Journal of Community Medicine*, 34(1): 15–18.

Lim, Stephens S., Lalit Dandona, Joseph A. Hoisington, Spencer L. James, Margaret C. Hogan and Emmanuela Gakidou. 2010. 'India's Janani Suraksha Yojana, a Conditional Cash Transfer Programme to Increase Births in Health Facilities: An Impact Evaluation. *Lancet*, 375(9730): 2009–23.

Lo Vuolo, Rubén. 2010. 'Las perspectivas del ingreso ciudadano en América Latina' (The Perspectives of Basic Income in Latin America). http://www.sinteseeventos.com.br/bien/pt/papers/LoVuoloLasperspectivasdelICenAL.pdf (accessed 10 March 2018).

Malini, S., R. M. Tripathi, P. Khattar et al. 2008. 'A Rapid Appraisal on Functioning of Janani Suraksha Yojana in South Orissa'. *Health and Population: Perspectives and Issues*, 31: 126–31.

Maluccio, J. A., and R. Flores. 2005. 'Impact Evaluation of a Conditional Cash Transfer Programme: The Nicaraguan Red de Proteccion Social'. IFPRI

Research Report no. 4, International Food Policy Research Institute, Washington, D.C.

Mohapatra, B., U. Datta, S. Gupta, V. K. Tiwari, V. Adhish and D. Nandan. 2008. 'An Assessment of the Functioning and Impact of Janani Suraksha Yojana'. *Health and Population: Perspectives and Issues*, 31(2): 120–25.

Molyneux, M. 2006. 'Mothers at the Service of the New Poverty Agenda: Progresa/Oportunidades, Mexico's Conditional Transfer Programme'. *Social Policy and Administration*, 40(4): 425–49.

Morris, S. S., P. Olinto, R. Flores, E. A. Nilson and A. C. Figueiro. 2004. 'Conditional Cash Transfers Are Associated with a Small Reduction in the Rate of Weight Gain of Preschool Children in Northeast Brazil'. *Journal of Nutrition*, 134(9): 2336–41.

NAC (National Advisory Committee). 2011. 'National Food Security Bill, 2011'. http://www.righttofoodindia.org/data/right_to_food_act_data/July_2011_national_food_security_bill_final_draft_nac_version_6_july_2011.PDF (accessed 10 March 2018).

Paul, Vinod. 2010. 'India: Conditional Cash Transfers for In-Facility Deliveries'. *Lancet*, 375(9730): 2009–23.

Powell-Jackson, T., J. Morrison, S. Tiwari, B. Neupane and A. Costello. 2009. 'The Experiences of Districts in Implementing a National Incentive Programme to Promote Safe Delivery in Nepal'. *BMC Health Services Research*, 9: 97.

Raman, Anuradha. 2011. 'Then there were three'. *Outlook*, 15 August.

Right to Food Campaign. 2008. 'Supreme Court Orders on the Right to Food: A Tool for Action'. Right to Food Campaign, Secretariat, New Delhi.

Saxena, N. C., and Harsh Mander. 2010. 'Response to Application of Ministry of Health and Family Welfare, Government of India (IA no. 92) for Directions regarding Implementation of Janani Suraksha Yojana', 5 April. http://www.sccommissioners.org/Correspondence/Commissioners_Response_JSY_050410.pdf

Standing Committee on Food, Consumer Affairs and Public Distribution. 2013. 'The National Food Security Bill, 2011: Twenty Seventh Report'. Department of Food and Public Distribution, Ministry of Consumer Affairs, Food and Public Distribution, Government of India. http://www.righttofoodindia.org/data/right_to_food_act_data/official_documents/Standing_Committee_Recommendations_on_NFSB_2011_January_2013.pdf (accessed 10 March 2018).

Tripathi, S. P. 2013. 'The Effectiveness of National Maternity Benefit Scheme in Selected Districts of Madhya Pradesh'. *International Journal of Advancements in Research and Technology*, 2(4): 140–54.

UNFPA (UN Population Fund) and CORT (Centre for Operations Research and Training). 2007a. 'Assessment of ASHA and Janani Suraksha Yojana in Rajasthan'. http://www.cortindia.in/RP%5CRP-2007-0302.pdf (accessed 18 September 2017).

———. 2007b. 'Assessment of ASHA and Janani Suraksha Yojana in Madhya Pradesh'. http://www.cortindia.in/rp/rp-2007-0301.pdf (accessed 18 September 2017).

Victora, C. G., J. P. Vaughan, F. C. Barros et al. 2000. 'Explaining Trends in Inequities: Evidence from Brazilian Child Health Studies'. *Lancet*, 356: 1093–98.

Vyasulu, Vinod. 2010. 'Brazil's "Fome Zero" Strategy: Can India Implement Cash Transfers?' *Economic and Political Weekly*, 50(26–27): 93–95.

Yanes, Pablo. 2011. 'Mexico's Targeted and Conditional Transfers: Between Oportunidades and Rights'. *Economic and Political Weekly*, 46(21): 49–54.

7

Cash Transfers or Transfers in Kind

Ashwin Parulkar

Introduction to Cash Transfers in India

One of the more controversial debates that emerged around the provisions of the NFSB concerned the possible use of cash transfers in place of foodgrains through the PDS. Given the known problems of corruption and leakages that have plagued the PDS, questions over whether the bill should dispense cash to the poor in lieu of foodgrains were rooted in divergent views held by policymakers and academics on both the existing state of India's social security system, and ideas on the extent of the government's role—and responsibility—in securing the basic needs of its poorest citizens vis-à-vis public institutions or the private sector.

Cash transfers refer to the direct transfer of money from governments to people for a specific purpose (Ghosh 2011). Jayati Ghosh argues that in the Indian context, cash transfers could be used to support a range of public services that already exist, and the two options—cash and service delivery—do not have to be mutually exclusive. Cash transfer schemes in Brazil, aimed at ensuring a minimum income and standardised public services, and in Mexico, designed to increase school enrolments and public health access, proved to be successful in reaching poor people who had previously been systematically excluded from social services (Bastagli 2011). These programmes, however, did not drastically reduce poverty in these countries (ibid.; Lindert et al. 2006; Skoufias and Di Maro 2007). A 2009 World Bank study showed that the impact of CCTs on the headcount poverty ratio was 8 per cent in Mexico and 2 per cent in Brazil (where the income transfer as a share of beneficiary income was lower), while the poverty gap reduced by 29 per cent in Mexico and by 15 per cent in Brazil (World Bank 2009).

Indian policymakers cited the experience of these countries as the inspiration for the possible use of cash transfers in the NFSB, but many argue that there are several nuances to these countries' experiences that have been overlooked.[1] Jean Drèze and Amartya Sen (2013) have noted that the successes of CCTs in Brazil and Mexico in positively impacting the lives of the poor were due to the simultaneous efforts of those governments in ensuring excluded groups' access to needed services such as healthcare and school enrolments. In Brazil, key dimensions of basic healthcare such as immunisation, antenatal care and skilled attendance at birth are 'virtually universal' (ibid.). In the absence of comparable health infrastructure and outreach in India, the authors warn that CCTs in the Indian context may not have similar impacts on poverty reduction, and that this policy move alone cannot 'fill the gap' (ibid.). Sen and Rajasekhar (2012) argue that while the rationale for social protection programmes evolved in similar phases in India and Latin America—from a focus on organised workers during import substitution models of industrialisation in the 1960s to market-based programmes in the 1980s and 1990s—the scale and scope of development challenges in India should caution policymakers from adapting Latin American models of cash transfers too readily. The major differences between India and Latin American countries, they claim, is that 'India's per capita income (and hence, overall financial capacity) is much lower, its poor population much larger, and its levels of absolute deprivation much higher.... the proportion of unorganised workers is significantly higher in India (over 90 percent of all workers) as against smaller proportions in Latin American countries' (ibid., 95).

Still, several forms of cash transfers in the context of food security already exist in India—pensions for the elderly, widows and persons with disabilities and maternity entitlements for lactating women. The National Family Benefit Scheme is a programme authorised under the Supreme Court's orders in the right to food case that provides cash to BPL families who suffer the premature death of a primary breadwinner between 18 and 65 years of age (Right to Food Campaign 2008). These cash benefits supplement other entitlements related to food security, such as the right of BPL households to purchase subsidised rations under the PDS, and the right of every rural household to obtain 100 days of manual work a year at minimum wages under MGNREGA. In other words, cash transfers are currently one of several existing policies in place for enhancing food security in India, and achieving overall social and economic goals, by targeting the multiple vulnerabilities poor people face.

Yet, a furore emerged over the possibility that cash transfers would supplant direct disbursements of food through the PDS. Cash transfers in lieu of the PDS would involve the transfer of money into the bank accounts of BPL cardholders. Recipients would withdraw this money to buy food from the market instead of going to their local ration shop to purchase grains at the subsidised rate. The amount of cash transferred would be sufficient in that it would enable the recipients to buy the same quantity of rations that they were previously buying from FPSs. Suppose a household was entitled to 35 kilograms of rice a month at the rate of Rs 3 per kilogram under the PDS. Assuming that the market rate of rice is Rs 15 per kilogram, the implicit subsidy from the PDS for this household would be Rs 420 a month.[2] So, the monthly cash transfer amount for this household would be Rs 420.

In late 2012, the debate intensified when the Government of India announced plans to implement cash transfers in January 2013 across 29 major development programmes in 51 districts with the help of the Unique Identification (UID) or Aadhaar scheme.[3] The Ministry of Finance recommended the release of funds to the various ministries for transferring subsidies and benefits under the various programmes, which were to be transferred from the government to Aadhaar-enabled bank accounts of recipients through a central payment platform. The recipients would then withdraw the money either through bank branches or through business correspondents equipped with micro-ATMs. Most of the newly proposed cash schemes were to be implemented in scholarship, healthcare and pension schemes. Still, food rights advocates grew concerned that cash transfers may replace direct food distribution since this provision was included in the government's 2011 draft NFSB, and pilot cash transfer schemes for non-food PDS items like kerosene were already under way in select districts. In some cases, the pilots fared poorly. For instance, in December 2011, a cash for kerosene scheme began in the Kotkasim region of Rajasthan state. The state government agreed to deposit cash in the amount of the difference between the PDS amount (Rs 15) and the market rate (about Rs 50) every three months in accounts at a local bank that agreed to host the programme. Instead of buying their kerosene from the local grain shops at the PDS amount, beneficiaries were to purchase it in the market under the agreement that the government cash deposits would subsidise their expenses. By December 2012, findings from the pilot revealed that the overall consumption of kerosene had fallen because either too few bank accounts were opened for the intended recipients, or the funds were not deposited in the accounts as promised. The problems in the rural banking infrastructure raised concerns among food rights advocates over the efficacy of cash for goods transfers in place of public goods at a time when the NFSB, which had provisos for cash transfers in the food system, awaited parliamentary debate and a vote.

While the draft bill prepared by the NAC was silent on the matter of cash transfers, the 2011 government draft mentioned cash transfers in two sections. In the section that detailed reforms in the PDS, the bill stated that cash transfers, food coupons and other schemes may be used in lieu of foodgrain entitlements, while a 'food security allowance' was also included in Clause 13 of the bill, which guaranteed cash to beneficiaries 'in case of non-supply of the entitled quantities of foodgrains or meals to entitled persons'. The stipulation of cash transfers in the draft bill was a catalyst for heated debates that followed in the months before its final passage. While the Standing Committee opposed the introduction of cash transfers on the basis of inadequacy of the banking infrastructure needed to successfully implement such a move, the final NFSA retained the provisions for cash transfers, food coupons and the food security allowance from the 2011 government draft. Some of these recommendations and parliamentary debates will be highlighted in further sections.

Types of Cash Transfers

Conditional Cash Transfers

Conditional cash transfers are safety net programmes that transfer cash to people on the condition that they fulfil certain requirements, which usually pertain to education or health. Conditional cash transfers have been used in several developing countries to reduce poverty, and more specifically to increase the levels of education and people's access to healthcare facilities (Narayanan 2011).[4] India's JSY programme is a CCT scheme for institutional deliveries, which was introduced in 2005 by the Government of India as a response to the slow and varied progress in improvement of maternal and neonatal health in the country. The JSY is the largest CCT programme in the world in terms of the number of beneficiaries (Lim et al. 2010).[5]

Unconditional Cash Transfers

Unlike CCTs, where the population for which the transfers are intended have to first fulfil certain behavioural requirements, unconditional cash transfers do not require the recipients to meet any conditions before receiving the transfer amounts. Unconditional cash transfers have typically been intended as safety net interventions, such as old age pensions in South Africa and Namibia. These programmes have contributed to a reduction in poverty in the short run and have enabled recipients to cope with HIV/AIDS while also enabling human capital investments, such as in grandchildren (Narayanan 2011). Another example from South Africa is the Child Support Grant, which had a positive impact on school enrolment (Case et al. 2003; Miller et al. 2011; Narayanan 2011).

Arguments in Favour of Cash Transfers in the Indian Context

The prominent supporters of cash transfers in India included many in the government, such as the former prime minister Manmohan Singh, who was also the chairperson of the Planning Commission, the then chairperson of the NAC Sonia Gandhi, former finance minister P. Chidambaram, then deputy chairperson of the Planning Commission Montek Singh Ahluwalia, and the former chief economic advisor to the prime minister, Kaushik Basu. The prevailing argument from the government was that providing subsidies in the form of cash directly to the poor would enable them to access goods currently denied to them due to defunct social programmes beset by corruption. Kaushik Basu recommended providing food coupons that would allow recipients to choose from a range of options in local markets. The World Bank, as well as other international agencies, also supported experimentation with cash transfers in lieu of the PDS.

Cash Transfers: A More Relevant Form of Social Protection

Guy Standing (2008) urges Indian policymakers to view cash transfers as the lynchpin of a new kind of social protection system in India, arguing that globalisation has led to increased vulnerabilities in developing countries like India that older models of service delivery—farm subsidies and social insurance regimes, for instance—are not equipped to address. He argues that India's social and economic system will continue to change in accordance with the dictates of globalisation, so policymakers should respond to these changes by creating new social security measures in line with those needs, instead of reforming older models with proven deficiencies that are not applicable to the changing world.

Santosh Mehrotra, similar to Standing, believes that India requires cash assistance for the vulnerable based on the large share of India's workers engaged in the unorganised sector, which is characterised by low and irregular incomes (Mehrotra 2010; NCEUS 2008). He also cites the need for a 'social contract' based on international examples of now-developed countries that transitioned to that state after World War II. Mere policies and institutions intended to provide access to basic services to the poor, he argues, do not fulfil the objectives of a 'social contract' if proper implementation is constantly undermined by corruption and malfeasance.

Standing suggests that policymakers should evaluate all social protection measures—cash transfers and similar forms of assistance—to gauge not only how effective the policy is in reaching the beneficiaries, but in actually leading to well-being in line with the goals of economic and social development. He argues that if properly executed, universal unconditional cash transfers satisfy these objectives; conversely, older models of service delivery, where the intended beneficiary's ability to access social support at times depends on the politics of the local bureaucracy, is disadvantageous to livelihood, development and the well-being of the most vulnerable.

Cash Transfers Will Reduce Large-Scale Leakages and Inefficiencies in Social Security Programmes

In general, those in favour of cash transfers argue that they will help reduce the fiscal deficit by curbing the amount of expenditures earmarked for poverty reduction programmes that are siphoned off through corruption. The programme under most scrutiny and debate is the PDS, routinely criticised for leakages and inefficiencies. Proponents of cash transfers thus argue that the system should be dismantled and replaced by direct deposits of money into the bank accounts of the poor. According to estimates by the Planning Commission, in 2004–05 the monthly PDS subsidy to an average poor household with a BPL card was only Rs 81.5. Based on NSS (2007) data, Svedberg (2012) calculates that of the total Rs 175 billion in subsidies to the PDS in 2004–05, only 11 per cent (Rs 19 billion) actually reached poor households. Svedberg argues that with this scale of diversion, the PDS has almost no effect on reducing poverty and malnutrition. In the same year, the PDS budget accounted for almost 3 per cent of the government's total expenditure and nearly 1 per cent of the net national product. Indeed, several government-appointed commissions have explored the causes of the level of mismanagement and inefficiencies in all right to food programmes (Planning Commission 2005; Supreme Court Commissioner's Office 2004, 2005, 2007; Swain and Kumaran 2012; Wadwa Commission 2009). A survey of the commissions' reports by Swain and Kumaran found that the reports point to corruption and discrimination against people on the basis of caste, gender and religion as major reasons for poor performances in the food security system, among other reasons:

1. A poor monitoring system resulting in diversion/leakages that prevent resources from reaching the service delivery points.
2. Local elite control over service delivery, which leads to overt corruption by service providers.
3. Unaccountable service providers who do not regularly open the service points, or open them at inconvenient times; distribute less than the stipulated entitlements in terms of quantity and quality; charge unwarranted prices or higher-than-stipulated prices to the beneficiaries; and mislead beneficiaries with false information.
4. Direct discrimination on the basis of caste, gender and religion.
5. Non-transparency on the part of the service providers and monitoring bodies.

According to Planning Commission estimates in 2008, of the 13.53 million tons of food purchased from FPSs in 2004–05, only 4.6 million tons were bought by poor households (Planning Commission 2008).

In the same year, 54 per cent of the grain earmarked for the PDS was leaked before reaching these shops, almost double the proportion of the leakages in 1993–94 and 1999–2000, estimated at 28 per cent. However, according to 'small' expenditure surveys conducted in 2007–08, the leakages have reduced to 43 per cent (Himanshu and Sen 2011; Khera 2011). A lengthier discussion on the reduction of leakages after 2004–05 will follow later.

In this context, proponents of cash transfers argue that providing the subsidy directly to the poor would both bypass agents and brokers as well as reduce holding costs of grains in government silos that accrue due to the lack of local storage transportation facilities. According to Svedberg (2012), if cash transfers are introduced, central storage of grain (amounting to 30–35 million tons a year) will not be required, and the remaining food security buffer stocks (20–30 million tons) could be stored in high-quality godowns, estimated to almost completely eliminate storage losses. Private agents would take over most of the transportation of grain which is currently carried out by the FCI; they are already providing such services at considerably lower prices (Planning Commission 2005). Cash transfers would also remove the need for FPSs, as rations would then be supplied through commercial shops, which would eliminate leakages at the retail level.

In parliamentary debates, Rangasayee Ramakrishna and Jayshreeben Patel, members of the BJP, favoured cash transfers for the reasons mentioned above. 'In my view it would be better to put the prescribed amount in the accounts of the beneficiaries and ask them to get to fend for themselves, but instead of that we continue with the existing system with all its infirmities,' Mr Ramakrishna said.[6] Ms Patel said, 'If money in cash is given instead of foodgrains then each person could be given double the amount of money in place of subsidy being provided through foodgrain distribution. This would make the FCI and PDS redundant.'[7]

Cash Would Enable People to Buy Food of Their Choice

A common argument in favour of cash transfers is that they would enable people to buy food of their choice from the open market (Basu 2011), and not be restricted to items sold in the PDS, which are often inferior in quality. It is also argued that cash transfers would enable people to buy food which is of better quality, as the food sold in the open market is sold at market prices and subject to competition.

Reforming the PDS a Costly Option

Himanshu and Sen (2011) note that the various reforms in the PDS are quite recent, and have so far been implemented only on a pilot basis in a few states. Although these reforms have led to improvements in operations and reductions in leakages, they have led to a greater subsidy burden on the state governments which are financing them. These state government expenditures on PDS reforms are in addition to the central government's PDS subsidy. Svedberg (2012) uses this information to argue that PDS reforms are a costly and hence somewhat unviable option.

Arguments against Cash Transfers in the Indian Context

While the government is keen to experiment with cash transfers, surveys have revealed that many of those who would receive these transfers in place of welfare programmes are opposed to this idea. In a survey that covered more than 12,000 rural households in nine states in May and June 2011, on the question of their choice of subsidised grain or cash, 67 per cent of respondents preferred the former. The percentages were especially high in states with well-performing food distribution systems (I will expand on this in the next section), such as Andhra Pradesh (91 per cent, where respondents were receiving 99 per cent of their grain entitlements), Chhattisgarh (90 per cent, where respondents currently received 95 per cent of grain entitlements) and Odisha (88 per cent, where respondents currently received 99 per cent of grain entitlements) (Khera 2013). The most common reasons for preferring grains over cash were 'food security', 'protection from exploitation' and 'self-control', where some respondents mentioned that cash may be used by some members of the family on wasteful expenses, like alcohol.

Improvements in PDS in Recent Years

As discussed earlier, those who argue that the PDS suffers from large-scale leakages use the NSS data collected in 2004–05 or earlier to support their stand. However, more recent studies reveal that there have been significant improvements in the PDS in several states. Between 2004–05 and 2007–08, the number of households receiving cereals from the PDS increased from 27 to 35 per cent. This proportion increased by more than 10 percentage points in several states, from 6 per cent to 18 per cent in Uttar Pradesh, from 22 per cent to 45 per cent in Odisha, from 25 per cent to 47 per cent in Chhattisgarh, and from 40 per cent to 59 per cent in Kerala (Khera 2011). This phenomenon can be largely attributed to reforms in the PDS, such as moving towards universalisation, lowering the price and increasing the quantity of PDS rations, giving the responsibility of running FPSs to self-help groups, and use of technology, such as computerisation of databases and using SMS and GPS services to track the movement of PDS rations. Critics of cash transfers thus argue that to enhance food security, the PDS should be strengthened and reformed.

The strongest case against cash transfers in lieu of food has arguably come from nationwide surveys by Jean Drèze and Reetika Khera (2013), who have drawn attention to significant improvements in the PDS, particularly in states with high poverty ratios that have had previous histories with corruption and leakages. They have shown that purchases increased by 50 per cent between 2004–05 and 2009–10 due to states like Andhra Pradesh, Chhattisgarh, Himachal Pradesh, Kerala and Tamil Nadu expanding the system to all residents, thus doing away with the divisive politics of poverty surveys.

A comparison between Chhattisgarh, which has a universal PDS, and Bihar, where the system is still in disrepair, with regard to the impact of food distribution on hunger is of note, since both states had a similar poverty 'headcount ratio' in 2009: 56 per cent of people in Chhattisgarh were poor compared to 55 per cent in Bihar (Drèze and Khera 2013). The authors' survey of households in each state showed that 70 per cent of respondents in Bihar claimed to have skipped meals in the three months before the survey, compared to 17 per cent of respondents in Chhattisgarh. This 'highlights not only the substantial impact of the PDS on rural poverty in states with a well-functioning PDS but also the misleading nature of poverty estimates that effectively ignore the PDS factor' (ibid.).

Using the national poverty line of consumption expenditure of Rs 673 per month, Drèze and Khera (2013) also show that due to the recent improvements in the system across certain states, the PDS reduced poverty by 11 per cent at the national level, with the understanding that there are large variances in the state-level impact on poverty due to continued problems in Jharkhand and Bihar. The best-performing states in terms of the reduction of poverty are Tamil Nadu (61 per cent), Andhra Pradesh, Karnataka and Kerala (between 33 to 41 per cent), Chhattisgarh (39 per cent), and Himachal Pradesh and Jammu & Kashmir (35 per cent) (ibid.).

Exclusion Errors Will Remain, Inclusion Errors May Increase

One of the main shortcomings of the TPDS is the large-scale exclusion errors that result from the process of selecting 'deserving' households for subsidised rations. According to NSSO data, about 50 per cent of poor rural households in the country did not have a BPL card in 2004–05, with rates as high as 80 per cent in Bihar and Jharkhand.[8] The system of cash transfers, however, would not remove this large-scale exclusion of food-insecure households from receiving assistance, and these excluded households would continue to be denied their right to food even if the PDS were to be replaced by cash transfers.

Currently, there is some level of 'self-selection' in areas where the PDS is universal, such as Tamil Nadu. This means that some non-poor households who are eligible to buy subsidised rations choose not to do so, because of the effort involved in the process, such as standing in queues at FPSs and/or because of being able to purchase superior-quality food items from the open market. This process of self-selection reduces the quantity of resources required for the PDS. However, if the PDS is replaced by cash transfers, the phenomenon of self-selection is expected to disappear, as no one is likely to forego the opportunity of obtaining money.

Leakages and Irregularities Will Remain

The main argument in support of cash transfers is that they would bring about drastic reductions in leakages in welfare programmes. However, irregularities in existing cash transfer programmes such as pensions, maternity entitlements and the National Family Benefit Scheme and bank payments of MGNREGA wages show that this claim is highly exaggerated, if not altogether untrue. These irregularities include delayed and/or incomplete payment of the cash entitlements.

Inadequate Banking and Market Infrastructure

The cash transfer system would require opening bank accounts for all the recipients of cash and going to markets to buy the food, although many villages, especially those which are far from block headquarters, are often several kilometres away from banks and markets. If the cash transfer system is introduced without expanding the banking system and markets, it would result in people having to travel long distances (and maybe even spend money) to reach these institutions. As the banks serving remote villages are likely to cater to a large population, people might also have to wait in long queues to withdraw money, which might even require foregoing a day's work (and that day's wages) to withdraw money before going to the market to buy food. Currently, many people, especially those who are unable to read or unable to travel to the bank because of old age, disability, long distances or a multitude of other reasons, are dependent on middlemen to withdraw money from banks. These touts in turn take a part of the pensions or wages which they help in withdrawing. Thus, unless the banking system is streamlined and made easier for people to use, those receiving cash transfers would pay substantial 'transaction costs' to access and utilise these transfers.

In the Kotkasim case mentioned earlier, the drop in consumption of kerosene from 84,000 litres a month for nearly 25,000 cardholders to between 1,000 to 5,000 litres a month was largely attributed to the failure of banks to open accounts for beneficiaries. Just above 14,000 cardholders had had accounts opened at the end of the first year. Further, while the government was supposed to deposit the first payments three months before the programme was scheduled to start in December 2001, only 6,000 accounts had been opened by that time.

The Standing Committee on Food, Consumer Affairs and Public Distribution (2013) also opposed the introduction of cash transfers in the NFSA on the basis of the country's poor banking infrastructure. 'Before introducing direct cash transfers in lieu of food entitlements', the committee wrote, 'the government should ensure that adequate banking infrastructure is made accessible throughout the country' (ibid.). While most of the right parties were in favour of cash transfers, several members of Parliament voiced disapproval of cash transfers. Arguments were based on beliefs that cash payments would not resolve problems in the access of poor people to the existing public food distribution system, and would further endanger the lives of farmers (Smt. Harsimrat Kaur Badal).[9] In the absence of concomitant reforms of the food delivery system, cash payments would also leave the poor largely reliant on the free market to secure food (Shri Thol. Thirumavalavan)[10] and, by undergirding public policy with an expansion of neoliberal reforms, would further exclude the poor from state-backed social protection measures (Prof. Sk Saidul Haque).[11]

Fungibility of Cash Will Undermine Food Security

Proponents of cash transfers highlight the fungibility of cash, i.e., the ability to use cash to buy whatever people want, as one of the biggest advantages of the cash transfer system. However, the fungible nature of cash also makes it possible to spend the money on non-food items, which would decrease the amount of money left for buying food.

Loss of Protection from Inflation

The PDS provides rations at a constant rate, irrespective of fluctuations in the market prices of these rations. Hence, the implicit subsidy provided by the PDS increases with the increase in market prices of the goods sold in the FPSs. So, apart from contributing to food security, the PDS also provides implicit income support by giving some relief from inflation. Although theoretically the cash transfer amounts can also be increased commensurately with inflation, the inability of the government to do this with the existing cash transfers—such as pensions, maternity benefits and the National Family Benefit Scheme—

does not give reason to believe that the cash transfers for food will be increased with increase in food prices.

No Protection from Arbitrary Increase of Food Prices during Scarcity of Food Supply

Cash transfers in lieu of the PDS are based on the assumption that at a given point in time, prices of food items will remain constant. This assumption is flawed, as in instances of scarcity of food supply to the markets, the cash transfer system would not be able to prevent shopkeepers from arbitrarily increasing the prices of rations as per the 'market forces of supply and demand'. This would decrease the real value of the cash transfers, resulting in people being able to buy lower quantities of food with their given cash amounts.

Adverse Impact on Agriculture

The PDS requires the government to procure food from farmers. However, it is feared that the system of cash transfers would remove this obligation of the government, which would have an adverse impact on agriculture and farmer protection. Indeed, the guarantee of MSP purchase by the government for wheat and rice is the most important instrument for the protection of farmers' income in India, and this would become unfeasible if the government could not offload a lot of this grain back into the market through the PDS. The fear is that cash transfers may roll back MSP and PDS, and through that farmers' income protection.

Cash Transfers Mark a Trend towards Privatisation

It is argued that cash transfers are merely another example of the government's insistence on furthering the neoliberal economic agenda by 'washing its hands away from its responsibility of providing basic services and social security to its citizens.'[12] Sheikh Saidul Haque of the CPI(M), for instance, opposed cash transfers in parliamentary debates, saying:

> I strongly object to these new provisions which will reduce the entire bill into a platform to push through non-liberal reforms with legal

sanctions which are against the interests of the people and which will lead to further exclusion. In a blatant violation of federal reforms, the Central Government reserves the right to notify the date of PDS reforms which will be mandatory for all the State Governments.

Alternative to Cash Transfers: Strengthening and Reforming the PDS

In the PDS survey mentioned earlier, most respondents in states where the PDS is functional did not want cash transfers. Preference for cash was more prevalent in areas where the PDS is dysfunctional, particularly in Bihar. Along with the reasons for opposition to cash transfers, this finding builds a clear case for strengthening and reforming the PDS. Reforms which have succeeded in improving the PDS in some states should be implemented elsewhere as well. These reforms are discussed below.

Moving towards Universalisation

For some time now, Tamil Nadu has had a universal PDS. Himachal Pradesh too has a universal PDS; although APL cardholders pay a higher price than BPL cardholders, they get the same rations in the same quantities. Andhra Pradesh and Chhattisgarh have a 'quasi-universal' PDS—80 per cent of the population of these states is entitled to PDS rations. These states also have a better-functioning PDS, as greater coverage under the programme increases people's stake in it, which in turn leads to better implementation and monitoring of the programme.

Reducing Prices of PDS Grains

Along with increasing the number of households covered under the PDS, some states have also reduced the issue prices of PDS rations below the centrally fixed issue prices for BPL households—Rs 4.65 per kilogram for wheat and Rs 6.15 per kilogram for rice. In Tamil Nadu, 20 kilograms of rice are provided to every household free of cost. In Jharkhand, rice is sold at Re 1 per kilogram and is free for households belonging to the 'particularly vulnerable tribal groups'.

In Chhattisgarh, Rajasthan, Odisha and Andhra Pradesh, PDS grain costs Rs 2 per kilogram. Reduced prices of PDS grains increase the implicit subsidy provided by the programme, which plays a crucial role in providing protection from hunger and poverty. Just like the move towards universalisation, reduction in PDS prices also increases the welfare obtained from the programme, which encourages people to ensure that it is implemented without irregularities.

Increase in the Variety of Commodities Sold in the PDS

Some states have added more nutritious items to their basket of PDS commodities. In Himachal Pradesh, for instance, all ration card holders are entitled to at least 1 kilogram of dal and 1 kilogram of oil a month. Larger households can purchase up to 3 kilograms of dal and 2 kilograms of oil. Dal and edible oil are provided in Tamil Nadu and Andhra Pradesh as well, with pulses having been provided intermittently in other states. This makes access to nutritious food more affordable, which is crucial for good nutrition.

Making Fair Price Shops Viable

The financial viability of FPSs depends on the quantity of rations handled by them, and the official commissions earned on these commodities. In the past few years, the official commissions have increased in several states, which has contributed to improving the viability of the shops. Due to PDS expansion in Chhattisgarh, Odisha and Rajasthan, the number of BPL cards handled per FPS has also increased. More viable FPSs reduce the pressure on ration dealers to indulge in corruption, and PDS beneficiaries are given their rations more regularly. Giving reasonable salaries to salesmen employed by cooperative societies that run FPSs also helps in curbing corruption at the retail level.

'Doorstep Delivery' of Grain to FPSs

Most of the PDS grain is leaked after it is removed from FCI godowns and before it reaches the FPS. The chances of this diversion are highest when the transportation is undertaken by the PDS dealers. To prevent this, the food departments in several states have started making arrangements to transport PDS rations to the FPSs themselves. This has resulted in ration card holders receiving a higher proportion of the quantity of rations to which they are entitled. Further ways of reducing leakages during transportation include the computerisation of the PDS (Department of Food and Public Distribution 2009). In Chhattisgarh, trucks carrying PDS grain are clearly marked and fitted with GPS to monitor their routes and unloading places (ibid.).

Elimination of Fake Ration Cards

To reduce leakage of grains at the retail level through fake ration cards, digital ration cards that are difficult to duplicate and can be used to document sales electronically are being used at some places.

In light of the various objections to cash transfers, the Right to Food Campaign emphatically supported an overhaul of the PDS, suggesting that the NFSB should do away with APL and BPL categories and universalise access to the system. It was also proposed that the PDS should include other nutritious staples such as pulses and edible oil, and finally, that the NFSB should make provisions for the deprivatisation of FPSs, doorstep delivery of food commodities and end-to-end computerisation, and institute a system of transparency, accountability and grievance redressal (Right to Food Campaign 2011). It is important to stress that the opposition has only been to cash as a substitute for food transfers. Cash transfers are supported for maternity benefits to pregnant and nursing mothers, and for social security pensions for the aged, single women and disabled. The argument is that transferring food, as grain or cooked food, directly to persons who may be food insecure is a more effective and reliable instrument than substituting this food with cash.

Notes

1. Whereas there are many important lessons from Brazil's Bolsa Família programme and Mexico's Oportunidades programme of CCTs (see Chapter 6), this chapter will focus on the experience and debates within India.

2. Total cost of purchasing the amount of ration entitled under the PDS from the open market *less* total cost of purchasing the amount of ration entitled under the PDS from the fair price shop = (35 × 15) – (35 × 3) = 420.

3. Current beneficiaries of the NFSA are required to possess an Aadhaar card to access grains entitlements—5 kilograms of foodgrains per person per month for priority households and 35 kilograms per household for families deemed by local officials to be the poorest of the poor, or Antyodaya families. Surveys in 2017 revealed that because eligibility to receive grains now depends on a person's Aadhaar numbers being located in the PDS database on hand at local ration shops, poor people are routinely denied their foodgrains quotas. Nayak and Nehra's (2017) study of Delhi shows that the current Aadhaar-enabled PDS has created two main barriers to access. People without Aadhaar cards are excluded from the system. Second, beneficiaries are penalised for faulty technology—when either the system does not recognise the beneficiary's Aadhaar number or the machine does not work due to poor electricity connections in ration shops. In their survey of 320 households in Delhi, nearly half (48 per cent) of households had at least one person missing from their ration card because of the absence of an Aadhaar number, while more than half (52 per cent) of the families who used ration shops had problems getting their grains because the system did not recognise their fingerprints or did not work because of a poor internet connection (ibid.).

4. For a detailed discussion on the rationale behind conditionalities and the dangers associated with conditional assistance, see Chapter 6 on 'Conditionalities'.

5. For more details on JSY and its conditionalities, see Chapter 13 on 'Gender-Just Food Entitlements'. See also Chapter 6 for a detailed discussion of conditionalities.

6. Message from Lok Sabha, Statutory Resolution Disapproving the National Food Security Ordinance (no. 7 of 2013) and the National Food Security Bill, 2013, 2 September 2013.

7. Lok Sabha, Synopsis of Debates (Proceedings Other Than Questions & Answers), Monday, 26 August 2013.

8. For a detailed discussion on targeting in the PDS, see Chapter 4, 'Universal or Targeted Entitlements'.

9. Lok Sabha Debates, Part I—Proceedings and Answers with Questions, Monday, 26 August 2013, pp. 168, 360.

10. Ibid., 312.

11. Ibid., 168.

12. Ibid.

REFERENCES

Bastagli, F. 2011. 'Conditional Cash Transfers as a Tool of Social Policy'. *Economic and Political Weekly*, 46(21): 61–66.

Basu, K. 2011. 'India's Foodgrain Policy: An Economic Theory Perspective'. *Economic and Political Weekly*, 46(5): 37–46.

Case, Anne, V. Hosegood and F. Lund. 2003. 'The Reach of the South African Child Support Grant: Evidence from KwaZulu-Natal'. Research Programme in Development Studies, Working Paper no. 224, Princeton University, New Jersey.

Cunha , J. 2011. 'Testing Paternalism: Cash vs. In-Kind Transfers'. *American Economic Journal: Applied Economics*, 6(2): 196–230.

Cunha, De Giorgi, and Jayachandran. 2011. 'The Price Effects of Cash versus In-Kind Transfers'. Working Paper no. 17456, National Bureau of Economic Research, Cambridge, MA.

Department of Food and Public Distribution. 2009. *Annual Report 2009–10*. New Delhi: Government of India.

Drèze, Jean, and Amartya Sen. 2013. *An Uncertain Glory*. London: Allen Lane.

Drèze, Jean, and Reetika Khera. 2013. 'Rural Poverty and the Public Distribution System'. Working Paper no. 235, Centre for Development Economics, Delhi School of Economics, New Delhi.

Ghosh, J. 2011. 'Cash Transfers as the Silver Bullet for Poverty Reduction: A Special Note'. *Economic and Political Weekly*, 46(21): 67–71.

Himanshu and Abhijit Sen. 2011. 'Why Not a Universal Food Security Legislation?' *Economic and Political Weekly*, 46(12): 38–47.

Kapur, D. 2011. 'The Shift to Cash Transfers: Running Better But on the Wrong Road?' *Economic and Political Weekly*, 46(21): 80–85.

Khera, Reetika. 2011. 'PDS: Signs of revival'. *The Hindu*, 12 June.

———. 2013. 'Cash vs. In-Kind Transfers: Indian Data Meets Theory'. Working Paper no. 325, Institute of Economic Growth, New Delhi.

Lim, Stephens S., Lalit Dandona, Joseph A. Hoisington, Spencer L. James, Margaret C. Hogan and Emmanuela Gakidou. 2010. 'India's Janani Suraksha Yojana, a Conditional Cash Transfer Programme to Increase Births in Health Facilities: An Impact Evaluation'. *Lancet*, 375(9730): 2009–23.

Lindert, K., E. Skoufias and J. Shapiro. 2006. 'Redistributing Income to the Rich and Poor: Public Cash Transfers in Latin America and the Caribbean'. Social Protection Working Paper no. 0605, World Bank, Washington, D.C.

Magen, B., C. Donovon and V. Kelly. 2009. 'Can Cash Transfers Promote Food Security in the Context of Volatile Commodity Prices? A Review of

Empirical Evidence'. International Development Working Paper no. 96, Michigan State University.

Mehrotra, Santosh. 2010. 'Introducing Conditional Cash Transfers in India: A Proposal for Five CCTs'. Institute of Applied Manpower Research, New Delhi.

Miller, Candace M., Maxton Tsoka and Kathryn Reichert. 2011. 'The Impact of the Social Cash Transfer Scheme on Food Security in Malawai'. *Food Policy*, 36(2): 230–38.

Narayanan, S. 2011. 'A Case for Reframing the Cash Transfer Debate in India'. *Economic and Political Weekly*, 46(21): 41–48.

Nayak, Nandini, and S. Nehra. 2017. 'Accessing the Right to Food in Delhi'. *Economic and Political Weekly*, 52(23).

NCEUS (National Commission of Enterprises in the Unorganized Sector). 2008. *Social Security in the Unorganized Sector*. New Delhi: Government of India.

NSS (National Sample Survey). 2007. *Public Distribution System and Other Sources of Household Consumption 2004–05*. Report no. 510. New Delhi: Ministry of Statistics and Programme Implementation, Government of India.

Planning Commission. 2005. 'Performance Evaluation of Targeted Public Distribution System (TPDS)'. Programme Evaluation Organisation, Planning Commission, Government of India, New Delhi.

———. 2008. *Eleventh Five-Year Plan*. New Delhi: Government of India, vol. II.

Right to Food Campaign. 2008. 'Supreme Court Orders on the Right to Food'. Right to Food Campaign Secretariat, New Delhi.

———. 2011. 'Replacing PDS with Cash Transfers'. http://www.righttofoodindia.org/data/May_2011_Draft_note_on_replacing_PDS_with_cash_transfers_English.pdf (accessed 10 March 2018).

Sen, Gita, and D. Rajasekhar. 2012. 'Social Protection Policies: Experiences and Challenges', in R. Nagaraj (ed.), *Growth, Inequality and Social Development in India* (Basingstoke: Palgrave MacMillan), pp. 91–134.

Skoufias, E., and V. Di Maro. 2007. 'Conditional Cash Transfers, Adult Work Incentives and Poverty'. *Journal of Development Studies*, 44(7): 934–60.

Standing, Guy. 2008. 'How Cash Transfers Promote the Case for Basic Income'. *Basic Income Studies: An International Journal of Basic Income Research*, 3(1): 41–48.

Standing Committee on Food, Consumer Affairs and Public Distribution. 2013. *The National Food Security Bill, 2011: Twenty-Seventh Report*. http://www.prsindia.org/uploads/media/Food%20Security/SC%20Report-Food%20Security%20Bill,%202011.pdf (accessed 10 March 2018).

Svedberg, Peter. 2012. 'Reforming or Replacing the Public Distribution System with Cash Transfers?' *Economic and Political Weekly*, 47(7): 53–62.

Swain, B., and M. Kumaran. 2012. 'Who Do ICDS and PDS Exclude and What Can Be Done to Change This?' *IDS Bulletin*, 43(1): 32–39.

World Bank. 2009. 'Conditional Cash Transfers: Reducing Present and Future Poverty'. Research Report, World Bank, Washington, D.C.

Part III

Food Entitlements and Special Groups

8

Ways by Which States Can Ensure Food Security

Ankita Aggarwal

In any country drafting a right to food law which includes obligations for public provisioning of food, the drafters need to address the question of what kinds of food provision duties the law should create for the state.[1] Access to food can be ensured by one or more of the following three alternatives: through direct provisioning of food; by providing the means to procure food; and by enabling the production of food. While everybody's access to food may be ensured by providing them with food, this is not the case with the other two options; for instance, young children would neither be able to procure nor produce food. Direct provisioning of food can take two forms: through subsidised or free food rations and through cooked meals. Means to procure food can be ensured by providing employment opportunities, or through social security cash transfers such as pensions, maternity benefits and unemployment allowances. Support for food production can be provided in a large number of ways; these include giving people rights over agricultural land, subsidising agricultural inputs such as seeds and fertilisers, making provisions for irrigation, sharing technical know-how with farmers, providing farmers with credit to buy agricultural inputs, and providing them with assured markets and MSPs for their crops to help ensure profitability. Supporting food production is required for ensuring not just the producer's but everybody's food security.

Apart from deciding on the method for ensuring people's food security, there is also a need to decide on the quantum of the support. This means deciding the quantities, types and prices of food rations; frequency, nutritional content and pricing of cooked food; and monetary values of wages, social security cash transfers, agricultural subsidies and MSPs for different crops.

FAO Guide on Legislating for the Right to Food on Types of Food Provisioning Entitlements

The FAO's *Guide on Legislating for the Right to Food* (2009) suggests that a right to food law needs to ensure that people have access to *adequate* food. While the precise meaning of adequacy will vary according to social, economic, cultural, climatic, ecological and other conditions, and the age, gender and nature of a person's work, it needs to take into account food safety, nutritional quality, quantity and cultural acceptability. In the end, decisions about adequacy are both normative and technical: nutrition experts estimate how much food a person needs for an active and healthy life, but the decisions about how much food is adequate is also influenced by cultural norms about human dignity, equality and choice, and political decisions.

The FAO guide goes on to state that while the magnitude and content of concrete entitlements that a person can rely on under the right to food law will vary according to the specific circumstances of that individual and the extent of economic development of the country, it must never fall below the entitlement to the minimum amount of food consistent with the imperative of human dignity. To operationalise the right to food, the substantive provisions of the law should be divided into the following two categories: (*a*) the right to adequate food; and (*b*) freedom from hunger; and should clarify their content and main components. The minimum level of the realisation of the right to food—freedom from hunger—requires that a minimum standard of satisfaction of nutritional needs is determined by the law itself.

Right to Adequate Food

The FAO guide suggests that in order to realise the right to adequate food, the law could state, for example, that every person has the right to live in conditions that enable her or him to:

- either feed herself or himself directly from productive land or other natural resources or rely on well-functioning distribution, processing and market systems, or both;
- be financially able not only to acquire a sufficient quantity and quality of food, but also to satisfy her or his other basic needs;

- be safe from the risk of losing access to food as a consequence of sudden shocks (e.g., an economic or climatic crisis) or cyclical events (e.g., seasonal food insecurity);
- have the opportunity of good food utilisation through access to an adequate diet, clean water, sanitation and healthcare to reach a state of nutritional well-being where all physiological needs are met; and
- access foods or diets that are the most appropriate under given circumstances in terms of their nutritional value and cultural acceptability.

The guide notes that young children, pregnant women and lactating mothers are particularly affected by malnutrition. Hence, it suggests that the law should therefore also include a separate, express provision stating that:

- Every girl and boy has a right to food and nutrition adequate for her or his age, allowing her or him to grow and develop.
- Every woman has a right to food and adequate nutrition during pregnancy and lactation.

To realise the right to food, the state needs to provide food (e.g., by rationing of essentials) or the means for its procurement to persons who cannot take care of their own needs, due to reasons beyond their control, in particular for children whose parents die or disappear or otherwise no longer take care of them. The state also needs to act towards creating conditions that enable people to produce food, for instance, by ensuring producers a fair price (Eide 2002). Realising the right to food also requires more far-reaching measures aimed at removing the causes of hunger and poverty, eliminating inequalities, improving people's means to ensure their livelihood and access to knowledge, resources, skills and opportunities needed to provide food for themselves.

The exact measures to be taken depend on the situation in the particular country (FAO 2009). The ICESCR suggests that state parties to the covenant should improve methods of production, conservation and distribution of food by making full use of technical and scientific knowledge, by disseminating knowledge of the principles of nutrition, and by developing or reforming agrarian systems in such a way as to achieve the most efficient development and utilisation of natural resources.[2] Other possible measures include land reform and other means to improve access to natural resources.

Realising the above-mentioned objectives in turn requires expanding the country's food production in an ecologically sustainable manner, strengthening production of healthy and nutritious food, organising training and education programmes on the advantages and importance of diet diversification, and distributing food to the most vulnerable through government social assistance programmes through exclusive local procurement. Other possible initiatives can include establishing incentives for microcredit institutions or adopting measures to support domestic production of certain basic crops.

Whenever individuals are unable to feed themselves, either because they have lost their pre-existing sources of food security due to a sudden and unexpected event or because they cannot ensure their food security due to sickness, disability or other reasons beyond their control, public authorities are obliged to take care of them directly. In such situations, the state should provide food of a quantity and quality that ensures food and nutrition security.

Right to Freedom from Hunger

Under this right, state authorities must ensure that no one is purposefully deprived of food or left to starve by the actions or omissions of public officials; they must take positive measures to protect persons suffering from hunger or those who are at risk of suffering from hunger. Because freedom from hunger is an immediate obligation, the framework law should also establish the specific entitlement to a minimum amount of food that persons have and that the state is required to provide.

Once the scope of a right to food law has been determined (as indicated in Chapter 3), a state should decide whether to provide the minimum amount of food in kind (i.e., by providing food products), through cash-like instruments (food stamps, coupons), through cash transfers, or through a combination of these options. Whatever means is chosen, it must be adequate to relieve persons from hunger and undernutrition.[3]

In the remaining sections of this chapter, we will look closely at the debates in India about ways in which public provisioning of food should be best organised.

Direct Provision of Food

Provision of Subsidised Food Rations

In the context of India's NFSB, the widest consensus was that the principal instrument for public provisioning of food security should be through reaching subsidised foodgrains to households. This was the approach of the central government, state governments, the Right to Food Campaign, the left and other political parties; indeed, virtually all except those who advocated transfer of subsidy instead of food through conditional or unconditional cash transfers, or those who oppose the idea of the food security law altogether. In fact, in the president's address to the Parliament in 2009,[4] when the government's resolve to enact a national food security law was first officially announced, the idea of the law was restricted exclusively to providing 25 kilograms of highly subsidised cereals every month to all families identified by the government to be BPL. This was also what was promised by the Congress Party in its manifesto in the general elections of 2009 (the only other component of food security which found mention in the party manifesto was community kitchens providing affordable food to urban homeless people and migrants). The extent of political consensus that had developed around this is reflected in the fact that all major national parties—the Congress, the BJP and the CPI(M)—made similar pledges and commitments to food security in their manifestos in 2009 (Raman 2009).

This wide consensus on using the PDS for delivering food security in India derives from several decades of experience with the food rationing system, which has remained the paramount instrument for ensuring an MSP guarantee to farmers for food production (mainly cereals), for price stabilisation by moving grains from surplus areas to those with shortages, and for ensuring cheap grain to millions of households. The PDS has its origins in the rationing system for urban areas, introduced by the British during World War II. However, in post-independence India, the scope of the PDS was gradually expanded to meet multiple objectives of price stability, support for farmers and providing affordable foodgrains for the population. The PDS was universal until 1992, when the revamped public distribution system was introduced in 1,775 tribal, arid, hilly and remote blocks of the country

(Khera 2011a). Thereafter, in 1997, the TPDS was introduced in an attempt to rein in food subsidy costs. Under the TPDS, households were classified as APL or BPL, and only households classified as BPL were eligible for subsidised foodgrains (Kochar 2005).

The PDS has been criticised for being highly mismanaged and ridden with corruption. By the government's own estimates for 2003–04, 58 per cent of the subsidised foodgrains issued from the central pool did not reach the BPL families because of identification errors, non-transparent operation and unethical practices in the implementation of TPDS (Planning Commission 2005).[5] Citing the high food subsidy costs associated with the PDS and its low effectiveness in ensuring food security for the country's population, some critics have called for a dismantling of the system in favour of cash transfers, food stamps and greater private sector participation (FICCI 2010). Such views are shared by some members of the government and the Planning Commission as well.[6]

On the other hand, there is evidence of significant improvements in the functioning of the PDS in states such as Himachal Pradesh, Tamil Nadu, Chhattisgarh and Andhra Pradesh, all of which have introduced reforms in the PDS. These reforms include expansion in coverage, reduction in prices, doorstep delivery of foodgrains, computerisation of stock management, increase in the variety of food rations provided at the FPS, etc. Results of a nine-state[7] survey of the PDS conducted in 2011 highlight that on an average, respondents received 84–88 per cent of their full PDS entitlements, with states that have brought about PDS reforms performing significantly better. Moreover, with the rapid rise in food prices, the effective subsidy being received through the PDS has risen significantly; in some states, it is equivalent to a month of wages that would be ensured by the MGNREGA programme (Khera 2011b). Such results highlight the important role being played by the PDS in contributing to food security.

While there is agreement on the principle that the primary instrument of food security should be subsidised foodgrains supplied through the network of FPSs, the specific details of how the PDS should be organised and constituted remain fiercely contested. Indeed, some of the most irreconcilable debates around the NSFB in India centred around the PDS; these debates related to: (*a*) coverage; (*b*) price; (*c*) types of rations; (*d*) quantities; (*e*) cash versus food; and (*f*) PDS reforms. We look at some of these debates more carefully in other chapters, especially those around universalisation and targeting, cash transfers instead of food, and PDS reforms. The other debates are reviewed briefly below.

Debates on the Extent of Coverage of the PDS

The debate on the coverage of the PDS was engaged by the left political parties and the Right to Food Campaign, among others, on the one hand, who were of the opinion that at the heart of any food security law, especially one focused primarily on guaranteeing access, there should be a universal PDS offering subsidised food rations to every household in the country; and on the other hand those who advocated, by conviction or compromise, limiting the coverage of households under the PDS. Even amongst this group of advocates, which includes members of the government, the Planning Commission, international financial institutions such as the World Bank and the IMF, and several academics, there have been large disagreements over what proportion of households should be covered under the PDS, and how they should be selected.

The NAC spent many months discussing the questions of coverage, price and quantities of rations which would be guaranteed by the law. One of its early suggestions was to cover all households in the 150 poorest districts in India. It reflected also on using transparent criteria to exclude around 10 per cent rural and 50 per cent urban households and to cover all others. The NAC suggested first an exclusion of 30 per cent rural and 50 per cent urban households, and extending thereby subsidised foodgrains to around 66 per cent of households, which was substantially higher than the coverage at the time. The remaining 70 per cent of rural and 50 per cent of urban households were to be given differential entitlements, both in terms of quantity and price. At least 46 per cent of rural and 28 per cent of urban households would be covered by 7 kilograms grains per person per month, at Rs 3 for rice, Rs 2 for wheat and Re 1 for millet. These would be designated 'priority' groups. The residual households (which were not 'priority' nor excluded) would be classified in the 'general' category and receive 4 kilograms per person per month, at a price not more than half the MSP at which the grain was purchased from farmers. It entitled single-member 'priority' households to a minimum of 14 kilograms of grains per month, and single-member 'general' households to 8 kilograms of grain per month at the same prices. The final NAC draft thus proposed covering 90 per cent of rural households under the PDS, in such a manner that not less than 46 per cent of all rural households would come under the 'priority' category, and the rest under the 'general' category. For urban areas, the draft proposed covering 50 per cent of the households under the PDS, with at least 28 per cent of all urban households getting 'priority' entitlements, and the rest coming under the 'general' entitlements.

The three-way division of households between 'priority', 'general' and 'excluded' categories was criticised for being complicated and lacking in transparency. It was argued that there was no clarity on how the different groups were to be identified, and there was no method of doing this in a practical, fair and effective manner (*Hindu* 2012a). The implementation of this proposal was also considered to be administratively difficult.

The draft bill submitted by the central government to the Parliament broadly followed this threefold division of households proposed by the NAC, but reduced the entitlement of the general category to 3 kilograms per person. The NAC draft had notified per capita entitlements of both categories to be a minimum, suggesting that state governments were free to raise these over time, and had laid down that prices could not be raised in the 'priority' category for at least 10 years. These provisions were also erased in the government draft, as well as the higher allocations to single-member households. Although it was unclear how the various categories would be identified, the 'priority' category seemingly compromised households falling below the Planning Commission's official poverty line, and an additional 10 per cent of this population, to provide for transient poverty and reduce exclusion errors. However, the official poverty ratios have been themselves hotly contested.

While the Lok Sabha Standing Committee on Food, Consumer Affairs and Public Distribution recommended folding the 'priority' and 'general' categories into a broader 'included' category, the final act opted to redefine the 'general' category to 'priority', and reintroduce the AAY scheme instead of the 'priority' category. The AAY was introduced in 2000 to focus on the 'poorest of the poor', and would affirm the belief that the former 'priority' category was aimed at BPL households. In their introduction into the act, Antyodaya households would receive the same entitlements that they had previously, 35 kilograms per household per month, at Rs 3 for rice, Rs 2 for wheat and Re 1 for millet, while 'priority' households would receive 5 kilograms per person per month, at the same prices as Antyodaya households. The final act would cover 75 per cent of the rural and 50 per cent of the urban population.

Various committees were assigned the task of calculating India's poverty ratio based on the NSS data of 2004–05. The Arjun Sengupta Committee estimated the country's poverty at 77 per cent. However, this estimate was based on a flawed methodology which did not take into account the varying prices across states and across rural and urban areas. The Tendulkar Committee estimated the country's poverty ratio to be 37.2 per cent. The Saxena Committee was of the opinion that the poverty ratio should be at least 50 per cent, but did not give any rationale for this figure (Himanshu 2010).

In March 2012, the Planning Commission revised the Tendulkar Committee poverty ratio based on NSS 2009–10 data. It declared that the all-India poverty ratio had declined from 37.2 per cent in 2004–05 to 29.8 per cent in 2009–10, with rural poverty declining from 41.8 per cent to 33.8 per cent and urban poverty declining from 25.7 per cent to 20.9 per cent (GoI 2012). The extent of poverty reduction declared by the Planning Commission was challenged by many, with an expert on poverty estimation noting that the commission had manipulated the consumption data for 2009–10 used for estimating poverty, resulting in an undercounting of 18 million poor in 2009–10 (Himanshu 2012).

While the issue of poverty estimation is fraught with contentions, there are wide disagreements on the methodology of identifying the poor as well. Governments have been singularly unsuccessful in identifying those who are poor, and several studies, including official ones, reflect high levels of exclusion and inclusion errors. Based on the NSS data from 2004–05, only about 36 per cent of poor households had a BPL or Antyodaya card (which is necessary to receive subsidised foodgrains through the PDS), while almost 60 per cent of such cards had been given to non-poor households (Planning Commission 2007). Similarly, the NFHS-3 conducted in 2005–06, and the India Human Development Survey of 2004–05, showed that about half of all poor families in rural India did not have a BPL card (Drèze 2012).

At the time of writing (in 2012),[8] there was a growing body of sympathetic analysts who suggested that if the government did not opt for a universal PDS, it should merely exclude the richest households by clear, transparent criteria, and cover all other households with uniform entitlements in terms of both quantity and price. For instance, in March 2012, a group of 36 development economists wrote an open letter to the prime minister, in which they urged the abolition of the distinction between general and priority households, and providing all households except those falling under the excluded category with uniform entitlements at uniform prices (*Hindu* 2012a).

Debates on the Types of Food Rations to Be Provided in the PDS

In a note written in March 2010, the Department of Food and Public Distribution of the Ministry of Food and Public Distribution expressed its opinion of including only rice and wheat as the PDS entitlements in the NFSB (GoI 2010). The Planning Commission in its presentation to the NAC in June 2010 also recommended providing legal entitlements for only rice and wheat under the PDS in the NFSB.

Despite the central government being seemingly content with only providing grains, the Right to Food Campaign was of the view that the NFSB should guarantee not only *food* security, but also *nutritional* security. It argued for not restricting PDS entitlements to rice and wheat, but also including millets, which have high nutrition value,[9] as well as pulses and oil. Pulses are the primary source of protein for a large section of the Indian population; as their prices had been hit hard by inflation, they were increasingly outside the reach of most poor families. The campaign argued that inclusion of millets, pulses and oils in the PDS would necessitate the government procuring the crops from farmers, which would boost the production of these crops and help in reducing the agricultural crisis. As these crops are dryland crops requiring less water than rice and wheat, their production assumes special importance in the context of increasing temperatures and decreasing availability of water because of climate change (Right to Food Campaign 2010).

In part due to this pressure from the Right to Food Campaign, the drafts of the NFSB gave legal entitlements for rice, wheat and millets.

Millets were added to PDS entitlements because of their high nutrition content and cultural preference in many areas. Further expansion of PDS entitlements to include pulses, oil and cooking fuel was included as a goal for progressive realisation in the NAC draft, and the Standing Committee recommended the inclusion of pulses, oils, fortified flour and sugar, as well as recommending that the bill promote local produce such as *madhuvi*,[10] ragi and suchlike. Despite this, as well as pressure from left parties in Parliament, the final act limited its provisions to rice, wheat and millets and has no clause promoting production of local grains.

Debates on the Quantity of Food Rations to Be Provided in the PDS

The various positions regarding the quantity of food rations that should be provided in the PDS were guided by concerns of availability of food crops, quantum of food subsidy and people's nutritional requirements. The Right to Food Campaign (2010) advocated for the provision of the quantity of food rations adequate for ensuring nutritional security for all people. The campaign quoted the ICMR, which stated that a moderately active male requires 2,700 kcal a day. This translates into 14 kilograms of cereals a month per individual (about 50 kilograms for a family of five members), 800 grams of oil a month per individual (about 2.8 kilograms per family of five members) and 1.5 kilograms of pulses a month per individual (about 5.25 kilograms per family of five

Table 8.1. Total monthly requirement for a five-member family, as per ICMR norms

Family member	Monthly requirement		
	Cereals (kg)	Pulses (kg)	Oils (g)
Man doing moderate work	14.4	2.7	1,050
Woman doing moderate work	10.8	2.25	900
1–6-year-old child	5	1.1	675
7–12-year-old child	9	1.8	750
Elderly person/third child	9	1.8	675
Total	48.2	9.65	4,050

Source: Right to Food Campaign (2010).

members) (ibid.). Table 8.1 provides the total monthly requirement for a family of five members as per ICMR standards.

However, as the government draft of the bill provided per capita entitlements, the Right to Food Campaign demanded that every person should be entitled to 14 kilograms of cereals, 1.5 kilograms of pulses and 800 grams of oil every month. Table 8.2 presents the total food requirement to meet the campaign's demand. The calculations are made based on a total population of 1,210,193,422 as per the census of 2011.[11]

Table 8.2. Total requirement of foodgrains, pulses and oil to meet the PDS demands of the Right to Food Campaign

	Total requirement (million tons)		
	Assuming 100% offtake	Assuming 80% offtake	Assuming 70% offtake
Cereals	169.4	135.5	118.6
Pulses	18.2	14.5	12.7
Oil	9.7	7.7	6.8

Source: Right to Food Campaign (2010).

Table 8.3 presents the total production of cereals and pulses between 2005–06 and 2009–10.

Table 8.3. Total production of cereals and pulses in India between 2005–06 and 2009–10

Year	Total production (million tons)			Total cereal availability (excluding 12.5 per cent for seed, feed and wastage
	Cereals	Pulses	Total foodgrains	
2005–06	195.21	13.39	208.60	170.8
2006–07	203.08	14.2	217.28	177.7
2007–08	216.02	14.76	230.78	189.0
2008–09	219.90	14.57	234.47	191.8
2009–10	203.45	14.66	218.11	178.02

Source: Directorate of Economics and Statistics, Department of Agriculture and Cooperation, Government of India.

The total requirement of cereals at 100 per cent offtake is 169.4 million tons. However, the campaign was of the opinion that due to self-selection in a universal PDS,[12] 70 per cent offtake of cereals is more realistic (Right to Food Campaign 2010), which would require procuring about 67 per cent of the total cereal availability in 2009–10.

Economists have argued that India does not produce enough grain to secure sufficient pulses to include in the PDS. The Right to Food Campaign disagreed, arguing that in the case of pulses, an offtake ratio of 80 per cent is realistic, yet this was close to the total production of pulses in 2009–10. The campaign added that due to lack of investment, the production of pulses was in serious crisis in India; if pulses were given a fair and stable procurement price, their cultivation was likely to increase. Also, the quantity of pulses supplied in the PDS could be gradually increased to 1.5 kilograms over a period of five years (Right to Food Campaign 2010). The campaign also stated that in the case of cooking oils as well, 80 per cent offtake was most realistic, which would require about 90 per cent of the net availability of edible oils from all domestic sources.

Meeting the campaign's demands would require the procurement of 70–90 per cent of the domestic production of cereals, pulses and oilseeds. This raised the question of whether governments should supply the entire food requirements of every household at subsidised rates. A more realistic aspiration was that 50 per cent of the food requirement should be offered by the state, and the remaining 50 per cent should be procured from the free market. In that sense, the provisions in the NAC and government drafts of 7 kilograms per capita seemed reasonable.

The group of development economists recommended providing every person (except those in the 'excluded' category) with 5 kilograms of foodgrains every month. The annual foodgrain requirement of this proposal was 51.5 million tons, marginally lower than the requirement of 52.4 million tons for the government's draft (*Hindu* 2012b). As the final act has 'priority' households receiving 5 kilograms of foodgrains per person per month, and Antyodaya households receiving 35 kilograms of foodgrains per person per month, the total foodgrain requirements would be around 60 million tons.

Debate on the Price of Food Rations Provided in the PDS

The NAC draft entitled every person in the 'priority' category to rice at Rs 3 per kilogram, wheat at Rs 2 per kilogram and millets at Re 1 per kilogram, with the prices locked in for at least 10 years from the date of notification of the act. 'General' category persons were entitled to foodgrains at a price not exceeding 50 per cent of the MSP for rice, wheat and millets. The government draft entitled foodgrains at the same prices, but did not have a clause which prohibited the increase of prices for the priority category. The Standing Committee recommended ensuring the price for 10 years, like the NAC draft; MPs also urged ensuring prices to prevent price hikes. Ultimately, the act ensured set prices for a mere three years from the date of notification.

The price of foodgrains for the 'priority' category in the NFSB draft was pegged so low as to be affordable even by most of the poorest and most destitute persons and households. A 'priority' category household of five members could buy its entire monthly quota of 35 kilograms of foodgrains with a single day's statutory minimum wage. The final act has kept the prices, although an Antyodaya household can receive a maximum of only 35 kilograms, regardless of the size of the household.

Some economists argued that these prices were too low, and a somewhat higher price would have been fiscally more attractive, and enabled coverage of still larger proportions of households, although the price was not the subject of much debate. The Congress manifesto and the president's address also promised rice and wheat at Rs 3 per kilogram. It was later felt that since wheat is cheaper than rice, its price should be lower. As millets are even cheaper than wheat, their price was fixed at Re 1 per kilogram. Some state governments, such as those of Tamil Nadu and Chhattisgarh, had been offering grain at similar prices for some years, which has not only contributed to the food security of poor people, but has also helped in winning votes.

The price for the 'general' category, fixed by the NAC draft at 50 per cent of the MSP, was much more contested. It was not only considered too high compared to the 'priority' category rates, but also as effectively pitting the impoverished consumer against the impoverished food producer. While it would be in the interest of the farmer to receive a higher price guarantee from the government, it would be in the consumer's interest to keep the price low, to ensure that the price of grains in the FPSs remained low. Otherwise, the proposed draft would have had the unintended consequence of increasing the unfair burden on the agricultural community, which would exacerbate the rural distress in the country. The final act instead replaced the 'general' category with the 'priority' category, and provided 5 kilograms per person per month at Rs 3 per kilogram of rice, Rs 2 per kilogram of wheat and Re 1 per kilogram of millets. This was in line with the Right to Food Campaign's demands that every person would be entitled to those prices. The group of economists also recommended this price for all those entitled to foodgrains.

Provision of Cooked Meals

Cooked food is typically provided to those who have special nutritional requirements—children, adolescent girls, pregnant women, lactating mothers and those who are unable to cook, such as the destitute and the homeless. There is no way of ensuring that the food rations bought by a family under the PDS are distributed equitably within a household. In this context, the ICDS and the MDMS assume importance as their entitlements of cooked food can only be utilised by those for whom they are intended. These direct feeding programmes for children, adolescent girls, pregnant women and lactating women assume even greater importance in the context of large-scale exclusion errors in targeting households for social assistance programmes such as the PDS and irregularities in the functioning of the PDS, which result in households not getting their full quota of food rations on time. The food provided by the ICDS and the MDMS is free of cost, as a large proportion of the users of these services would not be able to afford them if they were to be priced.

The specific entitlements of cooked food in the various drafts and the campaign's demands regarding these entitlements are detailed in the chapters on 'Children', 'Gender-Just Food Entitlements' and 'Vulnerable Groups' (see Chapters 10, 13 and 14). In the final act, the specific entitlements of cooked food were limited to children and women, with no specific entitlements for other vulnerable groups.

Provision of Means for Procuring Food

Creating Duties to Provide for Work

In India, there had not been a debate on creating duties in the NFSB to provide employment to people, so as to enable them to earn income to purchase food. This was because the Government of India passed MGNREGA in 2005, which guarantees every rural household 100 days of unskilled manual work a year. In case the government fails to provide employment within 15 days of application for work, it is legally obligated to pay an unemployment allowance. Although there are demands for increasing the number of days of wage employment guaranteed by MGNREGA and extending the employment guarantee to urban areas as well, these demands were outside the ambit of the NFSB.

Social Security Cash Transfers

There is a need for providing means for procuring food to those categories of people who are unable or have diminished capacities to work due to reasons beyond their control. These categories include pregnant and lactating women, single women, the aged, and persons with disabilities, among others.

Pensions

The NAC had, in its earlier discussions, considered including pensions for the aged, single women and persons with disabilities in the NSFB guarantees, as these are some of the most vulnerable groups that require additional support to meet their nutritional security. However, these were later dropped and shifted instead to the segment on 'Progressive Realisation', and in the final act moved to the non-legally-binding section on 'Provisions for Advancing Food Security'. The campaign demanded that single women (including widows, separated women, elderly women, divorced women and abandoned women), the aged and persons with disabilities should be entitled to an individual pension which was at least half of the minimum wage, equal to Rs 1,300 per month (at 2009–10 prices).

Maternity Entitlements

Maternity entitlements are a recognition of women's rights as workers and the need for them to rest during the last few weeks of their pregnancy, recuperate after delivery, and stay with their infants to exclusively breastfeed them for six months, which would usually require the mothers to stay away from work. There is a general consensus on providing maternity entitlements till six months after delivery as exclusive breastfeeding is recommended for infants till they reach 6 months of age.

The NAC and government drafts, as well as the final act, fixed maternity entitlements at Rs 1,000 per month for six months. The NAC draft stated that the amount of Rs 1,000 was to be revised every year based on an appropriate price index; however, the draft introduced in Parliament, nor the final act, did not have any such provision, which would lead to a reduction in the real value of the maternity benefits with increase in prices over time. The campaign was of the opinion that as maternity entitlements are provided as wage compensation, they must be at the minimum wage and must also be inflation indexed. It was also of the opinion that maternity benefits should be provided for a period of nine months; for three months before the date of delivery and six months after that date. The reason for demanding maternity entitlements before delivery is to support women in lieu of payment for work which they should not undertake to protect themselves and the foetus.[13]

Provision of Support to Produce Food

One of the important steps towards food security is ensuring adequate production of food crops in an ecologically sustainable manner. While food can be imported as well, this may not always be financially viable and threatens a country's food sovereignty. Providing support for increasing food production is essential for ensuring the food security of not just the producers, but of everybody else as well. Support for food production consists of a variety of measures, such as equitable distribution of resources such as land and water, subsidising agricultural inputs such as seeds and fertilisers, making provisions for irrigation, sharing technical know-how with farmers, providing farmers with credit to buy agricultural inputs, providing them with assured markets and procuring their produce at fair MSPs.

Legal entitlements supporting food production were largely kept outside the NFSB drafts and the final act. However, the NAC draft listed several goals related to agriculture for progressive realisation:

- Make efforts to revitalise agriculture and promote agrarian reform, through measures such as securing the interests of small and marginal farmers through remunerative prices, credit, irrigation, crop insurance and technical assistance, and focus on dryland regions.

- Prohibit unnecessary and unwarranted diversion of land and water from food production; and promote decentralised food production, procurement and distribution systems.
- Pay particular attention to small farmers, women and youth farmers who constitute the majority of the farming population.
- Progressively increase investments in agriculture, research and development, extension, micro and minor irrigation and rural power supply.

Most of these objectives were mentioned in the non-legally-binding provisions of the government draft and the final act, although the campaign demanded that these provisions should be given the status of legal entitlements.

The NAC, government drafts and final act recommended decentralised procurement as a means to reform the PDS, without giving as much regard to decentralisation as a way to protect food security at the state level, or of farmers. In 2012, government procurement was primarily restricted to Punjab, Haryana, Andhra Pradesh and parts of Uttar Pradesh. The campaign advocated procurement of foodgrains, including millets, pulses and oilseeds from all *krishi upaj mandis* (agricultural produce markets), so that all areas benefit from government procurement. It suggested further that the entire country should be divided into agricultural production-distribution zones, so that grain procured from a particular zone is distributed in that zone itself under the various schemes. In case of a shortfall in the availability of grain in a particular zone, required quantities of grain may be transported from other zones. To expand procurement, the campaign also recommended incentivising state governments to make timely payments to farmers.

The NAC considered but did not include MSP guarantees to all farmers as part of the NFSB, despite strong debates in Parliament in favour of MSP guarantees, with the Right to Food Campaign demanding that the NFSB should have had a provision that the government must undertake to buy any amount of grains that farmers were willing to sell at the MSP. As the NAC draft did not contain any MSP guarantees, subsequent drafts, Standing Committee recommendations and even the final act did not include any. The National Farmers Commission headed by Dr M. S. Swaminathan had suggested that the MSP should be at least 50 per cent more than the weighted average cost of production. The cost of production includes not only expenditures such as seeds, fertilisers, irrigation and wages of agricultural labour, but also the imputed value of owned assets including land and family labour. In 2012, the MSP of millets and pulses was very low, with the campaign advocating for a fair MSP not only for rice, wheat and millets, but also for pulses and oilseeds to boost the production of these crops, which were facing neglect due to uncertain prices and low levels of investment. An increase in the production of pulses and oilseeds could facilitate their inclusion in the PDS.

The exclusion of MSP guarantees from the legally enforceable provisions of the NFSA has serious implications, as Karat (2015) points out, as in 2014, the government reduced the rate of increase in MSP to only 3 per cent over the previous year, when input prices increased substantially. Further, when state governments tried to compensate, the central government decided to limit procurement of foodgrains from these states to the amount required for the PDS. Karat argues that as a direct consequence, states like Chhattisgarh and Andhra Pradesh have reduced procurement from individual farmers, leading to distress sales at low prices to private contractors.

Notes

1. This chapter does not enter the debate on the desirability of introducing cash transfers in lieu of subsidised food rations. That is discussed in Chapter 7.

2. UN Office of the High Commissioner for Human Rights, International Covenant on Economic, Social and Cultural Rights, http://www.ohchr.org/EN/ProfessionalInterest/Pages/CESCR.aspx (accessed 11 March 2018).

3. A more in-depth discussion on cash transfers can be found in Chapter 7.

4. 'My Government proposes to enact a new law—the National Food Security Act—that will provide a statutory basis for a framework which assures food security for all.' Address by the president of India, Smt. Pratibha Devisingh Patil, to the Joint Session of the 15th Lok Sabha, New Delhi, 4 June 2009.

5. There have been subsequent reductions in the scale of corruption in the PDS. According to NSSO data, PDS leakages came down to 42 per cent in 2011–12 at the all-India level (Khera 2017).

6. 'Cash transfer is the answer. There's a real problem in taking a system like PDS that's known to be really corrupt, giving it highly subsidised commodities and expecting the grain not to leak into the market.' Montek Singh Ahluwalia, deputy chairman, Planning Commission, as quoted in Mohan (2011).

7. The states surveyed include Andhra Pradesh, Bihar, Chhattisgarh, Himachal Pradesh, Jharkhand, Odisha, Rajasthan, Tamil Nadu and Uttar Pradesh.

8. The NFSA expanded the coverage of the PDS to 75 per cent of the rural population and 50 per cent of the urban population.

9. See Right to Food Campaign (2010, 11), for more on the nutrition content of millets.

10. A coarse grain eaten in India.

11. Census of India 2011, 'Table 1: Distribution of Population, Sex Ratio, Density and Decadal Growth Rate of Population: 2011', http://censusindia.gov.in/2011-prov-results/data_files/india/table_1.pdf (accessed 11 March 2018).

12. Discussed in detail in Chapter 4 on universal versus targeted entitlements.

13. The chapters on 'Gender-Just Food Entitlements' and 'Conditionalities' (Chapters 13 and 6, respectively), discuss maternity entitlements in greater detail.

References

Drèze, Jean. 2012. 'Just target the rich.' *The Times of India*, 4 February.

Eide, A. 2002. 'The Right to Food: From Vision to Substance', in M. Borghi and L. Postiglione Blommestein (eds), *For an Effective Right to Adequate Food*. Fribourg: University Press, pp. 27–53.

FAO (Food and Agriculture Organization). 2009. *Guide on Legislating for the Right to Food*. Rome: FAO.

FICCI (Federation of Indian Chambers of Commerce and Industry). 2010. 'Give "Food Stamps" instead of Price Subsidy to BPL Families for Targeted Food Distribution: FICCI'. FICCI, New Delhi, 22 May. http://www.ficci.com/PressRelease/601/press-may22.pdf (accessed 23 September 2017).

GoI (Government of India). 2010. 'Note for Empowered Group of Ministers'. Department of Food and Public Distribution, Ministry of Consumer Affairs, Food and Public Distribution, New Delhi.

———. 2012. 'Press Note on Poverty Estimates, 2009–10'. Planning Commission, New Delhi.

Himanshu. 2010. 'Demystifying Poverty: Which Estimate and Why'. *Mint*, 19 January.

———. 2012. 'India undercounts its poor'. *Mint*, 26 March.

The Hindu. 2012a. 'A simple proposal on food security', 12 March. http://www.thehindu.com/opinion/op-ed/article2985212.ece (accessed 23 September 2017).

———. 2012b. 'Simplifying the National Food Security Bill', 12 March. http://www.thehindu.com/migration_catalog/article13056807.ece/BINARY/Simplifying%20the%20NFSB (accessed 9 March 2018).

Karat, Brinda. 2015. 'Food insecurity acts'. *The Hindu*, 31 January. http://www.thehindu.com/opinion/op-ed/food-insecurity-acts/article6839851.ece (accessed 23 September 2017).

Khera, Reetika. 2011a. 'India's Public Distribution System: Utilisation and Impact'. *Journal of Development Studies*, 47(7): 1–23.

———. 2011b. 'Revival of the Public Distribution System: Evidence and Explanations'. *Economic and Political Weekly*, 46(44): 36–50.

———. 2017. 'Impact of Aadhaar on Welfare Programmes'. *Economic and Political Weekly*, 52(50): 61–70.

Kochar, Anil. 2005. 'Can Targeted Food Programs Improve Nutrition? An Empirical Analysis of India's Public Distribution System'. *Economic Development and Cultural Change*, 54(1): 203–35.

Mohan, Rohini. 2011. 'The Riddles of the Conscience Keepers'. *Tehelka Magazine*, 8(27), 9 July.

Planning Commission. 2005. 'Performance Evaluation of Targeted Public Distribution System (TPDS)'. Programme Evaluation Organisation, Government of India, March.

———. 2007. *Eleventh Five Year Plan 2007–2012*, vol. 2, Chapter 4. http://planningcommission.nic.in/plans/planrel/fiveyr/11th/11_v2/11th_vol2.pdf (accessed 23 September 2017).

Raman, Anuradha. 2009. 'Same Ol' Gruel: Rival Parties "Unite" on Food Security'. *Outlook*, 20 April.

Right to Food Campaign. 2010. 'National Food Security Act: An Introductory Primer on the Legal Guarantees Demanded by the Right to Food Campaign'. Right to Food Campaign Secretariat, New Delhi.

9

Starvation

Ashwin Parulkar

As mentioned in the Introduction, the catalyst for the human right to food in India was an episode of starvation deaths in Rajasthan that occurred at a time when public granaries held surplus stocks, and public food, nutrition and work programmes were found to be defunct. Reports on the travesty raised public awareness, and the petitions filed by the PUCL in the courts engendered debates on how and why poor people's access to food through the welfare system should be considered part of the constitutional right to life. It would seem logical that starvation and starvation deaths, matters of life and death which led to sustained public action through the democratic process, would be major issues pertaining to food security legislation. Considering the robust historical and contemporary literature on famine and starvation in India that should also inform policy decisions, it is surprising that this has not been the case. While interim court orders directed public officials to provide immediate relief to families of victims of starvation deaths through food and nutrition programmes, policymakers paid little attention to starvation response as a policy issue or starvation as a condition undermining food security in the country in the years preceding legislation.

One challenge policymakers face in tackling starvation is the task of defining the term. On the one hand, rigorous analysis is required on linkages between conditions of starvation (in terms of low food intake) and undernutrition (in terms of low nutrient absorption and lack of diversity of needed nutrients) that the poor in India experience. This also involves drawing clear lines between these conditions to establish methods of identifying people who suffer from particular variations of either or both of those conditions. Both are difficult tasks, yet doing so in the policy space is necessary to address the needs of people that a human rights based food law seeks to protect most. In the context of law, policymakers would have to agree on an accurate understanding of starvation that would mandate and enable public officials to address severe forms of hunger in line with legal duties placed on them to do so.

This involves a thorough analysis of the socio-economic and biological determinants of suffering that the destitute in India who experience chronic hunger as a daily reality face.

Currently, this is a major gap in Indian development policy. In popular understanding, starvation is understood as the complete absence of food intake. In *Hunger: The Biology and Politics of Starvation*, Butterly and Shepherd (2010) provide a comprehensive analysis of starvation relevant to the Indian context. They argue that the world's poorest people—particularly rural landless agricultural or daily wage labourers who live in unsanitary environments—experience a process of semi-starvation, or prolonged periods of low food intake, which compromises the ability of a person's immune system to prevent chronic and infectious diseases. Semi-starvation, especially in children, impairs intellectual potential and capacity, psychological and social well-being, and catalyses physical defects such as the stunting of growth and deformities that take shape early in life. Chronic low food intake of this nature is associated with malnutrition, which, over time, causes irreversible damage to the body's natural defence systems and leads to risk factors for diseases that, without proper public support measures like sanitation and healthcare, can lead to death. Some people who suffer such vicious forms of endemic hunger may not, over time, even feel hungry as a well-nourished person would. In the case of children, semi-starvation causes irreversible physical defects like stunting as well as chronic fatigue due to vitamin deficiency anaemia (Saxena 2008; see also Butterly and Shepherd 2010).

In *Starvation and India's Democracy*, Dan Banik (2007, 30) views starvation as 'a process resulting from a period of sustained undernutrition'. He goes on to state that 'during this period, the level of energy intake is below the level of energy expenditure such that there is a continuous loss of body weight which may ultimately lead to death' (ibid.). Banik carefully defines and differentiates variants of starvation and undernutrition in terms of how the duration and severity of low and insufficient food intake impact health and risk factors for diseases that lead to death. 'Short-term starvation', which can be seasonal or occur at 'regular times each year (for example, in months following a severe drought and destitution)', may last a few days or 'on and off for several weeks … and may not lead to death' (ibid.). Still, this form of starvation *can* 'result in increased morbidity and mortality among nutritionally vulnerable groups like children and the elderly' (ibid.).

'Starvation deaths', Banik argues,

> may result from a prolonged period of sustained undernutrition and the total absence of any forms of food for x days where x depends on various factors, including the general physical condition and nutritional status of an individual before the start of the starvation process, the rate of body weight loss, body size, metabolic changes, adverse climatic conditions and the level of physical activity undertaken during the period of starvation. (Banik 2007, 30)

The line Banik draws between undernutrition and severe undernutrition is instructive. While undernutrition is defined at the individual level as 'the consumption of incorrect types of food (quality) and/or inadequate intake of food (quantity)', severe undernutrition is characterised by a 'decline in the level of food consumption and high food insecurity among sections of the population'. Severe undernutrition 'leads to starvation, and the continuous loss of body weight can ultimately lead to death among certain individuals and groups more vulnerable than the rest of the population (e.g. children, pregnant women, nursing mothers, the elderly, the sick adults)' (ibid., 31).

The 2011 government draft food security bill and the final 2013 act defined starvation as *prolonged deprivation of food that threatens the survival of a person*. The NAC draft bill's definition included the phrase 'involuntary deprivation of food', alluding to contexts that drive many poor people's inability to access food, such as corruption, exploitation and discrimination in unregulated labour networks and social programmes. The bill also included a section on 'Rights of Persons Living in Starvation', which outlined rights of redressal, relief measures, and methods of identification of and investigation into starvation and starvation deaths. The most significant feature pertaining to starvation in the NAC draft was a starvation protocol, which was conceived as a legally enforceable code for identification and investigation of starvation. The Parliamentary Standing Committee did not provide any recommendations on the matter and there were no debates on this topic in Parliament when the bill was put to a vote in the fall of 2013.

While it is difficult to forecast how the current definition of starvation in the NFSA and the lack of guidelines for identification will address the needs of people who live in semi-starved and undernourished conditions,[1] the overall push to pass food security legislation is symbolic of civil society's call to reform the state's approach to endemic hunger, starvation and even famine.[2] In India, the government has largely dealt with starvation only after deaths are reported in the media, or when outbreaks of hunger are unmanageable, partly because, as noted earlier, policymakers have not addressed the public health and socio-economic dimensions of starvation. The lack of discourse on the matter is also partly why officials fail to identify or acknowledge starvation and act swiftly, with a political sensitivity in admitting to starvation deaths. Relief policies that actually exist in many individual states are also non-binding on officials. The questions before policymakers on issues pertaining to starvation and food security in legislation therefore include the following:

1. How should the state define starvation in the Indian context with due consideration to the problems of undernutrition and food access?
2. How should a food security law create legal duties on state officials to identify and respond to people who are starving?
3. Can the law create legal duties on the state to identify people who are starving?
4. How should those duties direct officials to provide people identified as starving their legal rights to food that are already on the books?

History of Starvation Relief in India

India's administrative problems in tackling starvation are rooted in the history of relief during the famine-plagued decades of British colonial rule in the nineteenth and twentieth centuries. In 1880, the British Government of India introduced the Famine Codes to respond to widespread hunger that had occurred in famines between 1858 and 1878. The writers of the codes acknowledged that ad hoc relief efforts were the cause, in part, of millions of unnecessary hunger deaths in rural India.[3] The Famine Codes ordered administrators to set up public works for the poor in famine-affected areas based on a set of conditions called 'tests of desperation'. The tests were used to decipher who among the poor were willing and able to work in famine conditions. There were three 'tests': first, work was offered far away from villager's homes; second, people were told they had to live at the work sites away from their families; third, work was to be monotonous, arduous and compensated with low wages.[4]

Jean Drèze (1988) argues that despite these stipulations, the Famine Codes proved a major step in official famine response, since the commission acknowledged the importance of 'diligence in starting relief operations' to prevent disaster. 'The great thing is to begin on time,' the 1880 Famine Commission Report states. The Famine Commission also identified poor agricultural labourers and artisans as the people driven most into starvation during drought and crop losses, while acknowledging that many poor lost their ability to secure food in terms of food surplus (ibid.). Later Famine Commissions (1898 and 1902) concluded that India had achieved large food surpluses in normal years, and multi-year storage facilities for foodgrains were also in operation in nineteenth-century India, evidence that periodic famines were not only due to overall food decline (ibid., 11).

The conventional belief at the time was rooted in Thomas Malthus's theory[5] on population growth, which held that famines were caused by population growth rates outpacing food production. Lord Lytton, India's viceroy from 1876 to 1880, supported Malthus's theory, arguing that famines were inevitable correctives to the food–population imbalance in countries like India, which faced droughts due to uncontrollable climatic shocks. 'India's population has a tendency to increase more rapidly than the food it raises from the soil,' he once wrote (J. C. Caldwell 1998, 683, cited in Davis 2002, 8). However, this was not true. The Famine Commission, assembled during Lytton's tenure just after the Madras famine, reported that rice and wheat production in the rest of India had been above average for the previous three years. The commission also concluded that food scarcity during times of drought and unemployment were heightened by exports of grain surpluses to England. Between 1875 and 1900, annual grain exports increased from 3 million to 10 million tons and, by the early twentieth century, India was supplying nearly one-fifth of Britain's wheat consumption (see B. Bhatia 1963, 38; R. C. Dutt 1904, 3–4; Dietmar Rothermund 1998, 36—all cited in Davis 2002, 31).

A century later, Whitcombe (1993) used district-wide data to show that the Madras famine victims were the rural destitute who lacked access to public works like irrigation. 'Malthus's predictions were not borne out,' she wrote. 'The coastal, intensively irrigated districts of Tanjore, Ganjam, Godavary and Kistna with populations of 540 square miles and above were prosperous and escaped while famine ravaged the up-country districts where population barely averaged 150 per square mile' (ibid., 1169). Moreover, in famine-hit regions like the Madras Deccan, the population rate had actually decreased 'more rapidly (23%) where districts were served by railways than where there were no railways (21 per cent)' (W. Digby 1900, 362–63, cited in Davis 2002, 7).

Several factors undermined the Famine Commission's efforts to target future famines and widespread hunger successfully.[6] First, while the commission agreed that famine occurred most when the rural poor could not secure employment in times of high food prices and drought, administrators did not take measures to control food prices. The British government feared that interfering with markets would hurt the inflow of cheap imports from England. Lord Lytton issued orders that 'there is to be no interference of any kind on the part of the government with the object of reducing the price of food' (R. Osborne 1879, 553; H. Srivastava 1968, 131; W. Digby 1900, 50–51; D. Steele 1999, 8—all cited in Davis 2002, 8). The Famine Commission's suggestion that government responses to starvation be swift and strong proved to be at odds with the Crown's grain export policy.

Second, the Famine Codes' warning signs used to diagnose starvation were not reliable indicators. Measures of rainfall, drought and crop prices were no different than those used in ad hoc responses to previous famines, and were only acted on after crisis had struck and lives were lost. The commission's acknowledgement of the need to identify famine victims for fast and effective responses was not followed by a revised method on how to accomplish this. In the Madras famine of 1876–78, which occurred before the Famine Commission assembled, the government relied on price signals to act, although this method led to a late response. Whitcombe (1993) argues that rising prices were not a reliable marker of distress because no indices of relations of prices to quantities of foodstuffs were created. The colonial government also never decided what the critical price point must be for a response to kick in, with the government not acting until 1877, a year after the famine struck, when rural workers were migrating in tens of thousands in a span of a few months. Medical inspections by William Robert Cornish that year showed that about 60 per cent to 70 per cent of the labouring poor in famine districts were suffering from scurvy, anaemia and wasting.[7] The birth rate in 1877 and 1878 were 25 and 50 per cent lower respectively than the previous two years. Meanwhile, the commission listed 'distress migration' as one of the key warning signals of famine, but the evidence from the Madras famine showed that local administrators encouraged migration because they saw it as a prospect of relief.

The third reason[8] that the codes failed to prevent subsequent crises was that they were not legally binding. The commission did instruct administrators to follow guidelines on how to recognise and declare a famine, but these instructions did not carry the force of law. In that sense, the ad hoc nature of famine relief was carried over, although when relief operations did begin, the codes did provide a set of guidelines to follow.

Fourth, the purpose of the Famine Codes was to restrict the death toll, not to restore the health and livelihoods of the people suffering from hunger. While the codes instructed local administrations to open up relief works and feeding programmes, action only took place after calamities struck, and the relief wages were also very low. Ultimately, these were relief efforts in the most minimal sense.

The latter two points are especially instructive for food legislation for policymakers in contemporary India seeking to assure access to the food system as a right in consonance with the constitutional right to life with dignity.

Need for a New Conception of Starvation Relief Policies

After independence, several Indian states—Orissa (now Odisha), Rajasthan and Andhra Pradesh, for instance—revised the Famine Codes to reflect government action required to respond to hunger under specific agricultural and climatic conditions. Dan Banik observes that in Orissa, despite the revision of the state relief code in 1996, meant to 'better enable Collectors and other district officials to "react" to press reports of starvation, the press reports of alleged starvation deaths in the years since has actually increased' (2007, 83). This is due, in part, to individual states' retaining the ethos of the colonial codes: they remained legally non-binding, were based on minimal relief efforts, and were usually enforced after hunger deaths were reported. Faster response in the revised Orissa codes was still understood in the context of reported deaths, rather than awareness of community-wide destitution. When the courts decided in 2001 that the right to food is inherent to the constitutional right to life, starvation was deemed a violation of that right: 'Whenever a report of death due to starvation is published and it comes to the notice of the Collector, he shall immediately cause an enquiry into the allegation' (ibid.). While the courts issued orders to provide immediate relief to people who had been denied access to food and nutrition programmes, there remains no system of identification and response that is legally binding to urge that action happen swiftly once starvation is detected.

Part of the problem in India lies in the fact that there is little acknowledgement of how chronic undernourishment and starvation are interrelated factors, despite numerous government surveys which have documented the large scale of undernourishment among India's most vulnerable people. National surveys from 1998–99 and 2005–06 show that malnutrition rates have not changed much in the last 14 years, raising questions as to why India's economic prosperity did not reach many of the country's poor and ailing. In the context of starvation response, the country's undernutrition problem sheds light on how it makes little sense to continue with a system of relief that acts only when deaths occur.

Most of India's workers are in the informal sector, predominantly farm labourers and manual labourers who have little social protection. These labourers, working across India at construction sites, brick kilns, farms and mines, expend more calories than people engaged in less physically demanding work, yet per capita calorie consumption in both urban and rural India has declined in the last 20 years. This pattern alone does not reveal much, and other factors, such as consuming the appropriate diversity of calories as well as having the capacity to absorb those calories, are as important to living a healthy life. Based on Deaton and Drèze's (2009) breakdown of NSSO (2001) data on cereal calorie consumption by income groups, Saxena (2008) observed that the poorest fourth of the population consumed about 30 to 50 per cent fewer calories than the top fourth.

Many in the informal sector toil in exploitative conditions marked by unequal labour relations, pushed to the brink of starvation because they have lost, or have never had, ownership or rights of access to resources such as land and water that could provide income, food and nourishment. The land and water they do utilise is often rife with chemicals or pollutants, and the wages they receive are often dictated by unregulated economies that thrive on cheap labour at rates that do not keep pace with rising food prices. Many poor people also live in unsafe conditions, with a lack of sanitation and clean drinking water and little to no access to healthcare.[9] Regardless of food intake, unsanitary conditions compromise the immune system and tear down the body's defences to disease-causing bacteria.

John Butterly and Vandana Prasad[10] argue that the combination of inadequate food and unsanitary conditions can lead to a state of malnutrition because the body is doubly vulnerable. The lack of nutritious food harms the body due to the inadequate intake of protein, energy and micronutrients, and exposure to unsanitary conditions makes it easier for viral or bacterial infections to occur.[11] Therefore, a deeper understanding of both the biological basis of starvation and the economic conditions in which the poor live is required to understand how policies should be implemented to secure rights to live free of hunger and malnutrition.

Butterly and Prasad also argue that addressing hunger is a public health concern. In their view, policymakers have to give more attention to the biological relationship between starvation and undernutrition. Based on current knowledge, low food intake over time, coupled with poor sanitation and unsafe water, weakens the body's defences to fatal diseases. Malnutrition in India, in other words, is both the result of low levels of food intake and also the major risk factor in infectious diseases.

The Planning Commission had acknowledged the need of policymakers to understand the continuum of hunger, poverty, malnutrition and disease. India's hunger problem, the commission said, was 'not one of hidden hunger; it is hunger for nutrient rich foods' (Planning Commission 2008). The Planning Commission based this claim on the exceptionally high numbers of Indian babies born with low birth weight, as well as nutrition indicators of SC and ST communities that were below already poor national means, and the high rates of energy and micronutrient deficiencies.[12] Still, current relief policies that lack proper definitions of salient terms and appropriate indicators to identify mutually enforcing conditions reveal a lack of understanding of the context behind endemic hunger, making adequate responses impossible. The Orissa Famine Code's chapter on health measures, for instance, states that 'cases of mortality may be high not so much due to direct starvation but due to increased incidence of diseases.'[13]

Policymakers will no doubt face challenges in drawing clear lines between starvation and undernutrition to target both endemic hunger and mortality-causing diseases, but some of the infrastructure to do this already exists, with numerous food and nutrition programmes under way in India. Indeed, the courts have decided that the programmes are legal rights, coming under the right to food, yet despite these rights, programmes such as the PDS and the ICDS are beset by problems in funding, high turnover, corruption and negligence. How will the law place duties on the state to identify and respond to the people who are denied their rights to these programmes?

We raise this issue in the final section of this chapter. For now, we note that if the new legislation includes methods of identification and diagnosis of starvation and undernutrition, officials in charge of food, health and nutrition programmes may be better served in streamlining their efforts to run development programmes efficiently. As of the final act in 2013, no methods of identification and diagnosis of starvation or undernutrition exist. The poor would be better served through legal writs and protection to access these programmes regardless of starvation protocols. If public debate emerges on the need, challenges and benefits of identifying the chronically undernourished, officials could also be made more aware of the socio-economic and public health dimensions of food security with an understanding that tackling hunger requires more than assuring grain quotas through the PDS, though it is no doubt an essential component.

The following sections deal with starvation from two perspectives. First, what is the biological relationship between semi-starvation, malnutrition and the onset of mortality-causing infectious diseases, and why are poor people, who experience long periods with inadequate nutritious food and unsanitary living conditions, more at risk? Second, what role have economic policies played since independence, and what role does India's legal framework play in preventing or failing to prevent starvation, starvation deaths and famine? I hope that understanding the latter will present policymakers with options to address both the economic and biological aspects of this complex problem. The remainder of the chapter will discuss each of these relationships.

Biological Basis of Starvation and Malnutrition

So far, I have mentioned that many poor people in India do not consume the required amount *and* diversity of nutrients to maintain normal health, which compromises their ability to absorb the nutrients they do consume, making them weaker and more prone to illnesses, such as mortality-causing infectious diseases.

I stress here the importance of the 'amount and diversity' of 'nutritious food' in drawing connections between food consumption and nutrition (e.g., semi-starvation and undernutrition). Here I explore why the amount of caloric intake does not in itself tell us anything useful about a population's food security and nutritional status. Nutrition is not merely a factor of *how much* one eats but how much nutrient-rich food one eats on a regular basis, in conjunction with other important factors such as a sanitary environment, access to affordable and good healthcare and clean drinking water. I will therefore explain the key factors linking food consumption to nutrition, while accepting that no conclusions can be drawn between consumption of food and nutritional status alone.

Deaton and Drèze (2009, 50) show that higher rates of child malnutrition actually exist in regions of India that have higher per capita calorie consumption. Kerala, for instance, has the highest prevalence of calorie undernourishment (28.6 per cent) but both the lowest proportion of underweight children under 5 (22.7 per cent) and lowest under-5 mortality rate (1.6 per 100) in India. Meanwhile, Uttar Pradesh has the lowest prevalence of calorie undernourishment (14.5 per cent) but the highest under-5 mortality rate, the fourth highest proportion of underweight children under 5 (42.3 per cent) and the highest under-5 mortality rate (9.6) (ibid.).

Per capita calorie consumption in India has also declined since the early 1980s while the overall nutritional status, although abysmal, has improved since the mid-1970s.[14] Deaton and Drèze (2009) cite Tara Gopaldas's (2006) work to argue that, in the context of food consumption and nutrition, the diversity of calories in diets such as essential vitamins and minerals is an important, and often overlooked, factor. Gopaldas argues that people in low-income groups are likely to be deficient in vitamins and minerals. She finds that on the basis of the National Nutrition Monitoring Bureau surveys, children in the 4 to 6 age group consume only 16 per cent of the recommended daily allowance (RDA) of vitamin A, 35 per cent of iron and 45 per cent of calcium. According to the Planning Commission, about 70 to 80 per cent of children's calorie intake, which they acknowledge is already low, comes from cereals and pulses. The heavy reliance on these foods comes at the expense of needed amounts of micro- and macronutrients such as proteins, fats, iron and vitamins. While the caloric intake of many poor children is often dangerously low, it is the vitamin and nutrient composition of their already deficient diets that puts them at risk—they do not consume enough nutrients on a regular basis to enable their bodies to absorb the nutrients they do get to grow and live healthy lives. Other factors such as unclean environments that are rife with disease-causing bacteria, zoonosis and unsafe drinking water also heighten their risk to disease and death.

Both points are important in the context of the NAC draft bill's definition of starvation as 'the prolonged involuntary deprivation of food that threatens survival of the person'. It may be more useful, as the Planning Commission stated, to view endemic hunger in terms of the lack of nutritious food over a period of time that threatens the possibility of living a healthy life.

Still, India's undernutrition problem is not a mere function of food options and choices that people make in their diets. N. C. Saxena (2008) argues that social practices which determine the low status of women are a major cause of poor nutrition in women and children.[15] Early marriages and low education levels, he argues, affect the amount and quality of Indian women's diets, hygiene, the number of children they bear, and their feeding and caring practices for the children they have. Saxena observed that low birth weights are linked to the poor health status of mothers, since underweight women give birth to smaller babies. The percentage of infants with low birth weight in 2006 was about 30 per cent, and Indian mothers put on less weight on average during their pregnancies (5 kilograms on average, compared to 10 kilograms for the average African or middle-class Indian woman) (ibid.). Improper feeding and caring practices of newborns also impair the development of the child. For example, only 23 per cent of Indian children under 3 are breastfed within one hour of birth, and only 46 per cent of children under 5 are breastfed at all (ibid.).

In starvation deaths, people do not merely die of zero food intake. It is true that many poor live on the razor's edge, with long periods without employment, defunct food and work programmes, and chronic illnesses resulting, in some cases, in going for days without food. These bursts of misfortune serve as the final blow to a person who has lived prolonged periods without essential calories, vitamins, minerals and proteins. The amount such people do get is, at times, already too low to maintain life itself, especially if they are plagued by disease and live in areas with no medical treatment.[16]

On the one hand, the conditions which lead to death are preventable. On the other, local governments routinely respond to hunger distress only when starvation deaths are reported in the media. While victims of starvation deaths suffered complications that would make it impossible to survive, living family and community members endure precisely the same conditions of poverty, malnutrition and denial of basic rights to social services. Such deaths and levels of destitution are indeed a function of unaddressed socio-economic inequalities and unexamined biological processes.

Infectious diseases are contracted through harmful bacteria, called microbes, which are passed from one organism to another. A well-nourished person who lives in a clean environment has built-in defences to these kinds of bacteria. Butterly and Shepherd (2010) explain that the skin and mucous membranes are the body's first line of defence to infection. They perform functions that maintain the body's immune system within the skin and blood cells. Humoral immunity, for instance, is the immune system in the blood maintained by the circulation of soluble proteins called B-lymphocytes. When foreign and potentially harmful bacteria, called antigens, enter the body, these complex proteins recognise the threat and produce their own antigens to destroy it. In a healthy person, B-lymphocytes reproduce themselves to kill bacteria and are able to respond quickly to similar attacks to protect the body against disease. Similarly, cellular immunity is maintained by

T-lymphocyte proteins, which involve the white blood cells directly in attacking foreign antigens. Like B-lymphocytes, they use their cellular memory to target repeated threats to the body.

Chronic undernutrition debilitates these immune responses: protein energy malnutrition and vitamin and mineral deficiencies destroy both the mucous membranes and the health of the skin. Butterly and Shepherd (2010) note that once these defences are compromised, the body's internal tissues become vulnerable to a range of infections that a normally fed and healthy person would avoid. Over time, undernutrition becomes, particularly for children who live in unclean environments, a major risk factor for infectious diseases, because the condition depletes the number of available complex proteins that ward off infection in the blood cells and weakens the efficacy of the ones that survive.

Overall, infectious diseases, including scarlet fever, diphtheria, dysentery, cholera, typhus and TB,[17] are more prevalent in populations with higher rates of hunger and malnutrition. Each of these diseases is prevalent in poorer communities in India that lack sanitation, basic healthcare, assured access to nutritious food and social security programmes.

Starvation Response in State Policy after Independence

So far, I have talked about how government responses to starvation spring from the colonial Famine Codes. Such responses have been out of step with reality because they stress minimal efforts to keep people from dying, and are usually only enforced after deaths are reported. After independence, while the states incorporated the codes into state policies, the new codes are still not in step with recent thinking established by the courts that the right to food is an inherent right to life—in other words, preventing death is not the same as protecting or enforcing one's right to life.

Response systems are still based on rainfall patterns and crop production, and often these signals do not paint an accurate picture of hunger and malnutrition distress. The Supreme Court has issued specific orders on starvation since the right to food case opened in

2001, with the 29 October 2002 order, for instance, stating that it is the duty of each state/union territory to prevent deaths due to starvation or malnutrition. If the commissioner reports and it is established to the satisfaction of the court that a starvation death has taken place, the court may be justified in presuming that its orders have not been implemented and the chief secretaries/administrators of the states/union territories may be held responsible for the same.[18] These orders also directed central and state governments to ensure that the AAY programme was extended to all destitute segments of the population. The orders were meant to provide people suffering from starvation access to programmes deemed as rights, but they still do not solve the problem of identifying starvation victims in a timely manner. They also do not address the basic problem of governance: local officials, who are often in charge of the food, work and nutrition programmes that are plagued by corruption, are reluctant to acknowledge that the people who are their constituents starve in part because they are denied access to right to food programmes.

An earlier section briefly raised the point that the process of semi-starvation and undernutrition leads to other complications such as deadly infectious diseases. This point was made to highlight the fact that it is necessary to identify people suffering from these complications before calamities such as drought or large crop failures do strike. Yet for that to happen, state relief codes, which rely on rainfall and crop production patterns as distress signals, have to be revised. Dan Banik's (2007) work on Orissa's starvation relief programme during the bad drought years of 1996 to 2000 gives us good insight into why this is so.

The Orissa Relief Code was drafted in 1975 with the aim of providing relief that was both 'preventative and curative'.[19] However, evidence of drought is necessary to open up relief works, with the codes possessing a range of warning signals to determine an emergency. The two main detection signals are rain gauges in each district after the rainy season and the estimation of crop losses. An 'analytical report' is also submitted every July to the district collector's office, which includes these signals as well as the stock of foodgrains available, the health situation, the functioning of the social security network and safety net oriented programmes. While this is a comprehensive list of indicators, drought is announced solely on the basis of the first two indicators.

The code goes on to instruct governments to open relief works 'only in the case of "widespread" calamities'. Banik (2007) observes that this caveat implies that the scale of distress has to be sufficiently large for the government to respond. The code's definition of calamity is quite narrow, since disease epidemics are only considered 'minor' forms of distress that do not require government intervention.

Banik shows that Orissa declared droughts in Kalahandi district every second year during the 1996 to 2000 period on the basis that low rainfalls occurred at times of paddy crop losses. He observed that rainfall may not be such a wise indicator of distress in this case, since the district had a higher average rainfall than most others in the state; it was equally important to know the 'timing' and 'distribution' of rainfall. Banik explains that since rain is most needed to grow crops in June, erratic rains in the monsoons can be as disastrous for crop growth as no rain at all. For instance, if the monsoon begins in mid-July and the total rainfall for the season is average or above average, farmers would still not benefit. His research highlights that distribution is also important, as in Kalahandi's drought years rainfall was much higher in areas that did not cultivate crops than in places that did.

The Kalahandi case also shows why crop production rates are not the sharpest indicator of distress. Drought did disrupt cultivation of the paddy crops in one season in the four-year period, but Banik finds that production 'once again passed 300,000 tons and surprisingly the drought in 2000 did not substantially reduce production [that year]' (2007, 57). Banik explains that other factors, such as high-yielding varieties of seeds, fertilisers, and surplus production in neighbouring blocks that received irrigation from local dams, made up for losses in the district alone. Overall, Banik writes, 'total food grain production in Kalahandi in the 1996–2000 period and per capita food grain production in the district—even in drought years—remained above Orissa and national averages' (ibid., 63). Despite this, 'large sections of Kalahandi's population were reported to have suffered from food shortage' (ibid.). Banik's research shows that even in the 1996 drought, which was dubbed 'the worst drought of the century', food was always available in nearby markets, but poor people could not buy it because they were suffering from unemployment and had no other sources of income. The second problem was that the district faced a storage problem, with the total storage space for the district at 40,000 tons, which was far lower than the requirement of 100,000 tons.

The Starvation Protocol

The previous sections highlighted the importance of revising government methods of identifying starvation victims based on a range of public health measures. The Supreme Court Commissioners devised a 'Starvation Protocol',[20] based on findings from a consortium of nutrition, economic, medical and social policy experts called the Hunger Watch Group, to identify and respond to starvation-related cases in a comprehensive fashion. The NAC draft implemented the protocol while the government draft and final act omitted it.

The protocol maintained, first, that 'starvation and malnutrition related deaths are public health problems requiring community diagnosis.' This understanding differs from the one currently held by many public officials, which is that starvation involves absolutely no food intake. However, it is not possible to definitively compare this definition with the government's understanding of a starvation death since the term is omitted from the act. The NAC draft defined a starvation death as a 'death of a person caused by starvation'.

Further, the protocol suggested understanding starvation not merely in terms of the immediate physiological causes of death, but also, and more importantly, the struggles faced by people living with the condition. It recommended a method called a 'verbal autopsy'—a two-part consultation that happens between public officials and people suffering from starvation and/or families of victims of starvation deaths. The first part is a discussion with the family of a victim of starvation death about the status of food and work access as well as periods of hunger. 'The idea', the protocol states, 'is not only to probe death and its causes but to understand the poverty and destitution faced by families and similarly affected people.'[21] The second part is a consultation with 'other members of the tribe, caste, class, gender or age group to which the affected people belong'. The aim, the protocol suggests, was to ask broader questions on food security that affected the whole community, such as the performance of food, work and nutrition programmes.

Third, the protocol states that findings from investigations should be used to provide relief in the form of legal entitlements. This is in stark contrast to the NAC draft's provision of two free meals for six months to persons identified as starving. As mentioned earlier, current court orders on starvation deaths direct local authorities to investigate causes, such as failed food and work programmes, and ensure that people who are starving and families of starvation death victims access entitlements to those programmes on a long-term basis. The protocol finally states that immediate relief should ensure 'immediate food availability to the family, free of cost for at least six months', but specifies that this should be converted into access to subsidised grains through the PDS in line with the AAY scheme, which is currently a legal entitlement for poor families who meet those criteria.

The Standing Committee did not comment on the protocol, but did suggest bringing people deemed to be starving into the ambit of the included category. The final act did not include the protocol and removed subsidies and targeted support for people living with starvation.

Notes

1. In a 1950 book called *The Biology of Human Starvation*, Ancel Keys broke new ground by providing a wealth of scientific data on biological changes that occur in the human body under conditions of semi-starvation. Keys was a researcher at the University of Minnesota and a consultant at the US War Department in the 1940s. Disturbed by reports of starvation of World War II victims living in famine-like conditions, he decided to recreate like conditions with willing volunteers to study discrete changes that occur in the body as a result of starvation. The experiment was known as the Minnesota Experiment and was based on observing 32 young men between the ages of 20 and 32 in three phases: a normal period of 12 weeks in which participants were fed a balanced diet at regular intervals; a semi-starvation period of 24 weeks based on a diet modelled on the food intake of prisoners of war; and a re-feeding period of 12 weeks, to restore the men to good health. Participants lost an average of 25 per cent of their body weight, which led to more obvious deprivations such as fatigue and vulnerability to serious conditions such as brain defects and oedema. Keyes found that severe weight loss induced by hunger can lead to serious health defects that require medical and nutritional rehabilitation due to the damage to the tissues, bones, muscle and skin.

2. Drawing clear lines between chronic malnutrition, hunger and starvation isn't straightforward. One international standard for measuring hunger is the body mass index, or BMI, which refers to an individual's weight relative to height—a proxy for body fat. The internationally acknowledged threshold for a 'normal' BMI is 18.5. An individual with a BMI of 17 to 18.4 is 'undernourished', someone with a BMI of 16.0 to 16.9 is considered 'severely malnourished', and a BMI less than 16 signifies a person is 'starving'. According to the WHO, a society with 40 per cent of its population registering a BMI less than normal (18.5) is in a state of famine. Based on this, the Planning Commission's (2011) *India Human Development Report* states, 'If India is not in a state of famine, it is quite clearly in a state of chronic hunger' (see also Parulkar 2012).

3. Estimates of nineteenth-century famines tally the dead between 5 to 12 million people, while estimates for the 1876–78 famine alone total 1.5 million deaths (see Davis 2002).

4. Jean Drèze (1988, 7–9) has said that while the tests of desperation were the foundation of the Famine Codes, 'gratuitous relief' was available for 'those unable to work—but only for them—in the form of doles or kitchens', which 'complemented the public works programme to 'form the core of relief measures'.

5. Malthus's theory rested on the belief that global population doubles every 25 years, unless checked, thus growing at a geometric rate (1, 2, 4, 8). Food production concurrently increases at an arithmetic rate (1, 2, 3, 4, 5, etc.). From this, he concluded that population would eventually outstrip food supply and that the poorest people would die as a corrective to this imbalance. One proposed solution was to keep wages at subsistence level. 'Besides this', Ancel Keys explains, 'he advocated, without much hope, the exercise of moral factors in advancing the marriage age and in the practice of continence.... towards the end of the nineteenth century Malthus's theory was much discounted because of its failure in quantitative prediction of population changes in that period and because of the increasingly obvious role of noneconomic, or at least nonfood, factors which affect the population level' (Keys 1950).

6. Three more major famines hit British-ruled India after the 1876–77 Madras famine. Drèze (1988) finds that the 1896–97 and 1899–1900 famines were brought on by crop failures across the country at a time when stocks were already diminishing, but also that the relief effort was bungled by faulty forecasts, confused information, no contingency planning, little motivation, and transport bottlenecks. There was also denial amongst administrators, as evidenced by the claim of the Famine Commission of 1898 that 'success actually attained in the relief of distress and the saving of human life was, if not complete, far greater than any that has been recorded in famines that are at all comparable with it in extent, severity, and duration' (ibid., 35). A brief discussion on causes of the 1943 Bengal famine will be provided later in this chapter.

7. Dr William Robert Cornish, sanitary commissioner, inspected famine conditions among 3,000 people in July 1877.

8. Reasons three to five are points raised in the writings of Jean Drèze (1988) and Harsh Mander (2010).

9. According to the WHO's *World Health Statistics* (2012), access to safe drinking water has improved in India, from 81 per cent in 2000 to 92 per cent in 2010, while access to sanitation remains poor, with only 34 per cent of the Indian population having access in 2010, a marginal improvement from 25 per cent in 2000.

10. Personal communication, John Butterly, January–February 2012; Vandana Prasad, February 2012.

11. Interviews with John Butterly, January–February 2012.

12. It was concluded that the high percentages of babies in India born with low birth weight was a problem that began in utero due to the mother's undernourishment. 'A mean deficit of 1.4 to 1.6 kg in weight at one year', the plan document stated, 'worsens to a deficit of about 9 kg at 10 years and 12–18 kg when adults.' Second, the mean heights and weights of children from vulnerable groups such as SCs and STs were below already poor national means. Third, in the case of energy and micronutrient deficiencies, there was an average of 500 to 600 kcal deficit in energy intake, which is about 40 per cent of the normal requirement, a 50 per cent deficit in the consumption of fat, calcium, iron, riboflavin and vitamin C. Fourth, diet surveys also showed that 70 to 80 per cent of calories consumed by children were derived from cereals and pulses, which means that 'children cannot consume more cereals to make up for the calorie deficiency because of its sheer monotony and lack of energy density.' And 'in the absence of fats, milk, eggs, and sources of iron, children starve themselves.' See Planning Commission (2008, 129–30).

13. Orissa Relief Code, para. 204, cited in Banik (2007).

14. The proportion of underweight children was 47 per cent in 1998–99 and 46 per cent in 2005–06. From 1974–75 to 2004–05, however, there was a reduction of around 50 per cent in the prevalence of severe undernutrition (measurement below 3 standard deviations of the median of the reference distribution). Reductions for weight-for-age in the child population were from 37 per cent to 18 per cent, while reductions in height-for-age were 53 per cent to 25 per cent. See Deaton and Drèze (2009, 50).

15. We understand that this claim may be viewed as a broad generalisation and, given that India is the largest democracy in the world, a bit hard to believe for people outside this country. However a 2011 TrustLaw survey on the status of women's rights in the world found that India was the fourth most dangerous country for a woman to live, behind Afghanistan, Congo and Pakistan. This was based on high rates of female foeticide, human trafficking and infanticide.

'In addition to sex slavery', the writers say, 'other forms of trafficking include forced labor and forced marriage, according to a U.S. State Department report on trafficking in 2010.' See Anderson (2011).

16. A Centre for Equity Studies study on starvation deaths in rural India observed these conditions in a study on government responses to starvation deaths in the poorest Indian states—Bihar, Madhya Pradesh and Jharkhand (see Parulkar 2012).

17. According to the WHO, India has the highest TB burden in the world, accounting for about 25 per cent of the world's cases. Government statistics indicate that nearly 40 per cent of India's population is infected with TB bacteria, the vast majority having latent rather than active TB. See 'TB Statistics for India—National & State Statistics', http://www.tbfacts.org/tb-statistics-india.html (accessed 24 September 2017).

18. Supreme Court of India, Interim Order of 29 October 2002, *People's Union for Civil Liberties v. Union of India & Ors*, Writ Petition (Civil) no. 196 of 2001, http://www.righttofoodindia.org/orders/oct29.html (accessed 11 March 2018).

19. Orissa Relief Code (1996), corrected up to 31/12/1996 by Board of Revenue (Special Relief), Revenue Department, Government of Orissa, Cuttack, cited in Banik (2007).

20. N. C. Saxena and Harsh Mander, 'A Protocol for Preventing Starvation', Submission as Special Commissioners of the Supreme Court in *People's Union for Civil Liberties v. Union of India & Ors*, Writ Petition (Civil) no. 196 of 2001.

21. Ibid., 6.

References

Anderson, Lisa. 2011. 'Trustlaw Poll—Afghanistan is most dangerous country for women'. Thomson Reuters Foundation, 15 June. http://news.trust.org//item/20110615000000-na1y8/ (accessed 24 September 2017).

Banik, Dan. 2007. *Starvation and India's Democracy*. London: Routledge.

Butterly, John, and Jack Shepherd. 2010. *Hunger: The Biology and Politics of Starvation*. Lebanon, NH: University Press of New England.

Davis, Mike. 2002. 'The Origins of the Third World: Markets, States, and Climate'. Corner House Briefing 27, December. http://www.thecornerhouse.org.uk/resource/origins-third-world (accessed 24 September 2017).

Deaton, Angus, and Jean Drèze. 2009. 'Food and Nutrition in India: Facts and Interpretations'. *Economic and Political Weekly*, 44(7): 42–65.

Drèze, Jean. 1988. 'Famine Prevention in India', WIDER Working Paper no. 45.

Gopaldas, Tara. 2006. 'Hidden Hunger: The Problem and Possible Interventions'. *Economic and Political Weekly*, 41(14): 3671–74.

Keys, Ancel. 1950. *The Biology of Human Starvation*. Minneapolis: University of Minnesota Press.

Mander, Harsh. 2010. 'Towards a Food Rights Code: The State, Food Denials and Food Rights'. Centre for Equity Studies Working Paper (unpublished).

NSSO (National Sample Survey Organisation). 2001. *Nutritional Intake in India 1999–2000*. New Delhi: Ministry of Statistics and Programme Implementation, Government of India, August.

Parulkar, Ashwin. 2012. 'Starving in India: The Forgotten Problem'. *Wall Street Journal*, 9 April.

Planning Commission. 2008. *Eleventh Five Year Plan 2007–2012*, vol. 2. http://planningcommission.nic.in/plans/planrel/fiveyr/11th/11_v2/11th_vol2.pdf (accessed 24 September 2017).

———. 2011. *India Human Development Report*. New Delhi: Institute of Applied Manpower Research, Planning Commission, Government of India, and Oxford University Press.

Saxena, N. C. 2008. *Hunger, Under-nutrition and Food Security in India*. New Delhi: Supreme Court Commissioners Office.

Whitcombe, Elizabeth. 1993. 'Famine Mortality'. *Economic and Political Weekly*, 28(23): 1169–73, 1175–79.

WHO (World Health Organization). 2012. *World Health Statistics*, Part III: *Global Health Indicators*. http://www.who.int/healthinfo/EN_WHS2012_Part3.pdf (accessed 11 March 2018).

10

Children

Ankita Aggarwal

Children are physiologically and psychologically dependent on adult protection and care, for meeting their right to food as well as a range of other rights. Children have special food requirements, and inadequate consumption of nutritious food, especially in the first three years of their life, can have devastating consequences for their health and future development. As small children are unable to grow or procure food, and infants are unable to communicate their needs, they are especially vulnerable to food insecurity that can lead to malnutrition and even mortality in extreme cases. Hence, a right to food law needs to have special entitlements for children, being particularly important in India, which has alarmingly high rates of undernutrition and where children's right to food is violated with impunity.

No mention was made of ensuring children's right to food in either the Congress Party manifesto in 2009 or the president's address to Parliament after the UPA government was elected to power, both restricting their pledges for the NFSB to cheaper grain entitlements under the PDS. The NAC draft, the government draft, as well as the final act, however, contained specific guarantees for children's right to food. There have been specific debates about what precise rights for children should be part of the law, and how these should be formulated.

This chapter begins with a discussion on the need for special entitlements for children in a right to food law, especially in the Indian context. The next section discusses the FAO's recommendations regarding provisions for children in right to food legislation. This is followed by a comparison of the various entitlements for children in the different drafts of the NFSB, as well as the final NFSA. Finally, these entitlements are compared with the FAO's recommendations and critiqued in detail.

Need for Special Provisions for Children in a Right to Food Law

Young children are particularly affected by malnutrition (FAO 2009), which can have devastating consequences. Malnourished children are more susceptible to illness and their brains do not develop properly. Malnutrition is responsible, directly or indirectly, for two-thirds of the deaths of children under 5 years of age in India, with two-thirds of those deaths taking place in the first year of a child's life. Insecurity about child survival causes families to have many children in succession, which further affects the health of women and children.

Malnutrition begins at birth, or even before, depending on the health of the mother; however, it intensifies sharply between the ages of 6 months and 3 years, with the first 1,000 days of a child's life being very crucial, as the damage done by malnutrition at this age is often irreversible. Nutritional interventions in this age group are most required to ensure adequate growth, cognitive development and good health. This is the period of most rapid development when the blood is formed, the neural channels of the brain are sculpted and personality is shaped.

Exclusive breastfeeding is recommended for children up to 6 months of age; beyond this period, the mother's milk alone is not sufficient for the growing child. In the first 1,000 days of a child's life, the child is also still helpless as s/he cannot feed her/himself or ask for more, and is also prone to more infections. A child at this age needs frequent meals of softened food that only an adult can give, along with continued breastfeeding (Right to Food Campaign 2007). The NAC draft of the NFSB, the government draft and the final act explicitly promoted exclusive breastfeeding for children under 6 months of age.

Many mothers are unable to take adequate care of their child as they are deprived of sufficient time, energy, resources, power and knowledge. Many of them have to work to earn a livelihood and have to juggle this with other tasks such as cooking, fetching water, cleaning, etc. Women often do not get enough support from other adults in the family, as taking care of a young child is regarded as the sole responsibility of the mother. A woman usually depends for support on her other children, sometimes as young as 4 to 5 years of age—young children who are in need of care themselves. In addition to this, the understanding of both the mother and other adults in the house regarding the child's nutritional needs at this stage may be poor.

All the issues discussed above highlight the necessity of giving special importance to children in a right to food legislation. The Right to Food Campaign too recognised the need for this emphasis, having hosted conventions on children's right to food and constituted a Working Group for Children under Six, and a joint Working Group with Jan Swasthya Abhiyan.[1] In this chapter, wherever the views of the Working Group for Children under Six are discussed, it is implicit that these are the views of the campaign as well.

The need for giving special attention to children in a right to food legislation assumes special importance in India, as children's right to food is violated with impunity in the country. As a result, India has intolerably high levels of child mortality, child malnutrition and other ailments amongst children arising due to the consumption of inadequate quantities of appropriate food. The following section discusses the status of child nutrition.

Need for Special Provisions for Children in India's Right to Food Law

This section highlights the appalling state of child well-being in India, which necessitates special attention to children's entitlements in the country's right to food legislation. Table 10.1 shows the trends in the state of infant mortality in India between 1988 and 2006.

The table not only highlights the alarming state of child well-being in the country, but also reveals the slow pace of improvement in malnutrition indicators. According to NFHS-3, conducted in 2005–06, 43 per cent of children under 5 years of age were underweight, 48 per cent were stunted and 20 per cent were wasted. The proportion of children in India under 5 years of age who are underweight is almost 20 times higher than would be expected in a healthy, well-nourished population, and is almost twice as high as the average percentage of underweight children in Sub-Saharan African countries (IIPS 2007a). In the 100 worst districts based on the child development indicators

Table 10.1. The state of infant nutrition in India, 1988–2006

Percentage of infants	National Family Health Survey data (NCHS[d] standards) Children under 3 years of age			National Nutrition Monitoring Bureau data[a] (NCHS standards) Children aged 1–5 years of age in rural areas			
	1992–93	1998–99	2005–06	1988–90[b]	1996–97[c]	2000–01	2004–05
Underweight (weight for age below 2 standard deviations cut-off)	52	47	45.9	69	62	60	55
Stunting (height for age below 2 standard deviations cut-off)	n/a	45.5	38.4	65	58	49	52
Wasting (weight for height below 2 standard deviations cut-off)	n/a	15.5	19.1	20	19	23	15

Notes: [a] These figures apply only to Andhra Pradesh, Gujarat, Karnataka, Kerala, Madhya Pradesh, Maharashtra, Odisha, Tamil Nadu and West Bengal.
[b] Data from West Bengal excluded.
[c] Data from Madhya Pradesh excluded.
[d] NCHS: National Centre for Health Statistics.
Source: Adapted from Deaton and Drèze (2009).

developed for the United Nations Children's Fund (UNICEF) India in 2009, 42 per cent of children under 5 years of age were underweight and 59 per cent stunted; of the children suffering from stunting, about half were severely stunted (HUNGaMA 2011).

According to NFHS-3, 70 per cent of Indian children between the ages of 6 and 59 months were anaemic; 3 per cent were severely anaemic, 40 per cent were moderately anaemic and 26 per cent were mildly anaemic (IIPS 2007a). About one-third of children do not get a full dose of diphtheria, pertussis and tetanus (DPT) immunisation and two-thirds do not have access to pre-school learning (Anon 2011). In 2010, according to World Bank data, India's under-5 mortality rate was 62.7 per 1,000 live births, while the global average was 57.89. In the same year, the country infant mortality rate was 48.2, as compared to the world aggregate of 41.

A 2003 *Lancet* study revealed that India accounted for more than 2 million child deaths of the global total of 10 million, concluding that underweight was a risk factor for several mortality-causing diseases (Black et al. 2003; see also Jones et al. 2003). The authors found that the fraction of deaths attributable to being underweight was 61 per cent for diarrhoea, 57 per cent for malaria, 53 per cent for pneumonia, 45 per cent measles, and 53 per cent for other infectious diseases. The effects of malnutrition in the first two years of a child's life are also irreversible, evidenced by lower educational levels and worse health conditions throughout life (Black et al. 2003). This is especially relevant in the case of India. A recent HUNGaMA study of child malnutrition, which covered about 20 per cent of the child population, revealed that 42 per cent of children under the age of 5, who lived in one of the 100 focus districts, were underweight and 59 per cent were stunted (HUNGaMA 2011). The report concludes that the prevalence of malnutrition is significantly higher among children of low-income families and that the condition starts very early in life (ibid.).

National surveys have also shown that the problem is especially acute in children of vulnerable communities such as SCs and STs. Among SC children, 47.9 per cent and 53.9 per cent are underweight and stunted respectively, whereas 54.5 per cent and 53.9 per cent of STs are underweight and stunted (IIPS 2007b). The children that elicit both risk factors are vulnerable to mortality-causing diseases while also showing irreversible physical effects because of their existing state of malnutrition.

Required Provisions for Children in a Right to Food Law

This section is drawn from the FAO's *Guide on Legislating for the Right to Food* (2009), which gives several recommendations on entitlements for children in a right to food law. An ideal right to food legislation should include a separate, express provision stating that:

- Every girl and boy has a right to food and nutrition adequate for her or his age, allowing her or him to grow and develop.
- As children are one of the groups that suffer most from discrimination, special measures are required to secure their right to food.

The right to food legislation should provide at least an amount of food entitlement which prevents children from suffering from hunger or undernutrition. In other words, the food entitlements should be adequate to enable children to function physically and mentally, maintain a normal resistance to illness, and to grow in a normal manner. The law should also specify the exact quantities of calories, proteins and micronutrients present in the food entitlements of children of different age groups: up to 6 months of age, between 6 months to 3 years, 3 to 6 years, school-age children, and adolescent boys and girls.

Under international human rights law, governments have a duty to prioritise the most vulnerable persons when the resources are limited. Such a list would include children, especially those living on the streets, in custodial institutions, working in hazardous industries, in conflict situations, belonging to refugees, with debilitating illnesses, of parents with debilitating illnesses, of parents from marginalised communities (such as Dalit and adivasi), etc. The advantage of such a detailed listing would be that it would require the competent state authorities to think of more specific entitlements for each of these categories. An ideal right to food legislation should place a public authority under a general obligation for persons who cannot take care of their own needs due to reasons beyond their control, in particular for children whose parents die or disappear or otherwise no longer take care of them (FAO 2009).

Provisions for Children in the Various Drafts of the NFSB and Their Critique

Provisions for Children under 6 Years of Age

For children in the age group of 0 to 3 years, the NAC draft provided for take-home rations and/or local and freshly cooked meals throughout the year through the local anganwadi. For all children between 3 and 6 years of age, the draft provided for a local and freshly cooked meal in the local anganwadi for at least 300 days a year. These meals for children in the age group of 0 to 6 years should include 500 kcal and 12–15 grams of protein per day. All children in the age group of 0 to 6 years are entitled to services which should at least include supplementary nutrition, immunisation, health check-ups, referral services, growth monitoring and promotion, and pre-school education.

While the NAC draft provided entitlements for children starting from birth, the draft NFSB introduced in Parliament stated that food entitlements for children would begin only from the age of 6 months, as exclusive breastfeeding would be promoted for children below 6 months of age. The government draft made no mention of the other ICDS services which should already have been considered legal entitlements as per the Supreme Court orders.

The final act does not differ much from the government draft, promoting exclusive breastfeeding for children below 6 months of age, meals for children from 3 months to 3 years of age in the form of take-home rations, for children between 3 and 6 years a morning snack and hot cooked meal, with a minimum of 500 kcal and 12–15 grams of protein per day. Children of age 6 months to 6 years who are malnourished are eligible for take-home rations containing 800 kcal and 20–25 grams of protein per day. Older children are eligible for MDMS in schools, but other ICDS services are found only in the non-binding 'Provisions for Advancing Food Security' section of the act.

Support for Breastfeeding

Until the age of 6 months, exclusive breastfeeding provides total food security to the baby, as it is a bacteriologically sterile source of nourishment which does not carry any risks of infection. Breastfeeding also provides the baby with immunity, and decreases the risk of a range of grave ailments.[2] For the mother, it suppresses ovulation and delays fertility, thereby increasing birth spacing. Women who breastfeed also have lower chances of contracting breast, uterine, endometrial and ovarian cancers.

Mothers need to be provided with support for breastfeeding. This includes giving them information about the value of breastfeeding, the need for exclusively breastfeeding the child until 6 months of age, the need for continued breastfeeding until 2 years of age and beyond, telling mothers about how frequently to breastfeed their child and how to do so, etc. The NAC draft provided support for practising exclusive breastfeeding for the first 6 months of a child's life through assistance at birth, skilled breastfeeding counselling and related assistance consistent with the provisions of the Infant Milk Substitutes, Feeding Bottles and

Infant Foods (Regulation of Production, Supply and Distribution) Act of 1992. Further, given the fact that many women are unable to provide enough breast milk because of being in poor health, supplementary nutrition during this time may be crucial. Studies have shown that breastfeeding combined with supplemental nutrition is more effective than exclusive breastfeeding, even for very young children (Anandaiah and Choe 2000), when mothers are undernourished. The Right to Food Campaign criticised the absence of any specific provisions of support for breastfeeding in the government draft. The final act also omits any specific provisions supporting breastfeeding, but does promote exclusive breastfeeding for children under the age of 6 months.

Nutrition Counselling

After 6 months of age, an infant needs other food apart from breast milk; however, the children's caregivers may not always know what type of food, in what quantities and at what intervals to give to young children. This necessitates the provision of counselling on optimal infant and young child feeding practices. The NAC draft had a provision for counselling on optimal infant and young child feeding to promote appropriate complementary feeding. However, neither the government draft nor the final act contained any such provisions.

Crèche Facilities

Crèches are critical for the right to food of children in many ways. However, none of the drafts, nor the final act, provided any legal entitlement to crèches, with only the NAC draft recommending progressive realisation of universal access to crèche facilities. The Right to Food Campaign demanded that crèches should be made available for children under 6 years of age, either provided through anganwadi-cum-crèche services or through other models operationalised by panchayats and/or women workers' cooperatives.

Midday Meal Scheme

The NAC draft entitled all children in the age group of 6 to 14 years studying in schools run by local bodies, government and government-aided schools up to Class 8 or beyond, *at least* one freshly cooked, nutritious midday meal every day of the year, except on school holidays. The draft went on to state that every school should have appropriate facilities for hygienic cooking and clean drinking water, and the food provided through the midday meal should be locally appropriate and nutritious. However, the government draft, as well as the final act, entitled *only one* midday meal a day, every day, except on school holidays, in all schools run by local bodies, government and government-aided schools, *only* up to Class 8. In both drafts and in the act, the nutritional standards for midday meals for children in primary classes was 450 kcal (including 100 grams of foodgrains) and 12 grams of protein, and for children in upper primary classes, 700 kcal (including 150 grams of foodgrains) and 20 grams of protein.

The Right to Food Campaign demanded that midday meals should be provided to all children below 14 years of age, including those who were out of school. Midday meals should also be provided at all alternative educational facilities and schools run by other bodies receiving grants from the government. The campaign also demanded that the midday meals should be provided every day, based on the Supreme Court order of 20 April 2004[3] according to which midday meals need to be supplied even during summer vacations in drought-affected areas.

Entitlements for Adolescent Girls

The nutritional security of adolescent girls is crucial to break the intergenerational cycle of malnutrition. The NAC draft, government draft and final act only mentioned nutritional, health and education support to adolescent girls in their list of enabling provisions, and not as legal entitlements. The Supreme Court orders of 28 November 2001 and 13 December 2006 legally entitle all adolescent girls to 500 kcal and 20–25 grams of protein per day in the ICDS. The Right to Food Campaign demanded inclusion of this legal entitlement in the NFSA in the form of hot cooked meals or take-home rations.

Entitlements for Children in the Age Group of 14–18 Years

As per the Juvenile Justice Act of India, individuals up to the age of 18 are considered children. However, the NFSB drafts and the final act gave legal entitlements to midday meals only to children up to 14 years of age, thereby depriving children in the age group of 14–18 of any legal entitlement to food. Only the NAC draft stated that the midday meal might be provided to children beyond Class 8, although the final say was left to the central and state governments. The Right to Food Campaign demanded that the NFSA provide legal entitlements to midday meals to children up to 18 years of age.

Special Provisions for Marginalised Children

The NAC draft entitled every child up to 14 years of age, including those who were out of school, to approach any feeding facility such as an anganwadi centre, school midday meals or a destitute feeding centre for a freshly cooked nutritious meal, and no such institution was to deny such a meal to any such child on any grounds. While the government draft and final act removed this entitlement, which may severely jeopardise the right to food of children in this age group who are unable to attend school and access midday meals, there are provisions for children suffering from malnutrition between 6 months and 6 years of age. Independent surveys such as the NFHS have shown that a greater percentage of children from SC/ST backgrounds are severely underweight, wasted and stunted compared to children from other backgrounds. Malnourishment is highest for tribal children, with 29.1 per cent severely stunted, 24.9 per cent severely underweight and 9.3 per cent severely wasted, followed by Dalit children with 27.6 per cent, 18.5 per cent and 6.6 per cent in the respective categories. Children from OBC communities showed similar levels of wasting to SC children (6.6 per cent) and slightly lower levels of stunting (24.5 per cent) and underweight (15.7 per cent). On some indicators, there exists a difference of over 10 percentage points among SC/ST children compared with children from other socio-economic backgrounds (IIPS 2007b, 270). The NFSA has no provisions for these severely malnourished categories of children. (For a detailed discussion on marginalised children, see Chapter 14 on vulnerable groups.)

Entitlements for Children with Malnutrition

For a detailed discussion on the definition, types, causes, prevention and treatment of malnutrition, and a critique of the provisions regarding malnutrition in the various drafts of the food security act, please refer to Chapter 11 on 'Malnutrition'.

FAO Recommendations on Children's Entitlements Compared with the NFSB Drafts

Table 10.2 offers a detailed comparison of the NFSB drafts with the FAO's (2009) recommendations regarding children's entitlements in a right to food law.

Table 10.2. Comparison between children's entitlements in a right to food law as recommended by FAO and those present in the NFSB drafts

Recommendations of FAO's Guide on Legislating for the Right to Food	NAC draft	Draft introduced in Parliament	Comments
Provide at least that amount of food entitlement which prevents children from suffering from hunger and undernutrition	Present	Present	
Law should specify exact quantities of calories, proteins and micronutrients present in the food entitlements of children of the following age groups: • Up to 6 months • 6 months–3 years • 3–6 years • School age • Adolescent boys and girls	• Provided a uniform nutrition standard for calories and proteins for children (separate standards for malnourished children) in the age group of 0–6 years. • Provided nutrition standards for calories and proteins for	• Did not specify any nutrition standards for children up to 6 months of age as exclusive breastfeeding is recommended for children of this age group. • Provided nutrition norms for children of different age groups which are	• Specifying exact quantities of micronutrients would 'disqualify' locally cooked food as it is not possible to prepare food locally which has specific amounts of micronutrients. This would be possible only in pre-packaged centrally produced food items. The chapter on locally

(contd.)

(Table 10.2 continued.)

Recommendations of FAO's Guide on Legislating for the Right to Food	NAC draft	Draft introduced in Parliament	Comments
	children of primary classes and for children in upper primary classes.	the same as those in the NAC draft. • Stated that take-home rations would comprise energy-dense food fortified with micronutrients and 50 per cent of the RDA. Both cooked meals and take-home rations should be prepared in accordance with the prevailing food laws.	cooked food versus ready-to-eat food lists the various advantages of the former over the latter. • The FAO guide does not specify the ages corresponding to 'school-age' children and 'adolescent boys and girls'. There is an overlap of some years in these two categories. The various drafts also do not specify the ages corresponding to 'adolescent girls'. • All the drafts of the NFSB had legal entitlements for children only up to the age of 14 years, whereas according to the Juvenile Justice Act, individuals up to the age of 18 years are considered children.
To prioritise vulnerable groups when resources are limited, the law should make an	None of the drafts or the act had such a list. In India, the BPL census is undertaken every five years or so to identify 'poor' households which are entitled to various social assistance services, not only regarding subsidised food but also for housing, pensions, health,		

(contd.)

(Table 10.2 continued.)

Recommendations of FAO's Guide on Legislating for the Right to Food	NAC draft	Draft introduced in Parliament	Comments
exhaustive list of such groups which would include children.	etc. For more on the process of BPL census, refer to Chapter 5, 'Selection for Targeted Entitlements'.		
Provide food or means for its procurement, particularly to children whose parents die, disappear or do not take care of them, especially those living on the streets, in custodial institutions, working in hazardous industries, in conflict situations, belonging to refugees, with debilitating illnesses, of parents with debilitating illnesses, etc.	None of the drafts or act provided any special entitlements for the various categories of children suggested by the FAO guide.		
The law can play a role in strengthening the link between the right to food and the right to education for children.	None of the drafts or act had any special provisions which contributed towards the realisation of children's right to education. However, the NAC draft stated that all children in the age group of 0–6 are entitled, inter alia, to pre-school education. All the drafts and the act made midday meals in government and government-aided schools a legal entitlement for children between 6 and 14 years of age, which gave children, especially those who are food insecure, an incentive to go to school; this not only increases enrolment rates, but also increases the retention of children in school, both of which in turn contribute to realising children's right to education. The draft introduced in Parliament, and retained in the final act, stated that the central government, state governments and local authorities shall, for the purpose of advancing food and nutritional security, strive to progressively realise education support to adolescent girls.		
Realisation of the right to food, particularly for	• Provided children in the age group of	• Provided for appropriate	

(contd.)

(*Table 10.2 continued.*)

Recommendations of FAO's Guide on Legislating for the Right to Food	NAC draft	Draft introduced in Parliament	Comments
children, also requires provisions for clean water, sanitation and healthcare.	0–6 years the entitlement to immunisation, health check-ups and growth monitoring and promotion. • Directed schools to make arrangements for appropriate facilities for hygienic cooking and clean drinking water.	facilities for cooking, drinking water and sanitation in schools and anganwadis. • Stated that the central government, state governments and local authorities should, for the purpose of advancing food and nutritional security, strive to progressively realise access to safe and adequate drinking water and sanitation, healthcare, nutritional and health support to adolescent girls.	
Competent state authorities should ensure that adequate legislation is in place to ensure that breast milk substitutes are not promoted but instead strongly regulated, and that the established rules are respected in practice.	India has legislated the Infant Milk Substitutes, Feeding Bottles and Infant Foods (Regulation of Production, Supply and Distribution) Act (no. 41 of 1992) which does this.		

Source: Prepared by the author.

Notes

1. People's Health Assembly, a respected association of public health activists in India.

2. Allergies, eczema and atopic illnesses, diarrhoea, Crohn's disease, childhood cancer, Hodgkin's disease, juvenile rheumatoid arthritis and many viral infections such as influenza and bacterial infections such as respiratory and middle-ear infections are prevented by breastfeeding. All these benefits of breastfeeding contribute to the health of the child, which in turn decreases the chances of malnutrition and mortality amongst children.

3. *People's Union for Civil Liberties v. Union of India and Others*, Writ Petition (Civil) 196 of 2001.

References

Anandaiah, R., and M. K. Choe. 2000. 'Are the WHO Guidelines on Breastfeeding Appropriate for India?' National Family Health Survey Subject Reports, Indian Institute of Population Sciences, New Delhi.

Anon. 2011. 'Malnutrition and ICDS: Pronouncements about the ICDS Scheme Have to Be Juxtaposed against the Grim Reality of Malnutrition'. *Economic and Political Weekly*, 46(12): 9.

Black, R., M. Saul and V. Jennifer. 2003. 'Where and Why Are 10 Million Children Dying Every Year?' *Lancet*, 361(9376): 2226–34.

Deaton, Angus, and Jean Drèze. 2009. 'Food and Nutrition in India: Facts and Interpretations'. *Economic and Political Weekly*, 44(7): 42–65.

FAO (Food and Agriculture Organization). 2009. *Guide on Legislating for the Right to Food*. Rome: FAO.

HUNGaMA. 2011. *HUNGaMA Survey Report*. http://hungamaforchange.org/HungamaBKDec11LR.pdf (accessed 25 September 2017).

IIPS (International Institute for Population Sciences). 2007a. *2005–06 National Family Health Survey (NFHS-3): National Fact Sheet*. Mumbai: IIPS.

———. 2007b. *2005–06 National Family Health Survey (NFHS-3)*, vol. 1. Mumbai: IIPS.

Jones, G., W. Richard, B. Robert, B. Zulfiqar, M. Saul et al. 2003. 'How Many Child Deaths Can We Prevent This Year?' *Lancet*, 362(9377): 65–71.

Right to Food Campaign. 2007. 'Anganwadis for All: A Primer'. Right to Food Campaign, Secretariat, New Delhi.

11

Malnutrition

ANKITA AGGARWAL, ASHWIN PARULKAR AND HARSH MANDER

Much of the discontent captured in debates on India's malnutrition burden in the years between the 2001 Supreme Court case and the passage of the NFSA in 2013 rested on the acknowledgement that India's high economic growth period in the early part of the decade of the 2000s had not led to a decrease in nutritional poverty. Angus Deaton and Jean Drèze (2009) showed that while per capita consumption expenditure increased from 2.2 per cent a year in the 1980s to 2.5 per cent a year in the 1990s and 3.9 per cent from 2000 to 2005, per capita calorie intake and nutrient consumption declined in the same period. Moreover, the proportion of underweight children under 3 remained high between 1998–99 and 2005–06, declining by 1 per cent from 47 per cent to 46 per cent.[1]

In the year preceding the passage of the food bill, one major point that framed the context of the severity of India's malnutrition burden was that compared to both BRIC nations (Brazil, Russia, India and China) with similar economic growth rates, and poorer South Asian neighbours with lower growth rates and less functional democratic institutions, India fared worse in terms of low birth weight, underweight and wasting among children.[2] The concern raised by Amartya Sen and Jean Drèze was not merely that India was lagging behind peer countries, but that the country had actually fallen behind in social indicators during the time of rapid economic growth. The comparison with Bangladesh is instructive. Between 1990 and 2010, Bangladesh had overtaken India in key social indicators, including life expectancy, child survival, fertility rates and immunisation rates, while still having only half of India's per capita income. In 1990, the child mortality rate in Bangladesh was nearly 24 per cent higher than India's, while 20 years later it was nearly 24 per cent lower (Drèze and Sen 2013).

At a time when major food security legislation was up for debate, the lack of improvement in nutrition and the relatively poor performance in social indicators in comparison to other countries with lower growth rates gave cause for nutrition and food security scholars to consider how law and policy should ensure legal entitlements to nutrition programmes, since the assurance of grains would have little bearing on the persistence of underweightedness, stuntedness and anaemia. In the chapter on starvation, we mentioned how undernutrition is an accompanying characteristic of endemic hunger in India, and explored how those who suffer from the condition face debilitating ailments: physical and cognitive impairment, an increased burden of disease and, in extreme cases, the loss of life. The debates we explore in this chapter can be read together with those in the starvation chapter, particularly concerning linkages between factors of food access and public health that, concomitantly and over time, debilitate the health and potential of an individual. Such debates shed light on the challenges policymakers face in ensuring additional infrastructure, service delivery, and practical and scientific expertise outside the scope of food distribution.

Debates have also called attention to the role of social inequities, such as discrimination against women and lower-caste people in education and access to basic services, which play a role in the undernutrition of vulnerable people. This point is particularly relevant in understanding the lack of improvement in nutrition indicators among historically marginalised groups who live in poorer states, such as ST and SC groups in Bihar, Madhya Pradesh and Jharkhand. As we have highlighted throughout this book, access to adequate food alone is not enough to tackle malnutrition. It is equally important, considering the scale of India's nutrition crisis, to consider legal mechanisms, institutional setups and monitoring systems of current social programmes required to ensure people's access to clean drinking water, healthcare and education, as well as specific nutrition services to target risk factors for diseases associated with chronic undernutrition. This begs the question of whether the recently passed NFSA has the potential to contend with major risk factors for malnutrition, or if additional policies in the context of food and nutritional security may be required to contend realistically with the problem.

This chapter begins with international definitions and norms of malnutrition for children and adults, followed by a discussion on malnutrition in the Indian context, and key scholarly debates on how and to what extent the state can address the numerous dimensions, causes and outcomes through law and policy. A subsequent discussion focuses on a subset of undernutrition, viz., severe acute malnutrition

(SAM), relevant to the large number of Indian children who face this condition. The final section critiques the entitlements under the NFSA and its prior drafts related to malnutrition.

Definition and Indicators of Malnutrition

The FAO guide (2009) states that it is essential for countries drafting a right to food law to define hunger and undernutrition to enable those suffering or at risk of these conditions to access necessary amounts of food, according to their age, sex, health status and occupation, in order to live with dignity. The guide recommends that in states where the immediate realisation of a comprehensive right to food is not a possibility, implementation can be phased, whereby eradication of hunger and malnutrition is a priority and removing the risk of these conditions is one of the next steps (ibid.).

Yet optimal definitions of malnutrition are still lacking in national public policies, mainly due to state governments not yet thoroughly examining the patho-physiological, public health, and other related dimensions required to wage appropriate responses (Meijers et al. 2010). Indian public nutrition programmes are largely guided by international norms set by the WHO to measure, diagnose and target malnutrition. The WHO understands malnutrition as a medical and social disorder rooted in poverty that is dependent on the interaction among the intake of food and overall state of health and the physical environment. It concerns conditions where a person has not enough or too much food,[3] and the body's response to a wide range of infections that result in malabsorption of nutrients or the inability to use nutrients properly to maintain health.[4]

People who experience malnutrition are vulnerable to infections due to poor intake of protein, energy and micronutrients.[5] In the Indian context, poor feeding practices during infancy and early childhood are also determinants of malnutrition (Planning Commission 2011). Scholars have argued that undernutrition, specifically, is the type of malnutrition associated with food insecurity resulting from the combination of poor intake of adequate nutritious food—a healthy balance of macronutrients and micronutrients—and unsanitary living conditions (IIPS 2007).

Devereux (1993, 13, cited in Banik 2007) draws a distinction between chronic malnutrition and undernutrition: 'Chronic malnutrition describes nothing more than a poorly balanced diet, lacking in essential nutrients, while undernutrition is the physical syndrome associated with prolonged food deprivation.' Banik (2007) defines undernutrition as the 'consumption of incorrect types of food (quality) and/or inadequate intake of food (quantity)' where 'an improvement in the nutritional status of the population is dependent on public action—including food subsidy programmes, direct food supplementation programmes, pension and labour-intensive programmes—and local coping strategies like seasonal migration, loans from moneylenders, and distress sale of assets.'

There are different indicators to measure malnutrition among children and adults. Among children, malnutrition can be measured by anthropometric indicators typically based on age, height and weight. Three standard indicators are 'height for age', 'weight for height' and 'weight for age'. Low height for age is often referred to as 'stunting', low weight for height as 'wasting', and low weight for age as 'underweight' (Deaton and Drèze 2009).

A child is deemed stunted, wasted or underweight if their score is at least 2 standard deviations below the median for the World Health Organization's (WHO) child growth standards.[6] In India, 45 per cent of children are stunted, which is an indication that they do not receive adequate nutrition from conception or birth and are at risk of recurrent and chronic illness (IIPS 2007). The first 1,000 days of life are a critical period for interventions in vulnerable populations whose children are prone to stunting, given that more than 15 per cent of children are estimated to be stunted at birth; this figure increases up to 58 per cent at 23 months of age, remaining stable thereafter (Cronin et al. 2016). Further, 23 per cent of Indian children are wasted, which indicates their current or recent failure to receive adequate nutrition; wasted children may be affected by frequent episodes of diarrhoea and other acute illnesses. Being underweight can result from either chronic or acute malnutrition, and is considered a comprehensive indicator since it encompasses aspects of stunting and wasting (ibid.).

Malnutrition in adults can be assessed using the BMI, which is defined as weight in kilograms divided by height in metres squared (kg/m^2). A BMI below 18.5 indicates chronic energy deficiency or undernutrition. Adults with BMI below 18.5 can be considered to be experiencing various grades of undernutrition including starvation, as mentioned in the chapter on starvation (see Chapter 9). Adults with a BMI of 25 or higher are considered to be overweight or obese. A normal weight for height is indicated by a BMI of 18.5–24.9. In India, 36 per cent of women and 35 per cent of men have less-than-normal BMIs (Planning Commission 2011).

In 1999, the WHO defined severe malnutrition in children to identify those in the world who were at higher risk of mortality and required intensive nutritional and medical treatment. Experts decided that severe malnutrition in children would be defined as a weight-for-height measurement of 70 per cent or less below the median—3 standard deviations below the mean NCHS reference values—and/or evidence of excessive swelling in body tissues, known as bilateral pitting oedema.[7] In 2005, a group of experts recommended that children in the age group of 6 to 60 months who have a mid-upper-arm circumference of less than 110 millimetres should also be diagnosed with severe malnutrition (WHO/UNICEF/WFP/SCN 2007).

In 2006, the WHO revised the terminology of severe malnutrition to 'severe acute malnutrition'. While the cut-off for the weight-for-height criterion for defining SAM remained the same, the cut-off for the mid-upper-arm circumference increased to 115 millimetres, to identify more children as suffering from SAM, and still having a high specificity of more than 99 per cent.

The NAC defined malnutrition in its draft as a 'condition that develops when the body, over a prolonged period of time, does not receive or absorb adequate and appropriate calories, proteins and other nutrients required for good health, growth and maintenance of the human body and mind' (NAC 2011). While the NAC draft defined severe malnutrition, SAM and severely underweight children as per the WHO norms, it did not refer to any norms for mild or moderate malnutrition. It may have been useful for the draft to define malnutrition as 'stunting', 'wasting' or 'underweight' for children and a BMI of less than 18.5 as malnutrition among adults. There was potential to utilise government surveys that have calculated the percentage of children and adults who suffer these conditions across a range of geographical and social groups. Still, this move is unlikely to occur in the near future since the government's 2011 draft of the bill and the final 2013 act did not define any grade of malnutrition, for children or adults.

Malnutrition in the Indian Context

Other chapters in this book discuss causes of undernutrition among vulnerable populations such as women and children in terms of declining consumption and access to food, social practices, intra-household power imbalances, and the lack of proper health, sanitation and water facilities. All are major factors in India's malnutrition burden. Malnutrition may be the result of one or more factors, with insufficient access to food an important determinant of malnutrition. A person may have inadequate access to food because of insufficient availability of food, inability to grow enough food, inadequate means to procure food and/or inability to access food-related assistance. Despite access to adequate quantity of food, a person may succumb to malnutrition if their food does not have adequate amounts of calories, protein and micronutrients such as vitamins. People may also fall prey to malnutrition despite being able to access adequate quantities of nutritious food; this occurs when they are unable to absorb the nutrients in their food, because of infectious diseases such as diarrhoea[8] and other infections. Malnutrition in turn makes the body more susceptible to infectious diseases, thereby triggering a vicious cycle, especially among children with unsanitary living conditions and unclean water leading to repeated infections that result in malnutrition (Ghosh 2006). Inadequate access to healthcare prolongs the duration and/or severity of the infections, which exacerbates the inability of the body to absorb nutrients. The combination of inadequate food and unsanitary conditions, according to John Butterly and Jack Shepherd (2010), can lead to a state of malnutrition because the body is doubly vulnerable—the lack of nutritious food rids the body of necessary protein, energy and micronutrients, while exposure to unsanitary conditions makes it easier for viral and bacterial infections to occur frequently.

Recently, scholars have stressed factors such as poor hygiene practices and sanitation facilities that impact undernutrition and require public health responses. Open defection is practised by 53 per cent of people in India, while 69 per cent of rural India lacks toilet facilities. The relationship between poor hygiene and the lack of sanitation facilities and its impact on child stunting has been established in studies of developing countries. A composite analysis of nine studies in multiple developing countries that researched the relationship between incidences of diarrhoea and child growth among 1,393 children found that the possibility of a child's stunting at 24 months increased by 2.5 per cent for each episode of diarrhoea. Further, one-fourth of all cases of stunting in 24-month-old children were related to having experienced five or more episodes of diarrhoea in the first two years of life (Checkley et al. 2008). This indicates the need for preventative measures such as proper health and sanitation infrastructure to protect against long-term deprivations associated with unclean environments, particularly to protect young children who have no choice in the environments in which they live.

In a study in Sudan, Merchant et al. (2003) also found that children from households that had access to proper sanitation facilities and water were about one-fifth less likely to become stunted, while children who were stunted but lived in households with access to such facilities had a 17 per cent greater chance of reversing stunting, compared to children who lacked such facilities. In India, Dean Spears has shown that access to appropriate facilities and healthy environments are stronger determinants for improving health and nutrition outcomes across communities than household wealth. In his analysis of the data on child height and defecation practices gathered by the *India Human Development Survey*, Spears found that, first, 80 per cent of children live in communities—primary sample units—in which more than half the people defecate in the open, indicating that children from both poor and wealthier families are exposed to the harmful effects of these practices. Spears finds that while poorer children are on average exposed more to open defecation, 'rich and poor children alike are threatened by disease externalities' (Spears 2016, 88).

These insights highlight neglected areas of public policy in tackling the complex problem of malnutrition. The problem is when interventions across sanitation, public health and food security are viewed as mutually exclusive determinants of hunger, malnutrition and poverty. Nutrition is not just about what people eat, but about how well the body is able to utilise it. The impact of environmental hazards—poor sanitation, lack of drinking water and open defecation—on malnutrition deserves as much attention in public policy as concerns raised by food security and food sovereignty advocates on the need to ensure food production systems and sustainable agricultural practices to address hunger.

N. C. Saxena has brought attention to the fact that the methodology used in official NSSO surveys of hunger in India—referred to as 'self-reported hunger', which involves asking people if they have had two square meals a day—has resulted in severe underestimates. Based on this, there was a 'drastic decline in self-reported hunger in India from 16.1 to 1.9 per cent' (Saxena 2008). Saxena points out that while this may reflect a significant improvement in the number of people who live with the most severe forms of hunger, the data does not include the conditions of people who live with chronic hunger that involves lack of adequate and nutritious food over time, required to live a healthy life. Saxena's analysis of the declining consumption of cereals between richer and poorer deciles shows that cereal consumption of the top decile of rural India declined between 1993–94 and 2004–05 due to higher levels of prosperity; whereas the decline in consumption of the poorer deciles should be considered a 'distress phenomenon, as with marginal increase in their incomes over time they are forced to cut down on their food consumption' (Saxena 2016).

The Centre for Equity Studies' own research on starvation and destitution in many parts of the country confirms that there are large populations who simply do not get enough to eat. In a study on government-reported starvation deaths across 10 villages in Bihar, Madhya Pradesh and Jharkhand, the study observed that in *all* of the nearly 30 cases, the demise of the victim was the result of multiple concurrent deprivations, including lack of access to regular employment, safe and quality food, adequate primary schools, clean drinking water, sanitation, and affordable healthcare to treat endemic infectious diseases, to name a few.[9] It is perilous therefore to reject out of hand feeding or subsidised food programmes, pitting them against public interventions for clean drinking water and sanitation.

While infected water and open defecation are important causes for malnutrition, they are not the only contributory factors. Many studies suggest that incorrect feeding practices are a common cause of malnutrition among infants. Directly or indirectly, studies suggest that they are responsible for more than two-thirds of the deaths caused by malnutrition (WHO 2003). Those children who are not breastfed as often as is recommended have a higher chance of becoming malnourished (Roy 2000). Thus, a lack of education about nutritious food, poor knowledge of the correct practices of feeding children, along with a lack of clean drinking water, sanitation and remedial action to be taken in case of infections, all contribute to malnutrition (ibid.).

Reduced levels of breastfeeding are frequently blamed by scholars and policymakers on irrational cultural practices and lack of education of mothers. What is frequently forgotten is that impoverished mothers often have to labour even just before and after pregnancy at work sites without crèches or childcare services. It is for this reason, and not irrationality, that they are unable to breastfeed the infant, and have to leave them to the care of older siblings, sometimes as young as 4 to 5 years of age—young children who are in need of care themselves. Due to their young age, these small children feed the infants in unsanitary conditions and with infected food and water. There is a need to recognise that many mothers are unable to take adequate care of their child as they are deprived of adequate time, energy, resources, power and knowledge. Many of them have to work to earn a livelihood and to juggle this with other tasks such as cooking, fetching water, cleaning, etc. Therefore, just as advocates of agricultural regeneration and sanitation may claim that they have the 'real' answer to the puzzle of malnutrition in India, it is equally reasonable for breastfeeding missionaries to claim that promoting and supporting breastfeeding and early childcare will do more than anything else to reduce malnutrition, and that therefore nutrition counselling, universal maternity entitlements and crèche and childcare services are the missing keys in the puzzle for ending malnutrition.

Inadequate public health services contribute to health crises and chronic poverty among poor people in India who are vulnerable to malnutrition. Government expenditure on health is about 1 per cent of GDP, much lower than in middle-income and industrialised countries. As a result, key health programmes for vulnerable populations are under-resourced or absent, such as maternal and child services. Balarajan et al. (2011) note that private expenditures on healthcare in India are among the highest in the world (71.1 per cent in 2004–05), and have increased more in the last decade among the poorest households. They find that health burdens and out-of-pocket expenses on health are contributing factors for more than half of households in India that fall into poverty.

In 2013 , the discourse on malnutrition shifted away from debates on factors that have led to, and adequately describe, India's malnutrition burden, to questions over the extent to which policymakers have an accurate understanding of the scale of the problem in the country. Economist Arvind Panagariya (2013), writing in the *Economic and Political Weekly*, argued that prevailing child malnutrition rates in India, which put the number of stunted and underweight children above those in Sub-Saharan Africa, were inflated, suggesting that uniform standards of child height and weight measurement, put forth by the WHO and promoted by the UN, were too rigid to account for the complex manifestations of malnutrition that may arise from 'genetic, environmental, cultural, and geographical factors'. Panagariya compared maternal and child health indicators between India and a range of Sub-Saharan African countries, noting that India fared worse in child weight and height for the relevant age than most of the 33 selected Sub-Saharan African countries, according to international data, but nearly all of the selected African countries registered higher rates of infant, under-5 and maternal mortality rates. He claimed that it was implausible for a society with high numbers of malnourished children to have lower death rates in comparison to poorer countries for each of these populations. Inaccurate methods of measurement, he concluded, were responsible for this inconsistency. Instead of height and weight measurements, policymakers should, he argued, require measurement of two components of malnutrition that can only be assessed by a thorough medical examination: protein energy malnutrition, 'which manifests in poor gains in height, weight, and head and mid-upper arm circumference', and micronutrient deficiencies, which 'result from low levels of iron, folate, iodine, and various vitamins' (see Gupta et al. 2013, 75).

These arguments led to a furore of rebuttals published in a later issue of *Economic and Political Weekly*. A group of academics and nutrition specialists associated with the Right to Food Campaign warned against dismissing height and weight measurements outright as appropriate indicators of malnutrition. 'Measuring heights and weights is not new and India has been measuring populations for nutrition status since the 1970s, and this includes height and weight' (Gupta et al. 2013,

74). The authors further mentioned that data on thorough medical check-ups of children already showed that 'more than 70 per cent of children are anaemic, more than 50–60 per cent have Vitamin A levels below the acceptable cut-off level of 20 ug/dl, and many have one more vitamin B complex deficiency' (ibid.). Most importantly, while the authors drew attention, as others had, to markers of social exclusion responsible for India's malnutrition problem—the unequal status of women and implementation problems in social programmes which prevent poor families from accessing food, to name two—they highlighted the discrete linkages between hunger and malnutrition, characterised by the poor's inability to access a range of nutrients and calories from the diverse range of foods required to live a healthy life. 'Nutrient deficiencies very rarely exist in isolation. The only solution is to make sure that people get their calories and proteins from diverse sources (or all the food groups)—cereals such as rice, wheat and millets, pulses, oils, potatoes, fruits such as bananas, fresh vegetables, milk, eggs, nuts, meat, and so on' (ibid., 75).

Malnutrition indicators are also far higher in communities that live in the poorer states of Bihar, Madhya Pradesh, Jharkhand, Odisha, Chhattisgarh and Uttar Pradesh, where 60 per cent of the country's poverty is concentrated. Each state fares worse than the abysmal national averages in nearly all nutrition indicators: height for weight and age of children, adults and children with anaemia, under-5 mortality rate, and the percentage of population with below normal BMI. With respect to most development indicators, at least one of these states ranks as the worst in the country. Madhya Pradesh, for example, has the most children under 5 who are underweight (60 per cent) and the highest under-5 mortality rate in the country (9.4 for every 100 children). The SC and ST populations in these states are, in terms of undernutrition, the country's most vulnerable people, experiencing even higher incidences of malnutrition and hunger than other groups in their state, SC/STs in other states, and other vulnerable groups in other areas of the country. For instance, in Bihar the under-5 mortality rate for SCs is 113 per 1,000 births, compared to the all-India SC average of 88, the Bihar average of 85, and the all-India average of 74. A large number of starvation and malnutrition deaths in India occur in these states, and nearly all of the cases occur in these communities.

In summary, it is clear that because of the extremely high incidence of malnutrition in India, it is imperative for the country to focus not just on food security, but also nutrition security, which comprises access to adequate amounts of nutritious food, clean drinking water, sanitation, maternity support, childcare services and public healthcare. While each of these interventions is vital, they do not exclude the need for simultaneous action on other fronts as well. An integrated, multi-view approach to malnutrition is a necessity, rather than each perspective furthering only its own agenda.

Preventing Malnutrition in the Indian Context

As the previous section concluded, the root causes of malnutrition lie in various structural problems. The debate on whether a right to food law should address these concerns is discussed in Chapter 3 on the scope of the food security law. This section discusses the prevention of the more immediate causes of malnutrition, such as inadequate access to nutritious food, clean water, sanitation, healthcare and education.

Improving the Implementation of Food and Employment Schemes

The various food-related schemes of the government, namely the TPDS, MDMS, ICDS, maternity entitlements and pensions for the aged, widows and persons with disabilities, all contribute to enhancing food security—either by directly giving food to people, or by providing them with money to buy food. By entitling every rural household to 100 days of manual work, MGNREGA contributes to the economic access of rural households to food. However, the implementation of these schemes and acts is far from perfect, and to alleviate malnutrition, there is a need to improve the implementation of these schemes.

Among all of the food-related schemes, the ICDS is the most important in tackling malnutrition. Under the Supplementary Nutrition Programme of ICDS, children up to 6 years of age, adolescent girls, pregnant women and lactating women are entitled to cooked food or take-home rations. Women in the age group of 15–45 years are supposed to get education on issues such as nutrition, infant feeding, family planning, sanitation, utilisation of health services and suchlike, which would enable them to look after their own and their family's nutrition and health needs, especially those of their children. The immunisation service of the ICDS plays a crucial role in providing protection against infectious diseases, where children under 6 are immunised against polio, DPT, measles and TB, and pregnant women are immunised against tetanus. The anganwadi workers at the ICDS centres are also supposed to regularly measure the height and weight of children to ensure that they are growing properly, and identify those who are malnourished. The referral services of the ICDS aim at linking malnourished children, along with other children requiring medical attention, with the public health care system (Right to Food Campaign 2007).

Despite the extreme importance of the ICDS in preventing malnutrition among children and facilitating their proper growth and development, the quality of these services is very poor in several areas. On 28 November 2001, the Supreme Court ordered the universalisation of ICDS services and made them legal entitlements. However, many settlements continue to not even have operational ICDS centres, leading to an urgent requirement to not only universalise but also improve the quality of ICDS services.

Focusing on Children in the Age Group of up to 2–3 Years

Child malnutrition sets in very early in life and increases rapidly until the age of 24 months. Thus, the crucial period to prevent malnutrition is the time from birth to 2 years of age. Prevention of malnutrition must therefore take place at the household level, and these efforts should also be supported by trained health and ICDS workers, as well as by experienced older women in the community who have successfully nurtured their children (Ghosh 2006).

Up to 6 months of age, children should be breastfed, starting almost immediately after birth, preferably even before the placenta is expelled and the mother cleaned up. Even though all mothers notionally breastfeed, the percentage of children exclusively breastfed drops steadily from 72 per cent for children under 1 month of age to about 20 per cent at 6 months (Ghosh 2006). Breastfeeding requires mothers to be with their children, which in turn requires maternity benefits and crèches at the place of the mother's work. Mothers who are in poor health themselves are unable to provide adequate breast milk to their young children. Breastfeeding, supplemented by additional nutritious food for children as young as even 2–3 months, may be more effective in preventing early malnutrition than exclusive breastfeeding, in this context (Anandaiah and Choe 2000). Pregnant and lactating women also need to be targeted and provided supplementary nutrition.

Children in the age group of 6 months to 2 years need to continue to be breastfed, but also need to be provided with other types of food, such as semi-solid foods, in the correct quantities and at specific intervals of time.

Other Measures

Other measures required for preventing malnutrition include ensuring access to adequate quantities of clean drinking water, sanitation, healthcare and education; this not only means physical access to these facilities, but also economic access, i.e., certain basic facilities should be free of cost.

Critique of the Provisions in the Drafts to Prevent Malnutrition

Both the government[10] and NAC (2011) drafts of the NFSB had a section on 'prevention and treatment/management of child malnutrition'. However, the provisions under this section were either for identification or for treatment/management of malnutrition, and not for prevention. So, even though none of the drafts gave any specific direction for preventing malnutrition, all the entitlements of the drafts that directly provided food or money to procure food and those that provided mothers with support for breastfeeding and counselling on child feeding practices, could be interpreted as measures that contributed to preventing malnutrition. However, the extent to which these entitlements would contribute to preventing malnutrition is debatable; for instance, the amount of subsidised cereals entitled under the drafts, especially for 'general' category persons, is far from adequate to meet people's nutrition requirements.

The drafts did not provide legal entitlements to pulses and oil, which are important sources of protein and energy respectively for a majority of Indians. The entitlements provided in both drafts and in the final act to cooked meals under the ICDS and the MDMS will play a crucial role in preventing malnutrition among children, pregnant women and lactating mothers. While the NAC draft provided legal entitlements for the other ICDS services as well, which are crucial in preventing malnutrition, the government draft and the final act did not, even though these services were given the status of legal entitlements by the Supreme Court in its order of 28 November 2001. Despite the importance of ensuring good health of adolescent girls to break the intergenerational cycle of malnutrition, neither draft, nor the final act, gave a legal entitlement to supplementary nutrition for adolescent girls through the ICDS, even though it is an existing legal entitlement as per the Supreme Court orders of 28 November 2001 and 13 December 2006. The provision of one free meal to people in destitution and affordable meals to people who are homeless may not be adequate to prevent malnutrition, but would contribute in preventing starvation, although this was excised from the final act. Maternity entitlement of Rs 1,000 per month for six months starting from the date of delivery of the child is another important provision that would contribute in reducing malnutrition among lactating women and newborn babies.

The Right to Food Campaign noted that poor working mothers lacked access to proper childcare support, which is crucial for preventing and treating malnutrition among children. The campaign demanded legal entitlements to crèches for all children under 6 years of age, suggesting that the ICDS should provide this service. This could be done in a phased manner, with 10 per cent of the anganwadis providing crèche services within the first two years of the act coming into force. There should also be a provision of 'anganwadi-cum-crèche' on demand, where there is such a demand from at least 20 mothers of children under 3 years in a habitation. The anganwadi-cum-crèches should have at least the following facilities:

- An additional room for children under 3 years of age.
- Additional wages for the two anganwadi workers and anganwadi helper for working full time in the crèche.
- An age-appropriate and nutritious morning snack and an evening snack for children less than 3 years, over and above the midday meal.

- Additional resources for health and medical supplies, sleeping and play equipment, cleanliness and hygiene, skilled care, etc.

The campaign suggested that other models of provision of crèche services through panchayats and/or women workers' cooperatives could also be worked out where there was a need for flexibility to suit the needs of working women. While the NAC draft mentioned universal access to crèche facilities as a goal for progressive realisation, the government draft did not do this, nor did the final act. The campaign also demanded that the goals of progressive realisation of access to adequate amounts of clean drinking water, sanitation and healthcare, which are crucial for preventing and treating malnutrition, be made legal entitlements.

Treatment of SAM

Children suffering from SAM can either be treated in their communities, or need to be admitted in an institution, depending on their condition. Children with SAM who develop medical complications—usually about 15 per cent of all children under 5 years of age suffering from SAM—require treatment in institutions. The rest, 85 per cent, can be managed through treatment in their communities (Sachdev et al. 2010). A combination of institutions and home-based management of SAM can also be adopted. For instance, SAM children with complications can be initially stabilised in a health facility and then rehabilitated by home-based management (Gupta et al. 2006).

Both in institution- and community-based treatment, ready-to-use therapeutic food (RUTF) and energy-dense local foods can be used for combating malnutrition. A discussion on institution- and community-based treatment of SAM is provided first, and then the debate on the preference between RUTF and locally produced energy-dense foods is taken up.

Institution-based Treatment of SAM

Institution-based care is the only option for treating SAM children with medical complications such as diarrhoea, acute respiratory infections, malaria and measles, which increase the risk of mortality among SAM children by 20 to 60 times (Gupta et al. 2006). However, there are several factors which hinder the success of this method of treatment. The capacities of institutions treating SAM, in terms of the number of inpatient beds and trained staff, are usually very low, compared to the number of children requiring treatment in them. Many have attributed the 'high' costs of running such institutions to their relative shortage, although what should be considered 'expensive' to save the lives of children can be debated (Collins et al. 2006).

Access to such institutions is also a problem, especially for people living in villages which are very far from such institutions and have poor connectivity (Damme and Boelaert 2002). The time taken by parents in deciding whether or not to take their SAM children to medical institutions, and the time spent in taking children to the institutions, may itself worsen their condition, or even prove fatal (ibid.). Immunosuppressed children with SAM also run a high risk of cross-infection in institutions treating SAM (Cook 1971).

Treatment in institutions requires the mother to stay with her child for several days (Briend and Collins 2010), which discourages many parents from taking their SAM children to institutions. It is not feasible for many mothers to stay with their children in the medical institutions, as that leaves their other children and household chores unattended to. Staying at the medical institution also prevents mothers from working and earning money, which is extremely important as most parents of SAM children are poor. Due to the relatively long periods of stay required in institutions to treat SAM, many parents take their children from the institutions midway into their treatment.

Very often, children who gain weight and recover in institutions fall prey to SAM again after discharge (Beaudry-Darisme and Latham 1973). Institution-based treatment also requires follow-up visits to ensure that the child is eating and growing properly and is free of any disease. However, often due to the various difficulties involved in taking children to these institutions, such as opportunity costs arising due to work that may have to be foregone, many parents are unable to take their children to the medical institutions for follow-ups.

In developing countries, even among SAM children who receive treatment in medical institutions, about 20 to 30 per cent are unable to survive. It is surprising that this case fatality rate has remained unchanged since the 1950s (Schofield and Ashworth 1996), given that since the 1970s, treatment techniques have been available that can reduce mortality rates among children with SAM to as low as 1–5 per cent (see Ahmed et al. 1999; Golden 1988, 1996).

In the past seven years, nutrition rehabilitation centres (NRCs) have been started in several states of India to provide institution-based treatment to children with SAM. Children identified by anganwadi workers as suffering from SAM are brought to the NRCs, where they are required to be admitted for a fortnight, along with their mothers. The SAM children are provided intensive care and their mothers are given training on nutrition. After the completion of their treatment in the NRCs, the children are required to be brought back to the NRC a few times for follow-up, to ensure that they are not slipping back into malnutrition.

The NRCs have been quite successful in treating children with SAM.[11] The Right to Food Campaign argued that they should be set up in every primary health centre where children should be given free treatment for as long as required to restore them to good health. It demanded that during the period when the child is admitted in the NRC, its mother or primary caregiver should also be entitled to free boarding and lodging at or near the NRC. For the proper functioning of NRCs, the campaign suggested convergence between the Health and the Women and Child Development Departments at all levels.

Community-based Treatment of SAM

The various limitations of institution-based treatment of children with SAM have contributed to innovations in community-based treatment of SAM. These involve the treatment of SAM children with special therapeutic foods such as RUTF or energy-dense locally produced foods, and careful monitoring of the children to ensure that they gain adequate weight (WHO and UNICEF 2007), as well as the counselling of their parents on health and nutrition. As mentioned earlier, community-based treatment is possible only for those SAM children who do not have medical complications. Hence, to enable community-based treatment of SAM children, they need to be identified in the early stages of SAM, as with time many develop medical complications (Briend and Collins 2010).

Despite the severe burden of malnutrition in India, the country did not have guidelines for effective home-based treatment of SAM children until November 2005, when a national workshop was organised to develop guidelines for effective home-based care and treatment of children suffering from SAM. The workshop was jointly organised by the Department of Human Nutrition, the All India Institute of Medical Sciences in New Delhi and the Indian Academy of Pediatrics (Speciality Chapter on Nutrition). The workshop decided on the types and frequency of foods to be provided to children with SAM and on guidelines regarding the monitoring of such children.[12]

Debates on the Use of RUTFs in the Treatment of SAM

What is RUTF?

Children with SAM need safe, palatable foods with a high energy content and adequate quantities of vitamins and minerals. Ready-to-use therapeutic foods are high-energy, fortified, ready-to-eat soft or crushable foods that can be consumed easily by children from the age of 6 months without adding water. At least half of the proteins contained in RUTF come from milk products, and RUTF has a similar nutrient composition to F100, which is the therapeutic diet used in hospital settings to treat SAM (WHO and UNICEF 2007). As RUTF does not require any cooking, cooling or heating prior to consumption, and can be safely stored at ambient tropical conditions for at least three to four months, its use is very easy (Manary 2005). Moreover, RUTF has very low water activity, which makes significant bacterial growth in it impossible (Briend 1997).

Efficacy of RUTF in Treatment of SAM

Ready-to-use therapeutic foods have been successful in treating SAM and lowering mortality rates among children suffering from SAM by facilitating rapid weight gain. They have also played a crucial role in enabling home-based management of SAM. For instance, in 2005 during a famine in Niger, Doctors Without Borders treated 60,000 children suffering from SAM with Plumpy'nut. Ninety per cent of the children recovered completely and case fatality was 3 per cent (Rice 2010). Apart from anthropometric recovery, RUTF also results in physiological and functional (including immunological) recovery (Sachdev et al. 2010).

Gera (2010) undertook a review of 18 published papers (two systematic reviews, seven randomised controlled trials, seven observational trials and two consensus statements) on therapeutic nutrition products for home-based management of SAM. The systematic reviews and randomised controlled trials showed RUTF to be at least as efficacious as F100 in increasing weight, and more effective in comparison to home-based dietary therapies, although locally made RUTFs were as effective as those that were imported (ibid.).

Due to its high rate of success in treating SAM, RUTF is recommended for the management of this malady by the UN and its agencies working in the field of malnutrition—the World Food Programme, WHO and UNICEF. The success of RUTF has inspired its manufacturers to produce more ready-to-use foods which can be used for preventing malnutrition. Its advocates compare the use of such foods with giving multivitamins to children (Rice 2010).

Use of RUTF in India

It is unclear to what extent RUTF is being used in India to treat SAM. Pilot studies in Bihar and Madhya Pradesh suggest that RUTF is effective for treating SAM and can also be scaled up (Sachdev et al. 2010). Plumpy'nut, a brand of RUTF manufactured by a French company called Nutriset, has been imported for distribution to SAM children in several states of India, including Bihar, Jharkhand, Madhya Pradesh, Maharashtra and Odisha, under the aegis of UNICEF and through the mechanism of NRCs. There is a proposal to make it the 'prescribed treatment for SAM'. However, if produced in India, Plumpy'nut would cost about US$40 or approximately Rs 2,500 per child per treatment.[13]

Production of RUTF in India

Ready-to-use therapeutic foods can be prepared by self-help groups and village industries. However, they need to be provided with the preground ingredients in a ready-to-mix form for hand mixing at home. As this mode of production would lead to the production of small quantities of RUTF in each batch, the cost of quality testing would be far higher as each small batch would need to be tested.

Ready-to-use therapeutic foods can be manufactured by dairy cooperatives as well, as they not only have all the necessary equipment and most (often all) of the ingredients required for producing RUTFs, but also have strong networks for distributing their products across the country. The many niche food companies in India could also produce RUTFs. For instance, a Norwegian company that manufactures therapeutic food products used by aid agencies has commenced production of RUTFs in India through its subsidiary, Compact India Pvt Ltd. Large-scale private companies can also produce RUTFs, and in much larger quantities than other companies (Beesabathuni and Natchu 2010). Ready-to-use therapeutic foods are already being manufactured in India, and there is a possibility of increasing the scale of RUTF production. In fact, some industrial units are already producing it for export (Sachdev et al. 2010).

The scale of operations and the costs of producing RUTF vary by the production model. Skimmed milk powder accounts for nearly 60 per cent of the ingredient costs; hence, milk cooperatives that produce skimmed milk powder are ideally placed to produce cheaper RUTF with lower ingredient costs. Fixed costs account for a small proportion of the overall cost of RUTF production in models with specialised or customised equipment capable of producing 15 tons of RUTF per day. Fixed costs also account for a small percentage (less than 3 per cent) of the cost of RUTF for smaller companies capable of producing about 1.5 tons per day of RUTF with less specialised equipment. Nearly 80,000 metric tons of RUTF would be required to treat SAM children in India with 300 grams of RUTF per day for 30 days.

Ingredients required for the manufacture of RUTF are already available in India, being the largest producer of milk in the world—over 1.6 million tons of milk powder is produced in the country every year—as well as the second largest producer of groundnuts. Easy domestic availability of these ingredients with pre-existing distribution channels and local vendors keeps the costs of these ingredients low, unlike in Africa where the bulk of ingredients for producing RUTF are imported. Bharat Immunologicals and Biologicals Corporation

Limited, a public sector company that manufactures vaccines and zinc tablets, has developed a composite mineral and vitamin formulation (Beesabathuni and Natchu 2010). This formulation, based on the WHO composition recommended for treatment of SAM, can be added directly as a premix to RUTF.

Concerns Regarding the Use of RUTF

In India, there has been a significant amount of disagreement over the use of RUTF to treat SAM children. In an attempt to resolve this disagreement, the Department of Human Nutrition of the All India Institute of Medical Sciences in New Delhi, in collaboration with the Department of Paediatrics and Clinical Epidemiology of the Sitaram Bhartia Institute of Science and Research in New Delhi, as well as the Speciality Chapter on Nutrition at the Indian Academy of Pediatrics, organised a National Consensus Workshop on Management of SAM Children through Medical Nutrition Therapy[14] in November 2009. One of the points that emerged in the consensus statement of this workshop, which was endorsed by the majority of the workshop participants, was that the special foods used for treating SAM should not be used in attaining food and nutrition security or promoting appropriate infant and young child feeding practices. In their view, these foods should only be used for treating SAM and not for preventing it (Sachdev et al. 2010).

The Working Group for Children Under Six and the International Baby Food Action Network (IBFAN) oppose the prevention as well as treatment of malnutrition with fortified commercial foods. They argue that the medicalised approach to malnutrition ignores the multiple underlying causes of the problem and destroys the diversity of potential solutions based on locally available foods (IBFAN 2012).

The Working Group for Children Under Six also highlighted that imported RUTF has been introduced in India without due process of discussion on its repercussions and implications. Investigation into the relative merits and demerits of various locally produced/procured and locally processed foods that have been successful in treating SAM has been completely ignored while deciding to introduce RUTF in India (Working Group for Children Under Six 2009). While IBFAN acknowledges that fortified commercial foods lead to weight gain in malnourished children, it argues that the efficacy of such foods is not comparable with improved feeding practices using local foods and support for breastfeeding (IBFAN 2012).

Both the Working Group for Children Under Six and IBFAN are of the opinion that commercially and centrally produced RUTF hinder self-reliance in food security and create unnecessary dependence on products over which families and communities have little control. The group goes on to note that there have been many instances in the past where commercial interests have tried to influence policy to serve their interests, even when they have been at the cost of children's health and well-being.

The concern that private companies may utilise the opportunity to produce and sell RUTF for commercial profits is expressed the world over. For instance, Nutriset, the private French company manufacturing Plumpy'nut, has the patent for the product until 2017, and the bulk of Plumpy'nut is produced in the company's factories in France. Critics of Nutriset say that the patent is so broad that it encompasses almost any kind of nut-based nutritional paste, and as a result the patent debunks the claim of the inventor of Plumpy'nut, Andrè Briend, that RUTF can be prepared anywhere. The objective of patents is to give incentives to innovate, by providing innovators with time-bound rights to exclusively reap profits from their ideas. However, many argue that life-saving products such as RUTF should be outside the ambit of patents. Nutriset defends itself by arguing that it needs the patent to protect its network of affiliates in countries such as the Democratic Republic of the Congo, Mozambique and Niger from big industrial companies (Rice 2010).

As most of Plumpy'nut is manufactured in France, it is exported to developing countries at high costs; this not only makes poor people dependent on therapeutic foods produced on other continents, but also makes the use of such foods very expensive. For instance, a two-month treatment of a SAM child with Plumpy'nut costs $60 (Rice 2010).

Treatment of SAM Using Local Energy-Dense Foods

The guidelines for community- and home-based treatment of SAM formulated by a large group of experts and supported by the Indian Academy of Pediatrics recommend the use of home-based food, modified from the family pot (Gupta et al. 2006). Several academic and medical institutions and NGOs have been engaged for many decades in community-based treatment of malnutrition using locally produced/ procured and locally processed foods along with intensive nutrition education (Working Group for Children Under Six 2009). Table 11.1 gives details of some of these mixes.

Table 11.1. Details of locally produced/procured and locally processed food used in India for treatment of SAM

Name of mix	Composition and calorific value	Developed by	Locally produced by	State
Davangere Mix	*Laddus* made of equal quantities of groundnuts, roasted Bengal gram, jaggery and ragi. 100 g gives 400 calories and 15 g protein.	Medical College, Davangere	Women's group	Karnataka
Shakti Nutrimix	Rice, wheat, whole gram (*chana*), groundnut, sugar, salt, cardamom, black pepper, vitamins and minerals. Each 100 g of mix provides 10.4 g protein, 5.3 g fat and 402 calories.	Shibpur People's Care Organisation, Shibpur, Dakshin Dinajpur	Women's groups	West Bengal
Nutrimix	Wheat (400 g), rice (400 g), gram (75 g), moong (75 g), groundnut (50 g), sprouted, dried, roasted and powdered. Two heaped spoons in glass of water or milk with sugar twice a day.	Development Research Communi-cation and Service Centre, Kolkata	Women's groups	West Bengal
Nutrimix	Wheat/rice and Bengal gram/ moong in ratio 4:1. used for treating SAM, for preparing f 75, f 100, as starter and catch-up foods. Each 100 g cooked provides 120–50 kilocalories and protein 2–3 grams. Can be made more	CINI (Child in Need Institute), Kolkata	Women's groups	West Bengal

(contd.)

(Table 11.1 continued.)

Name of mix	Composition and calorific value	Developed by	Locally produced by	State
	energy dense by adding seasonal fruits, and micronutrient-rich by adding electrolyte mineral solution.			
LAPSI	Green millet, peanut, jaggery. Successfully used for quick recovery from SAM.	Bharat Agro Industries Foundation and CAPART		Maharashtra
SAT Mix	Roasted and ground rice, wheat, black gram and sugar in ratio 1:1:1:2. Provides 380 calories per 100 g.	Sree Avittom Thirunal Hospital		Kerala
MIX		National Institute of Nutrition, Hyderabad		Andhra Pradesh
HCCM (high-calorie cereal milk)		Christian Medical College, Vellore		Tamil Nadu
Sattu maav	Wheat flour 42%, maize flour 10%, malted ragi flour 5%, Bengal gram flour 12%, jaggery 30%, vitamin premix 1% per 100 g. Provides 9–10% protein and 360 calories.	Nutrition Monitoring Programme (state programme)		Tamil Nadu

Note: Shelf life is not a necessary condition for these locally produced RUTFs as they are prepared in the quantities needed by local women's groups under the supervision of the respective hospital or NGO.
Source: Working Group for Children Under Six (2009).

Several ongoing experiments use modified family foods to treat SAM. Jodhpur Medical College has been using a mix of energy-dense *khichri*,[15] milk, dal (pulses), sugar, fruit, fruit juice and egg to treat SAM both in institutional and home settings, and is currently in the process of analysis and documentation. In Tamil Nadu, the Direct Nutrition Programme gives a mix of 80 grams rice, 10 grams dal, 2 grams oil, 50 grams of vegetables and condiments at a total cost of Rs 1.07 to each child between 2 and 4 years of age. This provides 358.2 kcal and 8.2 grams protein per child (Gupta et al. 2006). The *sattu maavu* listed in Table 11.1 is given as complementary food to children between 6 and 36 months of age and to pregnant and lactating women, and costs approximately Rs 15 per kilogram (Working Group for Children Under Six 2009). Other experiments by NGOs such as Mobile Crèches have used common foods including eggs, soya products and milk for demonstrable impact at a cost of Rs 8 per child per day for full day care nutritional facilities (Mobile Crèches 2008). Experiments in Gujarat, Maharashtra and West Bengal have shown weight gain in SAM children who have been given locally produced foods (Sachdev et al. 2010).

In an inpatient ward at a tertiary care government hospital in Hyderabad, 309 SAM children admitted to the nutrition ward were given energy-dense local foods along with multivitamin and multimineral supplements. While it was concluded that local energy-dense foods were found to be suitable for the nutrition rehabilitation of SAM children, the rate of weight gain (5 grams per kilogram per day) was moderate and the prevalence of morbidity was high, fever being the most common symptom (Mamidi et al. 2010).

A study was conducted in Delhi in which 26 SAM children were treated at home for 16 weeks using home-based diets. While half the children recovered completely, the other half recovered only partly. The study concluded that home-based management using home-prepared food and hospital-based follow-up was associated with slower recovery (Patel et al. 2010). It has been highlighted that use of locally produced food in India has only shown weight gain, and not increases in height or physiological or functional recovery (Sachdev et al. 2010).

The Right to Food Campaign suggested that for uncomplicated cases of SAM that did not require NRC-based treatment, after due examination by the medical staff, treatment should consist of giving the child locally produced therapeutic food, enhanced food rations under the SNP and proper counselling to the child's family on correct feeding practices.

Critique of the Provisions in the Drafts Regarding Treatment of Malnutrition

The NAC draft obliged state governments to identify children suffering from any grade of malnutrition, as well as those experiencing growth faltering or nutritional deterioration. Parents or guardians of every malnourished child were to be entitled to nutrition counselling for appropriate therapeutic foods, health check-ups and referral services. Malnourished children in the age group of 0 to 6 years were to be entitled to food containing 800 kcal and 20–25 grams protein every day. All severely underweight or undernourished children and children who were malnourished as well as sick, were to be entitled to appropriate therapeutic foods and special care, either in an NRC or in the community in which they normally resided. All these entitlements were also free of cost.

The provisions in the government draft for the identification and treatment of malnourished children were extremely minimal. The draft mentioned no special entitlements for children suffering from SAM. Both the 2011 draft and the 2013 act only stated that it is the responsibility of the state governments to identify malnourished children through anganwadis and provide them with free meals containing 800 kcal and 20–25 grams of protein every day. As mentioned earlier, the lack of definition of malnutrition may pose a problem in identifying children correctly. The draft recommends that the meals should be in the form of take-home rations; however, with take-home rations there is the danger of their getting consumed by other members of the family as well. Neither draft nor the final act had any provision for the identification or treatment of adults suffering from malnutrition.

The Right to Food Campaign demanded that state governments, through the anganwadis, identify, monitor and regularly track the growth of *all* children. The campaign argued that malnutrition could not be treated simply by the provision of an additional daily meal, and demanded that the act include all the provisions of the NAC draft regarding identification and treatment of malnutrition. However, it specified that the therapeutic foods should be locally produced.

Notes

1. Deaton and Drèze (2009) used National Center for Health Statistics (NCHS) standards, which were recommended by the WHO until 2006, to analyse NFHS data for these periods. Based on the new WHO standards, child undernutrition fell from 42.7 per cent in 1998–99 to 40.4 per cent in 2005–06. The authors caution against using this figure without 'other indicators of child nutrition, such as height-for-age and weight-for-height, since there was a decline in stunting (from 51 to 45 per cent), or height-for-age, and a small increase in weight-for-height (from 20 to 23 per cent).' The authors state, 'it is far from clear why wasting should increase while stunting goes down. Perhaps stunting in this period reflects continued improvements in maternal health.... The increase in wasting ... is quite puzzling, and it is difficult to come to any firm conclusion without a better understanding of these opposite trends.'

2. '43 per cent of under-five children in India were underweight during the period 2000–07, which is worst in South Asia and worse than the worst performer in the Sub-Saharan African region.... China had only 6 percent under-five children who were underweight during this period and Brazil only 4 percent.' See Planning Commission (2011). See also National Nutrition Monitoring Bureau, 'Salient Observations', http://www.nnmbindia.org/achievements.html (accessed 29 September 2017).

3. The manual does not discuss malnutrition caused by too much food, as that condition does not relate to the concern with the 'right to food'.

4. WHO, 'Water Sanitation Hygiene', http://www.who.int/water_sanitation_health/diseases-risks/diseases/malnutrition/en/ (accessed 11 March 2018).

5. Ibid.

6. WHO, 'WHA Global Nutrition Targets 2025: Stunting Policy Brief', http://www.who.int/nutrition/topics/globaltargets_stunting_policybrief.pdf (accessed 11 March 2018).

7. Statistical theory shows that in a well-nourished population, only 0.13 per cent of children have a weight-for-height measure of less than −3 standard deviations, giving this cut-off a specificity of more than 99 per cent (WHO 1999).

8. In diarrhoea, frequent stools prevent the adequate absorption of nutrients.

9. The 2011 Global Hunger Index Report places India among the three countries where, between 1996 and 2011, the index went up from 22.9 to 23.7, while 78 out of the 81 developing countries studied, including Pakistan, Nepal, Bangladesh, Vietnam, Kenya, Nigeria, Myanmar, Uganda, Zimbabwe and Malawi, succeeded in improving hunger conditions (Parulkar 2017).

10. 'National Food Security Bill, 2011', https://nls.ac.in/ccl/EGOM%20July_2011%20National%20Food%20Security%20Bill.pdf (accessed 11 March 2018).

11. For a more detailed account of the NRCs, see Citizens' Initiative for the Rights of Children Under Six (2006, 70).

12. The details of the workshop proceedings can be found in Gupta et al. (2006).

13. One sachet is 92 grams. Treatment for SAM requires 90 sachets per child of 10 kilograms. Nutriset, 'The Impact on Beneficiaries', http://www.nutriset.fr/index.php?option=com_content&task=view&id=30&Itemid=28 (accessed 30 September 2017).

14. Medical nutrition therapy is the therapeutic approach to treating medical conditions and their associated symptoms via the use of a specifically tailored diet.

15. A preparation of rice and dal.

References

Ahmed, T., M. Ali, M. M. Ullah et al. 1999. 'Mortality in Severely Malnourished Children with Diarrhoea and Use of a Standardised Management Protocol'. *Lancet*, 353(9168): 1919–22.

Anandaiah, R., and M. K. Choe. 2000. 'Are the WHO Guidelines on Breastfeeding Appropriate for India?' *National Family Health Survey Subject Reports*, no. 16, May. http://www.eastwestcenter.org/fileadmin/stored/pdfs/NFHSsubjrpt016.pdf (accessed 30 September 2017).

Balarajan, Y., S. Selvaraj, and S. V. Subramanian. 2011. 'Health Care and Equity in India'. *Lancet*, 377(9764): 505–15.

Banik, Dan. 2007. *Starvation and India's Democracy*. London: Routledge.

Beaudry-Darisme, M., and M. C. Latham. 1973. 'Nutrition Rehabilitation Centers: An Evaluation of Their Performance'. *Journal of Tropical Pediatrics and Environmental Child Health*, 19(3): 299–332.

Beesabathuni, K. N., and U. C. M. Natchu. 2010. 'Production and Distribution of a Therapeutic Nutritional Product for Severe Acute Malnutrition in India: Opportunities and Challenges'. *Indian Pediatrics*, 47(8): 702–6.

Briend, Andrè. 1997. 'Treatment of Severe Malnutrition with a Therapeutic Spread'. *Field Exchange*, 2, August.

Briend, Andrè, and Steve Collins. 2010. 'Therapeutic Nutrition for Children with Severe Acute Malnutrition: Summary of African Experience'. *Indian Pediatrics*, 47(8): 655–59.

Butterly, John, and Jack Shepherd. 2010. *Hunger: The Biology and Politics of Starvation*. Lebanon, NH: University Press of New England.

Checkley, W., G. Buckley, R. H. Gilman, A. M. O. Assis et al. 2008. 'Multi-country Analysis of the Effects of Diarrhoea on Childhood Stunting'. *International Journal of Epidemiology*, 32: 816–30.

Citizens' Initiative for the Rights of Children Under Six. 2006. *Focus on Children Under Six (FOCUS)*. New Delhi: Right to Food Campaign Secretariat.

Collins, Steve, Nicky Dent, Paul Binns, Paluku Bahwere et al. 2006. 'Management of Severe Acute Malnutrition in Children'. *Lancet*, 368(9551): 1992–2000.

Cook, R. 1971. 'Is Hospital the Place for the Treatment of Malnourished Children?' *Journal of Tropical Pediatrics and Environmental Child* Health, 17(1): 15–25.

Cronin, A. A., J. H. Rah, F. M. Ngure, F. Odhiambo et al. 2016. 'Water, Sanitation, Hygiene (WASH) and the Nutritional Status of Children in India: Understanding the Linkages and Structuring the Response', in S. Desai, L. Haddad, D. Chopra and A. Thorat (eds), *Undernutrition and Public Policy in India*. Oxon: Routledge.

Damme, W. V., and M. Boelaert. 2002. 'Therapeutic Feeding Centres for Severe Malnutrition'. *Lancet*, 359(9302): 260–61.

Deaton, A., and Jean Drèze. 2009. 'Food and Nutrition in India: Facts and Interpretations'. *Economic and Political Weekly*, 44(7): 42–65.

Drèze, Jean, and Amartya Sen. 2013. *An Uncertain Glory: India and Its Contradictions*. London: Allen Lane.

FAO (Food and Agriculture Organization). 2009. *Guide on Legislating for the Right to Food*. Rome: FAO.

Gera, Tarun. 2010. 'Efficacy and Safety of Therapeutic Nutrition Products for Home Based Therapeutic Nutrition for Severe Acute Malnutrition: A Systematic Review'. *Indian Pediatrics*, 47(8): 709–18.

Ghosh, Shanti. 2006. 'Preventing Malnutrition through Better Feeding at Home', in *Focus on Children Under Six*. Citizens' Initiative for the Rights of Children Under Six, New Delhi.

Golden, M. 1988. 'The Effects of Malnutrition in the Metabolism of Children'. *Transactions of the Royal Society of Tropical Medicine and Hygiene*, 82(1): 3–6.

Golden, M. H. N. 1996. 'Severe Malnutrition', in D. J. Weatherall, J. G. G. Ledington and D. A. Warrell (eds), *The Oxford Textbook of Medicine*, 3rd edn. Oxford: Oxford University Press, pp. 1278–96.

Gupta, A., B. Patnaik, D. Singh, D. Sinha et al. 2013. 'Are Child Malnutrition Figures for India Exaggerated?' *Economic and Political Weekly*, 48(34): 73–77.

Gupta, Piyush, Dheeraj Shah, H. P. S. Sachdev and Umesh Kapil. 2006. 'National Workshop on "Development of Guidelines for Effective Home Based Care and Treatment of Children Suffering from Severe Acute Malnutrition"'. *Indian Pediatrics*, 43(2): 131–39.

IBFAN (International Baby Food Action Network). 2012. 'Statement on the Promotion and Use of Commercial Fortified Foods as Solution for Child Malnutrition'. *Indian Pediatrics*, 49(4): 295–96.

IIPS (International Institute for Population Sciences). 2007. *National Family Health Survey 2005–06 (NFHS-3): India*. Mumbai: IIPS.

Mamidi, Raja Sriswan, Bharati Kulkarni, K. V. Radhakrishna and Veena Shatrugna. 2010. 'Hospital Based Nutrition Rehabilitation of Severely Undernourished Children Using Energy Dense Local Foods'. *Indian Pediatrics*, 47(8): 687–93.

Manary, Mary J. 2005. 'Local Production and Provision of Ready-to-Use Therapeutic Food for the Treatment of Severe Childhood Malnutrition'. Technical Background Paper. http://www.who.int/nutrition/topics/backgroundpapers_Local_production.pdf (accessed 11 March 2018).

Meijers, J., V. Marian, S. Joe, S. Peter and H. Ruud. 2010. 'Defining Malnutrition: Mission or Mission Impossible?' *Nutrition*, 26(4): 432–40.

Merchant, A. T., C. Jones, A. Kiure, R. Kupka et al. 2003. 'Water and Sanitation Associated with Improved Child Growth'. *European Journal of Clinical Nutrition*, 57(12): 1562–68.

Mobile Crèches. 2008. 'Tackling Malnutrition of Children under Six: Evidence from Two Microstudies'. Working Paper. http://docs.wixstatic.com/ugd/eec0d5_f26eb9c922d3411e95ff997b14c5f5f3.pdf (accessed 11 March 2018).

NAC (National Advisory Council). 2011. 'National Food Security Bill, 2011', July. http://www.indiaenvironmentportal.org.in/files/nfsb_final.pdf (accessed 11 March 2018).

Panagariya, Arvind. 2013. 'Does India Really Suffer from Worse Child Malnutrition than Sub-Saharan Africa?' *Economic and Political Weekly*, 48(18).

Parulkar, A. 2017. 'Born Busy Dying: Government Responses to Starvation Deaths in Bihar, Jharkhand and Madhya Pradesh', in A. Parulkar, S. Sharma, A. Shah, S. Sethia, R. John, A. Imaan and A. Baxi (eds), *Dispossessed: Stories from India's Margins*. New Delhi: Speaking Tiger Books, pp. 3–130.

Patel, Deepak, Piyush Gupta, Dheeraj Shah and Kamlesh Sethi. 2010. 'Home-Based Rehabilitation of Severely Malnourished Children in Resource Poor Setting'. *Indian Pediatrics*, 47(8): 694–701.

Planning Commission. 2011. *India Human Development Report*. New Delhi: Government of India and Oxford University Press.

Rice, Andrew. 2010. 'The Peanut Solution'. *New York Magazine*, 2 September. http://www.nytimes.com/2010/09/05/magazine/05Plumpy-t.html?pagewanted=1&_r=1&sq=peanut%20solution&st=cse&scp=1 (accessed 30 September 2017).

Right to Food Campaign. 2007. 'Anganwadis for All: A Primer'. Right to Food Campaign Secretariat, New Delhi.

Roy, Nikhil Chandra. 2000. 'Use of Mid-upper Arm Circumference for Evaluation of Nutritional Status of Children and for Identification of High-Risk Groups for Malnutrition in Rural Bangladesh'. *Journal of Health, Population and Nutrition*, 18(3): 171–80.

Sachdev, H. P. S., Umesh Kapil and Sheila Vir. 2010. 'Consensus Statement: National Consensus Workshop on Management of SAM Children through Medical Nutrition Therapy'. *Indian Pediatrics*, 47(8): 661–65.

Saxena, N. C. 2008. 'Hunger, Under-nutrition and Food Security in India'. Centre for Equity Studies Working Paper (unpublished).

———. 2016. 'Governance Challenges to Reducing Hunger and Malnutrition in India', in S. Desai, L. Haddad, D. Chopra and A. Thorat (eds), *Undernutrition and Public Policy in India*. Oxon: Routledge, pp. 193–211.

Schofield, C., and A. Ashworth. 1996. 'Why Have Mortality Rates for Severe Malnutrition Remained So High?' *Bulletin of the World Health Organization*, 74: 223–29.

Spears, Dean. 2016. 'Improving Sanitation in Rural India', in S. Desai, L. Haddad, D. Chopra and A. Thorat (eds), *Undernutrition and Public Policy in India*. Oxon: Routledge, pp. 79–100.

WHO (World Health Organization). 1999. 'Management of Severe Malnutrition: A Manual for Physicians and Other Senior Health Workers'. WHO, Geneva. http://www.who.int/nutrition/publications/en/manage_severe_malnutrition_eng.pdf (accessed 30 September 2017).

———. 2003. 'Global Strategy for Infant and Young Child Feeding'. WHO, Geneva.

WHO and UNICEF (United Nations Children's Fund). 2007. 'Community-Based Management of Severe Acute Malnutrition: A Joint Statement by the World Health Organization, the World Food Programme, the United Nations System Standing Committee on Nutrition and the United Nations Children's Fund'. http://www.who.int/nutrition/topics/Statement_community_based_man_sev_acute_mal_eng.pdf (accessed 11 March 2018).

WHO/UNICEF/WFP/SCN. 2007. 'Joint Statement: Community-Based Management of Severe Acute Malnutrition'. http://www.who.int/child_adolescent_health/documents/pdfs/severe_acute_malnutrition_en.pdf (accessed 30 September 2017).

Working Group for Children Under Six. 2009. 'Should India Use Commercially Produced Ready to Use Therapeutic Foods (RUTF) for Severe Acute Malnutrition (SAM)'. *Social Medicine*, 4(1): 52–55.

12

Cooked Food versus Ready-to-Eat Meals and Commercial Interests

ANKITA AGGARWAL

This chapter addresses the debate on whether children should be provided with freshly prepared and culturally appropriate hot cooked food or micronutrient-fortified[1] ready-to-eat food in children's feeding programmes, such as MDMS in government and government-aided schools and the SNP scheme of the ICDS. As ready-to-eat food is usually provided by private contractors, this option also leads to the debate on the desirability of introducing the private sector in the provision of food for children's feeding programmes.

The preference for cooked food versus ready-to-eat food was heavily debated among the government, policymakers, nutritionists, medical professionals, public health experts, scientists, educationists, academics, activists and civil society organisations, both before and during the drafting of the NFSB. In 2006, the Ministry of Women and Child Development (MWCD) proposed the introduction of micronutrient-fortified ready-to-eat food for tackling malnourishment among children. It substantiated its stand with several arguments, which were supported by private contractors who cited the advantages of packaged food over locally cooked food to support their stand. However, the ministry's proposal was rejected by Prime Minister Manmohan Singh, the Planning Commission, the NAC, the Planning Commission's steering committee on nutrition, civil society networks such as the Right to Food Campaign, Jan Swasthya Abhiyan, the survey conducted in 2004 by the Citizens' Initiative for the Rights of Children Under Six (2006; henceforth FOCUS survey), and individuals such as Amartya Sen, the noted economist and Nobel laureate. Opposition parties, including the left parties and some members of the UPA, also did not support the introduction of ready-to-eat packaged food (Sethi 2008). Two MPs expressed opposition to private contractors in debates that preceded the vote.[2] Most of the supporters of cooked meals also advocated that food for children should be produced locally, and not by private companies.

This chapter begins with a brief overview of the types of food that are provided in the MDMS and ICDS. It then describes the history of private contractors in these two programmes. The next section discusses the reasons provided by the MWCD and the private sector for supporting the provision of ready-to-eat food to children. Arguments opposing ready-to-eat food items and participation of private contractors in the supply of food for children's feeding programmes follow next. The penultimate section examines the discussion of this theme in the various drafts of the NFSB and in the final act, followed by the conclusion.

Current Food Provisions in MDMS and ICDS

As per Supreme Court orders, midday meals are supposed to consist of hot cooked food. Some of the common items provided in the midday meals are rice, dal, vegetables and khichdi. Some states serve 'special food' such as *poori*, *kheer* and boiled eggs once a week.[3]

The SNP arrangements under ICDS are generally different for children above and below the age of 3 years. Children in the age group of 3 to 6 years are generally fed 'on site' at the anganwadi. For younger children, however, on-site feeding is not practical as it requires mothers to come with their children to the anganwadi every day, something they may not be able or willing to do. Furthermore, this arrangement is not suited to the needs of young children, who require frequent feeding in small quantities over the day, rather than a hearty midday meal. As a result, for children in the age group of 6 months to 3 years, take-home rations are provided (FOCUS survey).

There are two broad types of on-site feeding arrangements for children in the 3 to 6 years age group: cooked food and ready-to-eat items such as *panjiri*[4] and *murmura* (puffed rice).

Supply of Food in MDMS and ICDS by Private Contractors

Legal Measures to Check Supply of Food for Children by Contractors in India

To prevent commercial interference in infant and young child feeding, India passed the Infant Milk Substitutes, Feeding Bottles and Infant

Foods (Regulation of Production, Supply and Distribution) Act in 1992, known as the IMS Act, for the 'protection and promotion of breastfeeding'. The act, amended in 2003, is one of the strongest pieces of legislation in the world that prohibits manufacturers of breast milk substitutes and baby foods from advertising or promoting their products for infants below 6 months of age, as this can interfere with exclusive breastfeeding. In particular, it prohibits sponsoring gifts, meetings, conferences, seminars, contests or giving funds for any other activity to healthcare workers and their associations. The IMS Act is the only legislation in India that explicitly prohibits commercial interference and influence with regard to food and nutrition (Right to Food Campaign n.d.).

The ICDS programme since 1985 has been the site for public–private partnerships, long before the notion gained the popularity it enjoys today. In virtually every Indian state, the SNP of the ICDS has been provided by contractors (FOCUS survey), but due to the poor experience with contractors in ICDS, the Supreme Court of India banned contractors in an interim order in the right to food case on 7 October 2004. The court directed that 'contractors shall not be used for supply of nutrition in anganwadis and preferably ICDS funds shall be spent by making use of village communities, self-help groups and *Mahila Mandals* for buying of grains and preparation of meals.'[5]

Apart from this Supreme Court order, there is no legal or policy framework for food fortification or micronutrient supplementation in India; nor are there any other guidelines on the kinds of public–private partnerships that can be entered into in the arena of food and nutrition. This policy vacuum is exploited by several groups, often those with profit motives, to enter the large-scale programmes tackling hunger and malnutrition such as the ICDS, MDMS and PDS.

Experiences of the Supply of Food for Children by Contractors in India

Despite the Supreme Court order, the presence of contractors in the provisioning of food for the SNP in the ICDS has continued. The office of the Commissioners of the Supreme Court, responsible for monitoring the compliance with orders, has been able to rid the SNP of contractors in barely half a dozen states, that too with enormous difficulty. Even today, more than 13 years after the Supreme Court order, private contractors are supplying food for the SNP of the ICDS in states such as Gujarat and Rajasthan. In some states, the contractor–politician–bureaucrat nexus has tried innovative ways to repeatedly circumvent the court's orders. The deep-rootedness of this nexus can be gauged from the responses of a few states to the commissioners. In Chhattisgarh, for instance, the state government wrote to the commissioners clarifying that they were not contravening court orders since they were procuring from 'manufacturers' and not contractors. This arbitrary distinction was raised by a number of state governments, and in other places where government orders to eliminate contractors were issued, subtle clauses to ensure their continuance crept in. Another example is Maharashtra, where orders to remove contractors and hand over preparation of food for the SNP to mahila mandals and self-help groups was accompanied by a clause allowing the 'co-operative federation' to supply those areas where such organisations were not present. Since the federation essentially sources all the supplies through private contractors, this allowed them a backdoor (re-)entry into the ICDS. This matter is presently sub judice in the Supreme Court.

States which were determined to comply with the court order saw lobbyists, employed by the contractors, attempt to place hurdles in its implementation. In New Delhi, a decision made by the Delhi government to hand over the SNP to NGOs led many contractors to register themselves as NGOs and apply under the revised scheme. Since the Delhi government had framed the scheme in consultation with the commissioners, a clause requiring the NGO to be registered for at least three years was subsequently included as a criterion for eligibility, and although the clause was subsequently challenged, opposition to the clause was turned down after a long-drawn-out legal battle, much to the contractors dismay.

States like Bihar, Jharkhand and Chhattisgarh have put in place alternative decentralised mechanisms for the provision of food for the SNP. Many other state governments have been pressured, particularly by the commissioners and the Supreme Court, to follow suit (FOCUS survey).

Attempts were made by biscuit manufacturers to introduce their products in the MDMS, with the Biscuit Manufacturers Association (BMA) sending a letter to MPs making a detailed case for biscuits as an alternative to cooked midday meals in primary schools. In the letter, the

BMA explained that glucose biscuit packets cost Rs 3.75 each, almost the same as a cooked midday meal, and allegedly contained all the nutrients that would be needed in the MDMS. The letter was signed by the BMA president, who was also an employee of Parle Products, which, incidentally, was the largest manufacturer of glucose biscuits in the country. Among other arguments, the letter stated that biscuits have a 'higher recall', which pointed to the real objectives of the BMA: the commercial jargon of 'recall' refers to the proportion of people who remember a particular brand, and recall data is used to track advertising effectiveness. An advertisement campaign based on giving every child the same packet of biscuits every day (at the expense of public money) would naturally have a high 'recall'.

Although it is not known how many MPs received these letters, 29 of them personally wrote to Minister of Human Resource Development (HRD) Arjun Singh and urged him to consider the proposal. These 29 MPs belonged to most of the major political parties (Congress, BJP, Rashtriya Janata Dal, Samajwadi Party and so on) except for the Bahujan Samaj Party and the left parties. Nine of these MPs represented constituencies in Maharashtra (which has a thriving biscuit industry), and six belonged to the Shiv Sena. It is interesting to note that the content of these letters was quite similar across signatories and parties, with, for instance, the letters signed by K. Natwar Singh (Congress) and Ramdas Athawale (Republican Party) being almost identical. Similarly, many MPs confidently cited that 'the ratio of carbohydrates, proteins, fats and glucose is quite balanced and beneficial to human health of all age group especially for children [*sic*]' (Drèze and Khera 2008).

The similarity in the letters arose because many of their signatories simply copied arguments from the BMA's promotional material verbatim. Even if some of the MPs actually believed that biscuits were better than cooked meals for children, it is unknown how much research they actually did on the benefits and disadvantages of the two options before expressing their support for the biscuits.

However, the Ministry of Human Resource Development (MHRD) strongly rejected the BMA's proposal after conducting a round of consultations with state governments. Apart from the state governments, Dr B. Sesikeran, director of the National Institute of Nutrition, shot down the proposal of replacing cooked meals with biscuits. Dr Sesikeran stated that the MDMS 'is supposed to provide one wholesome meal to schoolchildren and biscuits cannot replace it' (Drèze and Khera 2008). When this issue was discussed in the Lok Sabha, Mohammad Ali Ashraf Fatmi, the then minister of state for HRD, reiterated that the proposal of replacing cooked meals with biscuits did not 'fulfil the nutritional norms, dietary requirement and satiety of children and further it also deprives many intrinsic benefits that are being derived through present pattern of implementation' (ibid.).

In another instance, Real Contracts Private Limited also proposed to the MHRD that the freshly cooked hot midday meals should be replaced with 'ready to cook and serve hot' meals, which is dehydrated food that requires boiling before serving. Arjun Singh wrote a letter to Uttar Pradesh's chief minister Mayawati on this issue: 'We are, indeed, dismayed at the growing requests for introduction of pre-cooked foods, emanating largely from suppliers/marketers of packaged foods, and aimed essentially at penetrating and deepening the market for such foods' (Drèze and Khera 2008).

Having been unsuccessful in making inroads in the MDMS, the BMA then wrote to the MWCD and again to MPs, this time proposing the introduction of their products in the ICDS. They cited the reports of the Supreme Court Commissioners, which highlighted leakages in the supply chains of the MDMS and ICDS programmes, proposing a Swasth Bharat[6] brand of biscuits which matched the nutrition norms prescribed by the MWCD, supplied exclusively for the ICDS through its 'existing distribution and logistics network'. The manufacturers' association also suggested its own innovative ways of checking leakages of these biscuits in the ICDS—banning open market sale of the Swasth Bharat brand of biscuits and making it a crime that would incur penal action (Sethi 2008).

However, unlike the minister of state for HRD, the then minister of state for women and child development, Renuka Chowdhury, supported the replacement of hot cooked meals with ready-to-eat packaged food items in the SNP of the ICDS. In February 2008, she also proposed a public–private partnership for the provision of food for the SNP of the ICDS (Misra 2008). The next section discusses the reasons given by the MWCD and private companies for supporting the replacement of hot cooked meals with ready-to-eat food items in the ICDS.

In Andhra Pradesh, where the Naandi Foundation is in charge of the MDMS in Hyderabad and Secunderabad, children are given Britannia's iron-fortified Tiger biscuits in addition to cooked meals. While one study estimates the number of biscuits being given to every child as 16, the estimate of another study is exactly double at 32 (Jishnu 2012). It is interesting to note that Naandi was also involved in a recently concluded large-scale survey in India on hunger and malnutrition.

Arguments in Favour of Centrally Produced Ready-to-Eat Food in Children's Feeding Programmes

The MWCD pushed for ready-to-eat packaged food in place of hot cooked meals in the ICDS's SNP. The then minister of state Renuka Chowdhury advocated very strongly in favour of packaged food, and soon after NFHS 2005–06 revealed that little progress had been made in tackling malnutrition in the country, the MWCD issued a circular recommending the provision of micronutrients—vitamins and minerals—along with carbohydrates and protein to children covered under the ICDS to tackle malnutrition. The circular cited evidence which showed that a mild deficiency of micronutrients could adversely affect a child's development, immunity and growth, specifying the content of calcium, iron, zinc, iodine, vitamins A and B12, riboflavin, ascorbic acid and folic acid in the food, as well as a recommendation that micronutrient content be fortified in ready-to-eat foods and provided through 80 grams of supplementary nutrition or that raw food be given to meet micronutrient specifications (Misra 2008).

The MWCD, in an affidavit to the Supreme Court in the right to food case, argued that the mandate of the ICDS is to provide 'supplementary nutrition' and not 'meals' for supplanting the nutrition gap between the RDA and the actual average dietary intake. Hence, the nutrition and financial requirements under the ICDS are different from those of the MDMS, which is mandated to provide meals to children in the age group of 6 to 14 years at government and government-aided schools. While deciding on the nature of the food to be provided in the anganwadi centres and those who should prepare it, the logistical and financial support required for the preparation of hot cooked meals needs to be looked into. The ministry cited the 10th Five Year Plan

Document, vol. II: *Sectoral Policies and Programme*, section on Food and Nutrition Security (para. 3.3.8.7), on the following disadvantages of the provision of hot cooked meals in anganwadis:

- The children, especially those in the age group of 6 to 36 months, cannot consume the entire amount of food provided because of a smaller stomach capacity.
- Even if older children do eat the food provided in the anganwadi, this acts mainly as a substitute, and not an addition, to home food.
- The most needy segment, viz., children in the critical 6–36-month age group, and mothers may not be able to come to the anganwadi daily to receive the food.
- Cooking in poor hygiene conditions and keeping leftover food may result in bacterial contamination of food.
- Undernourished children, even those in the 3 to 6 age group, if given double rations, cannot consume all the food at one sitting in the anganwadi.

Apart from the justifications for providing ready-to-eat food items mentioned in the affidavit, Renuka Chowdhury also claimed that overworked anganwadi workers and helpers could not be expected to cook hot meals as well (Pande 2008). A senior official at the ministry claimed that apart from insufficient infrastructure and space for cooking and storing foodgrains, pulses and other ingredients, the ICDS programme was also ill-equipped to monitor the quality of food that was prepared and the water that was used for cooking, claiming that it would be easier to monitor the distribution of packaged food than to cook hot meals in anganwadis (Raman 2007). Renuka Chowdhury argued that there was no way of enforcing quality norms in cooked foods, expressing misgivings about rodents and lizards falling into meals cooked in anganwadis and food often being cooked with substandard ingredients in unhygienic conditions (Anon 2008a), with frequent reports of children falling ill after eating cooked midday meals (ibid.). She went on to cite the success of fortified snacks in African countries in a bid to push for their introduction in India (Raman 2008). Other supporters of packaged food also argued that the replacement of cooked meals in the ICDS and MDMS with ready-to-eat items would increase efficiency and help overcome fiscal and managerial constraints faced in the implementation of these schemes (Dasgupta and Baru 2008). Going against the Supreme Court order of 7 October 2004, which banned private contractors in the provisioning of nutrition in the ICDS, the ministry proposed centralised procurement of the fortified snacks through a chain of contractors (Raman 2008).

Companies producing micronutrient-fortified ready-to-eat foods saw the ministry's recommendations as an opportunity to increase the market of their products. For instance, Deepak Agarwal, executive director of Continental Milkose, a weaning food supplier in Uttar Pradesh, argued for the benefits of packaged food over cooked food: 'The standards set by the Ministry are such that they can only be met through pre-packaged food. We are equipped with proper provision for monitoring and an array of quality checks, which is not possible otherwise' (Misra 2008). V. K. Khandelwal, general manager of Suruchi Foods, another contractor supplying weaning food in Uttar Pradesh, said that although packaged food may not provide the variety children may want, it gives them their basic requirements. He went on to claim that packaged food is more hygienic than hot cooked food, arguing that because packaged food is replacing traditional food in urban India, there is no harm in emulating the trend in rural areas as well. He was of the opinion that junk food, which he termed as 'fun food', could be fortified with the correct proportion of protein and calories (ibid.).

Arguments in Favour of Locally Produced Hot Cooked Food in Children's Feeding Programmes

The proposal of the MWCD to introduce ready-to-eat food in anganwadi centres to tackle malnutrition was rejected by the Planning Commission after hearing the views of experts both in favour of and against ready-to-eat meals. The Planning Commission instead took the decision to support hot cooked meals prepared by women's groups and other members of the local community (Sarin 2008). In November 2007, the Planning Commission's steering committee on nutrition, comprising Biraj Patnaik, Jean Drèze, Vandana Prasad and Veena Shatrugna, wrote to Montek Singh Ahluwalia, deputy chairman of the Planning Commission, expressing the view that locally cooked meals, in accordance with the tastes and preferences of children, were the best means of providing nutritious food to children in the age group of 3 to 6 years (Raman 2007). The committee believed that the

MWCD's decision of introducing packaged food was influenced by commercial interests (Raman 2008). Amartya Sen wrote a letter to the prime minister to dissuade the MWCD from serving biscuits and other packaged foods in the ICDS. He asked the prime minister to prevent the introduction of packaged food in the MDMS (Menon 2008), stressing that private contractors, while pursuing their commercial interests, may jeopardise children's health (Anon 2008b). The left parties and some members of the UPA, including the prime minister and his office, were also not in favour of introducing packaged food. N. C. Saxena pointed out that, in light of the Supreme Court's interim orders, providing children ready-to-eat items such as biscuits would not only be blatantly illegal, but would also be construed as contempt of court by the Government of India (Sethi 2008). It is interesting to note that in a presentation made to the NAC regarding the NFSB in September 2010, the MWCD recommended that cooked food and take-home rations provided in the ICDS should be prepared locally and not by private contractors.

In light of the proposals of replacing cooked meals in the ICDS and MDMS with biscuits and other ready-to-eat food items, the University School Resource Network project at Jawaharlal Nehru University initiated a consultation (henceforth 'the JNU consultation') in early 2008, which reviewed the scientific evidence that favoured cooked meals over ready-to-eat food items in nutritional support programmes. The JNU consultation was attended by civil society alliances, professional networks like the Indian Association for Preventive and Social Medicine, as well as academics. Those present at the JNU consultation unanimously rejected the replacement of cooked meals with dry ready-to-eat items such as biscuits (Dasgupta and Baru 2008). The NAC and civil society networks such as the Right to Food Campaign, Jan Swasthya Abhiyan and the FOCUS survey of 2004 (Citizens' Initiative for the Rights of Children Under Six 2006) also supported the provision of locally prepared hot cooked meals.

Freshly Cooked Food More Nutritious than Micronutrient-Fortified Ready-to-Eat Food

The world over, there is a shift away from the daily consumption of processed and packaged food (Anon 2008c). Veena Shatrugna, a nutritionist and deputy director of the National Institute of Nutrition, and member of the Planning Commission's steering committee on nutrition, has listed the advantages of freshly cooked food over ready-to-eat food items. She says that processed and packaged food usually has high salt content (ibid.), and predominantly contains cereals and pulses with preservatives, hydrogenated fats, and potentially trans fats; these have to be dehydrated while packaging, in which many nutrients are lost (Misra 2008). On the other hand, cooking vegetables, spices and herbs enhances the taste and nutritional value of food. Shatrugna adds that micronutrients need to come from natural food and not from food that has been artificially fortified. While natural food ensures that it enters the food matrix, binds with it and slowly gets processed, chemicals in packaged food require lots of enzyme secretion for proper absorption, which can strain the liver enzymes in the long run. A fresh meal also offers a greater range of nutrients than ready-to-eat food items. While there are about 20 vitamins and minerals and almost 300 phytonutrients, antioxidants and flavanoids in natural food, not all of these are available for fortification (ibid.).

The members of the JNU consultation also concluded that the nutritional impact of dry snacks is questionable and that the impact would likely be much less than that of cooked meals. In the consultation, it was also highlighted that biscuits are processed foods which do not have much fibre, but are high in both monounsaturated and polyunsaturated fatty acids, which are important long-term risk factors for various ailments such as coronary heart disease and diabetes. A freshly cooked meal is also cheaper in terms of the per rupee nutrient yield than ready-to-eat food items (Baru et al. 2008).

Shatrugna adds that food deficiencies are being disguised as deficiencies in micronutrients; when children get adequate and appropriate food, the deficiencies in micronutrients disappear in more than 90 per cent of the children (Misra 2008). Calorie deficiencies, one of the biggest challenges in malnutrition, go away with wholesome balanced diets (Dasgupta and Baru 2008). The Right to Food Campaign was also of the opinion that wholesome, balanced, adequate and diverse food is the best solution for reducing hunger and malnutrition. If micronutrient deficiencies persist despite the consumption of enough and good-quality food, which is sometimes the case, supplements can be given to solve the problem.

Citing the examples of contaminated packaged water and worms in confectionaries, both produced by multinational companies, the notion that packaged food is more hygienic than freshly cooked meals has also been challenged (Anon 2008b).

Greater Acceptance and Attraction of Freshly Cooked, Culturally Appropriate Food

Ready-to-eat food items are not always culturally accepted by their recipients. For instance, there have been reports of ready-to-eat *daliya* (porridge) mixes provided as a take-home ration in the ICDS being used as cattle feed (Misra 2008). The FOCUS survey also suggests that children seem to like cooked food much more than ready-to-eat

Table 12.1. Effects of cooked food and anganwadi location on child attendance

Probit analysis of child attendance (dependent variable: dummy for 'regular attendance' [1 = regular, 0 = not regular], as reported by the child's mother)	
Independent variables	*Marginal effect*
Anganwadi provides cooked food	0.47 (3.41)**
Anganwadi is located in the same hamlet as the child's house	0.35 (5.16)**
State dummies	
Himachal Pradesh	–0.10 (0.63)
Maharashtra	–0.40 (2.91)**
Rajasthan	0.07 (0.43)
Uttar Pradesh	0.12 (.75)
Chhattisgarh	–0.47 (3.13)**
Number of observations	323

** Significant at 1% (z statistics in parentheses).
Notes: (1) This analysis focuses on children aged 3–6 years. (2) Tamil Nadu is the 'default' state (no dummy). (3) Socio-economic variables such as education, SC/ST status, owning a pucca house or having an APL card are not significant. The coefficient of the 'cooked food' variable (first row) suggests that the probability of regular attendance rises by 47 percentage points when cooked food is provided at the local anganwadi.
Source: Citizens' Initiative for the Rights of Children Under Six (2006).

items; mothers are also usually happier when cooked food is provided in the ICDS. These preferences of children and mothers are reflected in the fact that child attendance at the anganwadis seems to be much higher when cooked food is provided than when ready-to-eat items are distributed. In the FOCUS survey, sample mothers were asked whether their child attended the anganwadi regularly; statistical analysis of the responses indicates that the chance of an affirmative answer ('yes, the child attends regularly') is higher by almost 50 per cent when cooked food is provided at the local anganwadi. The details are provided in Table 12.1. As the table indicates, the probability of regular attendance depends overwhelmingly on two variables: provision of cooked food and location of the anganwadi. When cooked food is provided, and the anganwadi is located in the same hamlet as the child's house, the probability of regular attendance is around 80 per cent, irrespective of the socio-economic characteristics of the child's family.

Hot Cooked Meals Reach Intended Beneficiaries More

As it is easier to take ready-to-eat food items back home from the anganwadi or school than hot cooked meals, the former have a higher likelihood of being 'shared' with other family members and having a lesser impact on improving children's nutrition (Baru et al. 2008; FOCUS survey). For instance, in Rajasthan it has been often found that murmura was freely passed around the house and even the village, where it had become a common snack, defeating the purpose of providing 'supplementary nutrition' to vulnerable children. To some extent, the same can happen with cooked food if the SNP is a substitute for home meals. Despite this, if cooked meals served at the anganwadi are of good quality (as in Himachal Pradesh, Maharashtra and Tamil Nadu), children would benefit from 'qualitative supplementation' from cooked food even if there is some substitution for home meals. With ready-to-eat items, the supplementation effect is likely to be much lower to the extent that the food is shared (FOCUS survey). The experiment with the dry rations and biscuits which were part of the National Programme of Nutritional Support to Primary Education, prior to the Supreme Court order on cooked meals, showed that they were often not eaten by children (Baru et al. 2008).

Shorter Shelf Life and Poor Quality of Some Ready-to-Eat Food Items

Some ready-to-eat foods, such as the panjiri mixture, have a short shelf life and are often stale by the time they are distributed to children. There have been several instances of children falling seriously ill after consuming local panjiri (FOCUS survey). Ghosh (2006) highlights that ready-to-eat food is often poor in quality, and poor-quality packaged food also poses health risks (Raman 2008).

Success of Hot Cooked Meals in MDMS

Supporters of cooked meals also highlight the success of the MDMS, especially in states like Tamil Nadu and Karnataka (Raman 2008). In a study of the MDMS in select districts of three states (Drèze and Goyal 2003), districts in Karnataka were found to have the best quality and quantity of meals; children got a variety of food items, including vegetables, rather than solely grains and pulses. Second helpings were allowed as the food was cooked in adequate quantities. Enrolment jumped by 15 per cent in the sample villages of Karnataka. Districts in the two other states, Chhattisgarh and Rajasthan, showed increases of 11 per cent and 18 per cent respectively, even though the quality of meals was less than satisfactory. In Tamil Nadu, meals typically comprised 100 grams of rice, 50 grams of vegetables, 15 grams of pulses and 1 gram of oil, along with one egg/banana, which provided calories above the 450 kcal prescribed norm on an average (Government of Tamil Nadu 2013)—effectively replacing a meal at home (Samson et al. 2007). A hot, nutritious meal at school could therefore be crucial in realising children's right to food.

Social Benefits of Hot Cooked Meals

Opponents of ready-to-eat food items note that hot cooked meals also inculcate a culture among children to sit and eat together, which helps in destroying various taboos regarding caste that forbid people of some castes from eating with people of other castes, or eating food cooked by members of SCs and STs (Pande 2008). In an assessment of the MDMS in Karnataka, half the cooks in sampled villages were

Dalits. Most importantly, there did not seem to be resistance to this. Guidelines mandate the hiring of female cooks, with widows being given preference.

Micronutrient-Fortified Ready-to-Eat Food: A Sly Way of Introducing Private Contractors

The Right to Food Campaign and many nutritionists oppose the proposal of introducing ready-to-eat food items in the ICDS or MDMS and setting up public–private partnerships for the provision of such foods, as they see it as a move to serve the interests of private contractors and the food industry. Biraj Patnaik commented that the exact micronutrient specifications prescribed by the MWCD can only be met by packaged food and not in home-cooked food. The recommendations have been a sly way of introducing ready-to-eat food (Misra 2008). The FOCUS survey states that the 'technical' analysis of the preference between hot cooked meals and micronutrient-fortified ready-to-eat food overlooks the politics of the issue: ready-to-eat foods tend to be pushed by commercial lobbies that are often hard to resist or regulate. The problems associated with the supply of food, particularly for children, by private contractors are discussed below.

Concerns Regarding Public–Private Partnerships

Expressing concerns regarding the desirability of public–private partnerships, the Working Group for Children Under Six[7] noted that, considering the vast and diverse nature of the private sector that ranges from large corporate bodies to small charitable unfunded institutions, public–private partnerships themselves are poorly defined. There is also a complete absence of a regulatory framework that safeguards the interests of the public, when the latter might be in conflict with the interests of the private party in such a partnership, and also ensures that the costs to the public exchequer are justified.[8]

Potential for Conflict of Interest

The Working Group for Children Under Six argued that as the objective of the private sector is primarily to create profit for their shareholders, rather than perform a public good, there is a high probability of a conflict of interest, with a danger of the public interest being compromised in the process of commercial interests influencing public policy.

A number of national and international declarations which have received the endorsement of the Government of India also hold this view. The Global Strategy for Infant and Young Child Feeding, adopted by the World Health Assembly (WHA) Resolution in May 2002 and by the UNICEF Executive Board in September 2002, calls for implementing programmes on infant and young child feeding consistently with accepted principles for avoiding conflict of interest. Paragraph 44 of the Global Strategy for Infant and Young Child Feeding clearly delineates the role of manufacturers, limiting it to full compliance with the International Code of Marketing of Breast-Milk Substitutes and relevant WHA resolutions, and to meeting the quality, safety and labelling standards of Codex Alimentarius.

During one of the debates of the WHA in 2005, India took the position that 'commercial enterprises by definition are profit driven entities. It is neither appropriate nor realistic for the WHO to expect that commercial groups will work along with governments and groups to protect, promote and support breastfeeding' (INFACT Canada 2002). Following this, the WHA adopted Resolution 58.32, which further urged the member states 'to ensure that financial support and other incentives for programmes and health professionals working in infant and young child health do not create conflict of interest' (ibid.). The most recent WHA Resolution, 61.20, urged member states 'to strengthen implementation of the International Code of Marketing of Breast-Milk substitutes and subsequent relevant WHA resolutions by scaling up efforts to monitor and enforce national measures in order to protect breastfeeding while keeping in mind the WHA resolutions to avoid conflicts of interest.'[9]

To maximise profits, corporations often sell products that are not medically recommended, e.g., breast milk substitutes, biscuits and suchlike, and/or compromise on the quality of their products to cut costs and thereby increase profits. To sell their products, they also often mystify malnutrition. For instance, instead of acknowledging that malnutrition is a result of chronic deprivation of food, they reduce malnutrition to micronutrient deficiencies and hence justify the sale of processed foods fortified by micronutrients or micronutrient supplements. Each bout of diarrhoea in children under 6 months of age can lead to significant weight loss, which in turn can lead to malnutrition, and is often caused when breast milk is replaced by other foods, including water, which may be contaminated and carry infection. It is best prevented and treated with exclusive breastfeeding for infants below 6 months of age and oral rehydration solution for older infants. However, by linking diarrhoea prevention with treatment with vitamin A and minerals like zinc, corporations create a market for their products.

High Levels of Corruption among Private Contactors

The Planning Commission had opposed the centralised purchase of packaged food for the ICDS and MDMS due to complaints received from several experts, who argued that the centralised procurement of packaged food was difficult to monitor and check (Anon 2008a). In September 2007, the Supreme Court Commissioners wrote to the prime minister that the 'entire feeding programme was riddled with corruption and leakages since the supply of food, the ready-to-eat food powders, had passed into the hands of private contractors. Not only were these calorifically inadequate and culturally inappropriate, most of the time they never even reached the intended beneficiaries' (Misra 2008).

Until about 2005, the costs of the SNP were met entirely from state government funds, while all other costs of the ICDS, such as infrastructure, salaries and other administrative costs, were financed with funds given by the central government. It was therefore the state governments' prerogative to administer these contracts. The fact that the size of these contracts is often huge—ranging between Rs 250 million (in smaller states) to over Rs 2.5 billion (in larger ones)—makes them prone to corruption and diversion of funds. Over time, the tenders for these contracts have been drawn to favour key players, and irregularities remain the norm rather than the exception. In Chhattisgarh, for instance, the last contract was renewed despite the presence of a 'no renewal' clause in it. Complaints about the quality of the SNP and irregularity in delivery to the anganwadis were also overlooked by compliant officers.

In Uttar Pradesh, SNP contracts worth hundreds of millions of rupees have been doled out to a few private contractors without any transparency. According to a senior official in the MWCD, some of this money finds its way back to the politicians who pull the strings, and is used to buy votes at election time. Meanwhile, the SNP programme in Uttar Pradesh has become a health hazard, with stories of food poisoning appearing periodically in the local media (FOCUS survey).

Lack of Accountability of Private Contractors

In the past, the contractor system did not allow for monitoring to be decentralised, or for the community or panchayats to exercise any control on the nature and quality of food given at the anganwadi, or even whether it reached the anganwadi centre at all (FOCUS survey).

Supply of Food by Private Contractors Necessitates Centralised Food Production

The supply of food by private contractors necessitates centralised production, procurement and distribution of food, which, apart from often being more expensive, also detracts from local control and autonomy over diets. Sometimes it may even displace local livelihoods such as milling. Many, especially the Right to Food Campaign, have stressed the various advantages of food being produced by the community using locally available foodstuffs. This not only provides members of the community with employment opportunities, but also increases their knowledge about nutrition. Procuring foodgrains, pulses, eggs, milk and meat from local farmers and producers not only boosts the local economy, but also reduces carbon emissions. The campaign recommended that self-help groups could be involved in cooking food for the ICDS, and that they should be supported and their capacity should be enhanced for this purpose, with the ultimate goal of providing all the required infrastructure, including revolving funds, to prevent the disruption of local food supply.

Cooked Food versus Ready-to-Eat Food, Commercial Interests in the Supply of Food, and the NFSB

The NAC draft of the NFSB clearly stated that 'no private contractors shall be used for the production and processing of take home rations or freshly cooked meals under any maternal or child feeding programmes.' It defined a private contractor as 'any entrepreneur, commercial enterprise or company not owned or controlled or funded or aided by the Appropriate Government'. The draft stated that the food provided in the ICDS and MDMS should be local and freshly cooked meals.

The draft tabled in Parliament did not have any clause which prohibited the participation of private contractors in supplying food for maternal or child feeding programmes. The final act defines a meal as 'hot cooked meal or ready-to-eat meal or take home ration, as may be prescribed by the Central Government', which leaves the possibility open of introducing ready-to-eat meals in the future. The schedules of the draft described take-home rations in the ICDS as energy-dense food fortified with micronutrients as per 50 per cent of the RDA. This is clearly a clause enabling ready-to-eat food. It also mentioned that meals in the ICDS and MDMS should be prepared in accordance with the prevailing food laws.

Critique of Provisions in the Drafts Regarding Cooked Food, Ready-to-Eat Food and Commercial Interests

The Working Group for Children Under Six (a joint working group of the Right to Food Campaign and Jan Swasthya Abhiyan) heavily criticised the introduction of ready-to-eat food in the 2011 government draft. The Working Group argued that by introducing ready-to-eat food, the draft NFSB was more interested in creating markets and protecting corporate interests than looking after the interests of children.

The Working Group added that the requirement in the NAC draft and the government draft that take-home rations be fortified with micronutrients containing up to 50 per cent of the RDA, and that meals should be prepared in accordance with the prevailing food laws, opened the door to centralised production and distribution, as fortification cannot be done in a decentralised manner by the community, and food prepared locally by the community may not meet the standards set by the food laws. These provisions of the NAC draft, the draft introduced in the Parliament, as well as the final act contradict the Supreme Court orders for banning private contractors and encouraging decentralised production by self-help groups, women's groups, panchayats, etc.

Conclusions

Locally produced cooked food has been shown to be more nutritious and also preferred by children. The claims of the advantages of micronutrient-fortified ready-to-eat food cited by owners of food-processing companies have been challenged by many, and need to be examined in light of the large profits which such companies stand to earn if the policy of providing ready-to-eat food items in place of cooked food in the ICDS and MDMS is implemented.

Various shortcomings of the ICDS programme, such as inadequate space and infrastructure for cooking meals, lack of sanitation at the place of cooking, and shortage of manpower at the anganwadi centre for cooking food, have all been used for justifying the provision of ready-to-eat food items in the SNP of the ICDS. While there is no doubt that these are legitimate concerns regarding the programme, the solutions to these shortcomings do not lie in replacing cooked food by ready-to-eat food, but in addressing the root causes of these problems and implementing thought-out reforms in the ICDS, such as increasing the financial allocation for ICDS to improve the quality of food provided in the SNP, increasing the number of anganwadi workers per anganwadi centre, improving the monitoring systems to check the quality of services delivered in the programme and to reduce the leakages in the supply of foodgrains. Reform options for the ICDS are discussed in detail in Chapter 15, 'Reforming Delivery'.

Past experience has clearly demonstrated that private contractors who have supplied food for the SNP in the ICDS, in collusion with government officials at various levels, have not only indulged in blatant corruption and pilferage, but have also lobbied for the provision of unhealthy food items such as breast milk substitutes and biscuits to children. The fear of public interest being compromised when commercial interests influence public policy is also very real, as private contractors are not only unaccountable to the community, but the sale of their products is at the cost of possible employment opportunities for the community.

Due to the significant advantages of cooked meals over ready-to-eat meals and the undesirability of private contractors in the supply of food for public nutrition programmes, the NFSA needs to prohibit the supply of ready-to-eat food items and the entry of private contractors in the supply of food for such programmes. The NAC draft was the only one which mentioned that food provided in the ICDS and MDMS should consist of local and freshly cooked meals. It was also the only draft which clearly prohibited private contractors from supplying food for the feeding programmes.

Notes

1. Detailed discussion on micronutrient fortification, its benefits and harms is offered in Chapter 11, 'Malnutrition'.

2. Shri D. Raja (CPI): 'You must take steps to contain the private contractors who are entering into the Mid-Day-Meal Scheme or ICD Scheme.' Shri Bhartruhari Mahtab (BJD): 'The footnote in page 16 to the Schedule is to bring in contractors. This should be dropped. I am given to understand that the HRD Ministry is opposed to this footnote. The Ministry of Health and Family Welfare also supports HRD's position. Why then the WCD Ministry is hell-bent to introduce this extremely acute standard of micronutrients level in the food supplied under ICDS, which at the village level self-help groups or other village bodies can never certify leaving the market open for contractors and manufacturers to supply? I would urge upon the Government to drop these provisions that would open the ICDS scheme to a legally mandated monopoly for manufacturers and contractors.' Rajya Sabha, Supplement to Synopsis of Debate (Proceedings Other Than Questions and Answers), 2 September 2013, http://164.100.47.5/newsynopsis1/Englishsessionno/229/Supp.%20Synopsis%20_E_%20dated%202.9.13.pdf (accessed 11 March 2018).

3. *Poori* is a light, round, deep-fried Indian bread; *kheer* is an Indian rice pudding.

4. A dry mixture made from whole wheat flour fried in sugar and ghee, heavily laced with dried fruits and herbal gums.

5. Supreme Court order of 7 October 2004 in *People's Union for Civil Liberties v. Union of India and Others*, Writ Petition (Civil) 196/ 2001. A *mahila mandal* is a group of women formed for the purpose of credit, saving, carrying out economic activities, etc.

6. Literally, 'Healthy India'.

7. Joint Working Group of the Right to Food Campaign and Jan Swasthya Abhiyan.

8. Statement by Working Group for Children Under Six, http://www.righttofoodindia.org/data/public_private_partnerships_in_the_area_of_

infant_and_young_child_feeding_and_nutrition.pdf (accessed 1 October 2017).

9. Ibid.

References

Anon. 2008a. 'Contractors? That Means Pilferage'. *Outlook*, 31 March.

Anon. 2008b. 'Too many cooks'. *Business Standard*, 20 March.

Anon. 2008c. 'Cold comfort'. *The Times of India*, 18 March.

Baru, Rama V., Rajib Dasgupta, Mita Deshpande and Aparna Mohanty. 2008. 'Full Meal or Package Deal?' *Economic and Political Weekly*, 43(24): 20–22.

Citizens' Initiative for the Rights of Children Under Six. 2006. *Focus on Children under Six* [FOCUS survey]. New Delhi: Right to Food Campaign Secretariat.

Dasgupta, Rajib, and Rama V. Baru. 2008. 'No package deal'. *The Indian Express*, 18 March.

Drèze, Jean, and Reetika Khera. 2008. 'Glucose for Lok Sabha?' *Hindustan Times*, 14 April.

Drèze, Jean, and Aparajita Goyal. 2003. 'The Future of Mid-day Meals'. http://www.righttofoodindia.org/data/mdm.pdf (accessed 1 October 2017).

Ghosh, Shanti. 2006. 'Food Dole or Health, Nutrition and Development Programme?' *Economic and Political Weekly*, 41(34): 3664–66.

Government of Tamil Nadu. 2013. 'National Program of Mid-day Meals in Schools'. Department of Social Welfare and Nutritious Meal Programme. http://mdm.nic.in/Files/PAB/PAB2012-13/TamilNadu/1_St_Writeup_AWPB_2012-13.pdf (accessed 1 October 2017).

INFACT Canada. 2002. 'World Health Assembly 2002 Rejects Commercial Influence in Infant Feeding Programmes'. http://www.infactcanada.ca/WHA_2002_Rejects_Commerical.htm (accessed 11 March 2018).

Jishnu, Latha. 2012. 'Tiger Takes the Biscuit'. *Down to Earth*, 31 January.

Menon, Sreelatha. 2008. 'Don't serve biscuits in anganwadis: Amartya Sen'. *Business Standard*, 5 February.

Misra, Savvy Soumya. 2008. 'ICDS Gets Packaged Food for the Malnourished'. *Down to Earth*, 15 March.

Pande, Mrinal. 2008. 'Midday meals: welfare by rote, not reason'. *Mint*, 8 April.

Raman, Anuradha. 2007. 'Hard to Swallow'. *Outlook*, 10 December.

———. 2008. 'Biscuit Lunch'. *Outlook*, 31 March.

Right to Food Campaign. n.d. 'Commercial Interference in Tackling Hunger and Malnutrition'. http://www.righttofoodindia.org/data/pamphlet_on_

commercial_interference_in_tackling_hunger_and_malnutrition.pdf (accessed 1 October 2017).

Samson, M., C. Noronha and A. De. 2007. 'Toward More Benefit from Delhi's Midday Meal Scheme'. Collaborative Research and Dissemination, New Delhi. http://www.cordindia.com/images/Midday.pdf (accessed 1 October 2017).

Sarin, Ritu. 2008. 'Renuka pushes for pre-cooked meals, Plan panel says no, open to misuse'. *The Indian Express*, 17 March.

Sethi, Nitin. 2008. 'Biscuit firms tweak plan to sell wares'. *The Times of India*, 24 March.

13

Gender-Just Food Entitlements

Ankita Aggarwal

The world over, women are discriminated against in various ways. One of the consequences of the unequal status of women is their inability to access adequate food, education, healthcare, clean drinking water and sanitation required for their nutritional security. Pregnant and lactating women have enhanced nutritional requirements because of their physiological condition, and single women are much more food insecure than their married counterparts because of their extremely low status in Indian society. Hence, any right to food legislation needs to address the heightened vulnerability of women to food insecurity, especially that of pregnant, lactating and single women.

This chapter begins with a discussion on the vulnerability of women's nutritional security. It talks about the enhanced nutritional requirements of pregnant and lactating women and the additional discrimination faced by single women, which increases their food insecurity as compared to married women of the same economic category. The next section discusses the provisions for enhancing women's nutritional security in CEDAW and the FAO guidelines. This is followed by a summary of the provisions for women's nutrition security in the drafts of the NFSB and in the NFSA. The final section discusses the existing schemes and legislations in India which aim at enhancing women's nutrition security, and critiques the provisions of the NFSB drafts and the NFSA.

Vulnerability of Women to Nutrition Insecurity

Women play a crucial role in guaranteeing their families' nutrition security. Those who are able to access livelihood opportunities are more likely to spend a greater part of their income on the family's nutrition than men, who tend to spend more money under their control on alcohol, tobacco and suchlike (Haddad et al. 1997). However, because of the various forms of discrimination faced by females, such as those within the family (Choudhary and Parthasarathy 2007), in owning land and other means of production, and in accessing livelihood opportunities, a large proportion of women and girls are highly vulnerable to food insecurity (Krishnaraj 2005). Women and girls also face barriers in accessing education, healthcare, clean drinking water and sanitation, all of which are required for nutritional security. In families where males are given preferential treatment over females, especially those that are food insecure, women and girls often not only get less food to eat, but may also be forced to eat food which is inferior in quality and nutrient content (see Choudhary and Parthasarathy 2007; IFPRI et al. 1998; Mukherjee and Mukherjee 1994).

Studies since the 1980s have shown that across socio-economic groups, girls face feeding neglect. In Punjab, thrice the number of girls as boys were found to be malnourished in wealthy families and twice as many in less wealthy families (Cowan and Dhanoa 1983). A study of intra-familial food allocation in south-central India was interpreted by the researchers as a seasonal gender bias; in the lean season, boys were favoured over girls in the distribution of food (Beherman 1984). The HUNGaMA 2011 survey shows that the prevalence of stunting and low weight-for-age was higher for boys between 0 and 35 months, but this is reversed in the 36–59-month time period (HUNGaMA 2011, 13), indicating that girls seem to have a nutritional advantage at birth, which they lose in the early years. However, not all studies have been unequivocal in stating this gender bias (for instance, see Haddad et al. 1996). It has been noted that gauging food allocation at an individual level can be difficult; study results should be interpreted with caution and, consequently, policy that attempts to mitigate intra-familial inequities in food distribution should be formulated keeping this in mind (see for instance Harriss-White 1997). Policies can also have unintended consequences—for instance, a school meal programme that targets girls specifically can lead to parents giving lower quantities of food to them at home.

In 1980, the UN Women's Conference in Copenhagen stated that although women constitute 50 per cent of the world's population, they own only 1 per cent of the world's resources (UN 1980), as in most developing countries, land titles are in the name of men. Apart from facing barriers to owning land, women also often lack other resources required for growing food (World Bank 2001), and even when women are able to undertake food production, their returns are often not commensurate with their effort (Krishnaraj 2005).

There is a growing trend of the 'feminisation of poverty', i.e., poverty among women is rising faster than among men. This phenomenon is associated not just with higher incidence of poverty among women than men, but the *severity* of poverty is also higher among women (Commissioners of the Supreme Court 2008). Out of the 1.3 billion people considered to be living in poverty worldwide, over 70 per cent are believed to be women (FAO 1996).

Table 13.1 presents the proportion of adult Indian men and women who have a BMI of less than 18.5 (a standard cut-off conventionally associated with 'chronic energy deficiency') in the nine states selected by the National Nutrition Monitoring Bureau.[1] Except 1975–79, for every other year, the proportion of women suffering chronic energy deficiency is equivalent to or higher than that of the men in the same year. Also, the rate of decline in chronic energy deficiency from 1975–79 to 2004–05 among men is 10 percentage points higher than among women. The higher incidence of chronic energy deficiency in women is glaring evidence of their greater food insecurity.

Table 13.1. Nutrition status of Indian adults, 1975–79 to 2004–05 (BMI)

	Proportion (%) of adults with BMI below 18.5					% Decline (1975–79 to 2004–05)
	1975–79	1988–90	1996–97	2000–01	2004–05	
Men	56	49	46	37	33	41
Women	52	49	48	39	36	31

Sources: National Nutrition Monitoring Bureau (1999, 2002 , 2006). These figures apply to the nine 'NNMB states': Andhra Pradesh, Gujarat, Karnataka, Kerala, Madhya Pradesh, Maharashtra, Odisha, Tamil Nadu and West Bengal. Data for 1975–79 and 1988–90 exclude West Bengal; data for 1996–97 exclude Madhya Pradesh. See text for further discussion.

Table 13.2 compares the incidence of anaemia among Indian men and women in 2005–06. The significantly higher incidence of anaemia among women reflects higher iron deficiency among them—an indicator of the higher levels of nutrition insecurity faced by women.

Table 13.2. Incidence of anaemia in Indian men and women in the age group 15–49 in 2005–06

		Proportion (%) of men in the age group of 15–49 years	Proportion (%) of women in the age group of 15–49 years[a]
Incidence of anaemia	Mild	13	39
	Moderate	10	15
	Severe	1	2

[a] Excludes pregnant women and women who gave birth in the preceding two months.
Source: IIPS (2009).

Specific Vulnerabilities of Certain Categories of Women

Pregnant and Lactating Women

Pregnant and lactating women have enhanced nutritional requirements to facilitate the growth and development of the foetus and the infant, as well as for maternal metabolism and tissue development specific to reproduction. The total nutrient requirement of pregnant and lactating women is not necessarily the simple addition of those accumulated in maternal tissues, products of pregnancy and lactation, and those required by non-reproducing women, although this summation is sometimes used to arrive at estimates of recommended nutrient intakes.

Not all pregnant and lactating women in India are able to obtain the extra nutrition or maternity care required by them. This is highlighted in Table 13.3, which presents the status of maternity care and incidence of anaemia among pregnant women over three NFHS surveys. Access to maternity care has improved over time, increasing with the level of women's education, and is higher in urban areas. Similarly, the incidence of anaemia, although still very high, has reduced over the years, decreasing with the increase in the level of women's education, and is lower in urban areas.

Table 13.3. Access to maternity care and incidence of anaemia among pregnant women in India

	NFHS (2005–06)	Urban	Rural	No education[a]	<5 years complete[a]	5–9 years complete[a]	10 years complete and above[a]	NFHS (1998–99)	NFHS (1992–93)
Mothers who had at least 3 antenatal care visits for their last birth (%)	50.7	73.8	42.8	29.8	52.1	64.6	85.3	44.2	43.9
Mothers who consumed iron folic acid for 90 days or more when they were pregnant with their last child (%)	22.3	34.5	18.1	9.5	20.6	27.5	49.4	NA	NA
Births assisted by a doctor/nurse/lady health visitor/auxiliary nurse midwife/other health personnel (%)[b]	48.8	75.3	39.9	27.9	48.0	61.9	86.8	42.4	33.0
Institutional births (%)[b]	40.8	69.4	31.1	19.8	38.9	52.9	80.6	33.6	26.1
Mothers who received postnatal care from a doctor/nurse/lady health visitor/auxiliary nurse midwife/other health personnel within 2 days of delivery for their last birth (%)[b]	36.8	60.8	28.5	18.0	34.1	46.4	73.4	NA	NA
Pregnant women age 15–49 who are anaemic (%)	57.9	54.6	59.0	63.0	58.5	56.2	47.4	49.7	NA

[a] Education of women.

[b] Based on the last two births to ever-married women in the three years before the survey.

Source: IIPS (2007).

Pregnant and lactating women are particularly affected by malnutrition (FAO 2009). Nutritional deficiencies among pregnant women are a leading cause of maternal and child mortality (Nag 1994), and can also irreversibly impair the development of foetuses and infants (Ercilla and Chikoko 2006). More than 40 per cent of children in India are born underweight, primarily because of the poor nutritional status of their mothers (IIPS 2007). Low birth weight is a common cause of malnutrition among children, which in turn can have several debilitating impacts.[2]

Single Women

In predominantly patriarchal societies such as India's, women are viewed instrumentally as reproducers, caregivers, sexual outlets and facilitators of a family's prosperity. They are not treated as agents and ends in their own right, or as individuals complete in themselves; instead, their cultural acceptability is defined only in terms of their status as appendages of men, in the roles of wives, mothers, daughters and sisters. Hence, single women are characterised by the absence of male 'protection' in their lives, which in most instances is actually an absence of male 'control'. This erodes the social status of women who are widows, divorced, separated, abandoned or never married (Mander 2012).

A deep social stigma is attached to widows, who face formidable social barriers to remarriage. Although widows in India have a legal right to their deceased husbands' property, few of them are able to exercise this right; even in cases where widows are able to establish their rights, they are unable to exercise actual control over the property. Women who leave their husbands' families or are abandoned by their husbands are often not accepted back in their parents' houses, which leaves them with nowhere to go.

Because of the various social restrictions faced by most single women, it is extremely difficult for them to find gainful employment. Even when they are able to find jobs, they may be paid very low wages and forced to work for long hours. Single women are often unable to access credit, even if they have the same earning capacity as men or married women. To sustain themselves and their children, single women often have to mortgage or sell their assets in distress, resort to undignified options such as sex work, begging, and sending their children to work (Commissioners of the Supreme Court 2008).

As per the 2001 Census, 6.9 per cent of the women in India were widowed, 0.5 per cent were divorced or separated, and 1.4 per cent were above 30 years of age but unmarried. In total, 7.5 per cent of the women in India (39 million women) were single in 2001. Almost 58 per cent of the widows were more than 60 years of age, and hence faced the additional burden of old age. Low-income single women comprise the most vulnerable 10 per cent of the entire female population in India (Dhar 2011).

The additional discrimination and deprivation faced by a large proportion of single women make them much more food insecure than others belonging to the same economic category (Commissioners of the Supreme Court 2008). This is evident from the fact that single women consume far fewer calories than married women. As per the 61st Round of the NSS conducted in 2004–05, the average daily consumption of married women was 2,086 kcal, but the corresponding consumption of widows was 2,005 kcal and of divorced or separated women, 1,989 kcal.

Provisions for Enhancing Women's Food Security in CEDAW and the FAO Guidelines

Several initiatives have been taken globally to address women's heightened food insecurity. These include raising the issue of women's right to own land and other property, assisting women to grow food, advocating diversification of livelihood opportunities for women, and enabling women to increase their access to education, healthcare, sanitation and clean drinking water.

An international treaty to guarantee women the protection of their rights, CEDAW, was adopted by the UN General Assembly in 1979 and came into force in 1981. Of the 193 countries in the world, 187 have ratified CEDAW (Afsari 2010). The convention has specific clauses on the equal treatment of women in agrarian reforms and also on similar rights for both spouses in the ownership, management and disposition of property.

Article 14 of CEDAW has several provisions that promote the nutritional security of rural women. These include guaranteeing rural women access to adequate healthcare, direct benefits from social security programmes, all types of education and training, opportunities of organising self-help groups and cooperatives to obtain equal access to employment, access to agricultural credit, marketing facilities and appropriate technology, and access to adequate living conditions, including sanitation and water supply.[3]

In 1979, the FAO's World Conference on Agrarian Reform and Rural Development called for promoting ownership rights for women and repealing laws that discriminate against women regarding inheritance, ownership and control of property (Rao 2005). The 127th Session of the FAO Council in 2004 adopted voluntary guidelines to support the progressive realisation of the right to adequate food in the context of national food security. Table 13.4 provides excerpts of those guidelines that aim at promoting women's right to food and nutrition security.

Table 13.4. FAO's voluntary guidelines promoting women's right to food and nutrition

Guideline no.	Guideline
7.4	States should consider strengthening their domestic law and policies to accord access by women heads of households to poverty reduction and nutrition security programmes and projects.
8.6	States should promote women's full and equal participation in the economy and, for this purpose, introduce, where it does not exist, and implement gender sensitive legislation providing women with the right to inherit and possess land and other property. States should also provide women with secure and equal access to, control over, and benefits from productive resources, including credit, land, water and appropriate technologies.
8.10	States should take measures to promote and protect the security of land tenure, especially with respect to women, and poor and disadvantaged segments of society, through legislation that protects the full and equal right to own land and other property, including the right to inherit. As appropriate, states should consider establishing legal and other policy mechanisms, consistent with their international human rights obligations and

(contd.)

(Table 13.4 continued.)

Guideline no.	Guideline
	in accordance with the rule of law, that advance land reform to enhance access for the poor and women.
11.2	States should strengthen and broaden primary education opportunities, especially for girls, women and other underserved populations.
13.4	States may wish to give priority to channelling food assistance via women as a means of enhancing their decision-making role and ensuring that the food is used to meet the household's food requirements.
17.5	States should, in particular, monitor the food security situation of vulnerable groups, especially women, children and the elderly, and their nutritional status, including the prevalence of micronutrient deficiencies.

Source: FAO (2006).

The FAO *Guide on Legislating for the Right to Food* (2009) acknowledges the widespread discrimination against women and recommends that a right to food act should include a specific clause prohibiting discrimination against women in matters that affect the realisation of their right to food. The guide states that the framework law should oblige public authorities to 'eliminate and prevent all forms of discrimination against women with regard to the guaranteed right to food, including less favourable treatment of women for reasons of pregnancy and maternity, and to promote equality of opportunities between men and women' (ibid.).

The guide instructs governments to take immediate measures to ensure that particularly disadvantaged people can either directly access food or the means for its procurement. Debating the merits of a right to food act listing all 'vulnerable groups' as opposed to defining 'vulnerability', it states that women would certainly constitute one of the vulnerable groups. The guide reminds states that as per international human rights law, states are required to prioritise the needs of the most vulnerable persons when resources are limited (FAO 2009). The guide recommends that a right to food act should have an express provision stating that 'every woman has a right to food and adequate nutrition during pregnancy and lactation' (ibid., 87).

Provisions relating to Women's Nutrition Security in the NFSB Drafts and the NFSA

NAC Draft

The preamble of the NAC draft referred to Articles 12, 13 and 14 of CEDAW, which call for ending discrimination against women in healthcare and recognise the special needs of rural women. The draft designated the eldest adult woman of each household as the household head, unless there is no adult woman in the household, in which case it would be the oldest male member of the household. Ration cards for PDS, according to the NAC draft, were to be issued in the name of an adult woman member of the household, if any.

State governments were directed to provide take-home rations or nutritious and freshly cooked meals, free of charge, to women during pregnancy and six months after the delivery of the child, through the local anganwadi. Such meals are required to contain 600 kcal and 18–20 grams of protein.

The draft also directed state governments to provide nursing mothers with maternity benefits of Rs 1,000 a month, for a period of six months. The amount of Rs 1,000 was to be revised every year based on an appropriate price index as determined by the central government. Women who are regularly employed with the government, public sector undertakings or state public sector undertakings, or those receiving similar benefits under any law, would not be eligible for the maternity benefits under the draft.

Among other categories of vulnerable people, pregnant and lactating women affected by emergency or disaster were entitled to two freshly cooked meals every day, free of charge, for a period of three months after the disaster. If pregnant and lactating women were living in households, groups or communities threatened by starvation, they would be entitled to at least two free freshly cooked meals a day and Rs 2,000 as maternity benefits.

The draft urged the government to progressively realise universal access to adequate pensions for single women (and also for the aged and persons with disabilities) which were not less than the prevailing statutory minimum wages for unskilled workers. It also stated the desirability of nutrition, health and education support to all adolescent girls.

Draft Introduced in Parliament

The eldest adult woman in every 'priority' and 'general' household was to be considered the household head for the purpose of issuing ration cards. Only in the case of a household that did not have an adult female would the eldest adult male member of the household be considered the household head. If such households had girls less than 18 years of age, they would assume the status of household heads on becoming adults.

Every pregnant and lactating woman would be entitled to a free meal during pregnancy and until six months after birth, through a local anganwadi. Schedule II of the draft, which prescribed types of meals and nutritional standards for various categories of people, prescribed take-home rations for pregnant and lactating women that contained 600 kcal and 18–20 grams of protein.

Every lactating woman would be entitled to a maternity benefit of Rs 1,000 for a period of six months.[4] Women in regular employment with the central or state governments or public sector undertakings, or those who received similar benefits under any other law, were not entitled to these maternity benefits.

Schedule III of the draft listed access to nutritional, health and education support to adolescent girls and adequate pensions for single women as provisions for advancing food security.

Comparing Provisions in the NFSB Drafts with the Recommendations of the FAO Guide on Women's Food Security

The FAO guide recommends the inclusion of a specific clause in the right to food act that prohibits discrimination against women. Such a clause was absent from both drafts of the NFSB and the final act.

Although the NFSB drafts had some specific provisions for vulnerable groups, they neither gave a list of 'vulnerable groups' nor defined 'vulnerability', as recommended by the FAO guide, and the final act even omitted certain provisions for vulnerable groups except women and children. Although the NFSB drafts and the final act did not have an express provision stating that 'every woman has a right to food and adequate nutrition during pregnancy and lactation,' they did have special food entitlements for pregnant and lactating women.

Standing Committee Recommendations

The Standing Committee on Food, Consumer Affairs and Public Distribution spent many months collecting information and recommendations from various sources. They recommended that pregnant women should receive maternity benefits of Rs 1,000 per month after three months of pregnancy up until delivery, as well as 5 kilograms of foodgrains every month during pregnancy and up to two years after childbirth. While these recommendations were more wide-ranging than the content of the NFSB draft, the Standing Committee also recommended limiting the maternity benefit up to the birth of a second child, ostensibly to stabilise the population. Their further recommendation was for the maternity benefit amount to be indicated in a schedule and not the main body to ensure that it could be revised. They further recommended that the grievance redressal mechanism should have elected women representatives.

National Food Security Act

Despite the recommendations of the Standing Committee, the final act did not differ much from the government draft, having not implemented any of the Standing Committee recommendations on provisions dealing with women.

The eldest adult woman in every household is still to be considered the household head for the purpose of issuing ration cards. Only in case a household does not have an adult female would the eldest adult male member of the household be considered the household head. If such households have girls less than 18 years of age, they would assume the status of household heads on becoming adults. In a rare show of unity, the various parties in Parliament all agreed that the act would help empower women due to this clause.

Every pregnant and lactating woman is entitled to a free meal during pregnancy and until six months after childbirth, through a local anganwadi. The act prescribes take-home rations for pregnant and lactating women that have 600 kcal and 18–20 grams of protein.

Every lactating woman is entitled to a maternity benefit of Rs 1,000 for a period of six months. Women in regular employment with the central or state government or public sector undertakings, or those who receive similar benefits under any other law, are not entitled to these maternity benefits.

The state commissions that will be set up to monitor the implementation of the act should have at least two women as part of the committee, as chairperson, members or member-secretary.

Schedule III of the act lists access to nutrition, health and education support for adolescent girls and adequate pensions for single women as provisions for advancing food security, although this is provided under a legally non-binding section.

Enhancing Women's Nutrition Security in India

Women's Access to Property, Means of Food Production and Employment

In India, various initiatives have been in place to increase women's access to land and other means of production, and also to livelihood opportunities. After the publication of the *Report of the Committee on Status of Women in India* in 1974 (GoI 1974), and the women's movement in the country which represented the demands of peasant women,[5] the Sixth Plan (1980–85) recommended that all land distributed under the land reform initiative should be registered jointly under the name of both spouses (GoI 1980). This was India's first public document which recognised the importance of land for women (Rao 2005). The Sixth Plan was also the first plan to include a separate chapter on women and development, which recommended the utilisation of at least a third of the financial resources for rural development programmes for women by starting the Development of Women and Children in Rural Areas programme (GoI 1980). The Seventh Plan (1985–90) focused on enhancing women's skills in agriculture and allied sectors through training and increasing the participation of women in rural development programmes (GoI 1985). The Eighth Plan (1992–97) highlighted the need for increasing women's control over economic resources and services and recommended joint titles for married women for productive assets, houses and house sites. It recommended that state governments allot 40 per cent of the surplus land collected under the Land Ceiling Act to women, especially women-headed households, the rest being in joint titles (GoI 1992). The New Agricultural Policy in 2000 also recognised women's rights to land (Rao 2005).

Over the last few decades, the issue of women's right to property has also received legal recognition. The Hindu Succession Act of 1956 treats daughters and sons as equal heirs to the property of their parents, regardless of the marital status of the daughters. The act also gives widows legal rights over their deceased husbands' property.[6] Muslim personal law also allows women to inherit property from both their natal as well as their marital family, although their share is lower than that of their male counterparts.

Despite the various provisions for women's rights to land and other assets in the various plan documents and legislations, a large proportion of women continue to be alienated from their right to property (Thakur 2001).

The drafts of the NFSB, and the final act, are completely silent on the issue of land rights, nor do they provide any legal entitlements to other factors of food production. However, given the importance of equitable access to land, water and other inputs required for agriculture, and considering the agrarian crisis being faced by the country, a separate legislation guaranteeing these entitlements may be considered.

The MGNREGA programme has contributed to providing employment to women in rural areas (Khera and Nayak 2009). However, there is an absence of a similar legislation for urban areas.

Women as Household Heads

Before the introduction of the NFSA, adult men were considered household heads, and ration cards were issued in their names. Despite the recent change, in the eventuality that a married couple divorces or separates or a man abandons his wife, the ration card is usually retained with the man and the woman is unable to access subsidised rations from the PDS (GoI 2009). The change to making women the household heads was in part to prevent this disadvantage to single women.

Enhancing Nutrition Security of Lactating and Pregnant Women

Integrated Child Development Services

The ICDS, introduced in 1975, addresses the nutrition and health needs of pregnant and lactating women by providing them with supplementary nutrition, immunisation, health check-ups and nutrition and health education. The NFSB drafts, as well as the final act, provide legal guarantees only for supplementary nutrition and not for the other ICDS services.

The Supreme Court order of 28 November 2001 universalised the ICDS, which means that every hamlet should have an anganwadi and all pregnant and lactating women are entitled to ICDS services (Right to Food Campaign 2008). However, several settlements in the country are still not served by an anganwadi, and even in areas that are served by anganwadis, the utilisation of ICDS services is generally low. As per NFHS-3 conducted in 2005–06, only 21 per cent of pregnant women and 17 per cent of lactating women in areas served by an anganwadi received supplementary nutrition (IIPS 2009).

The order of 28 November 2001 also entitled all pregnant and lactating women to 500 kcal and 20–25 grams of protein for 300 days a year (Right to Food Campaign 2008). At most anganwadis, pregnant and lactating women are given dry rations such as rice and dal. Although this ration is intended only for the consumption of the woman, it usually gets shared among all the members of the family, and the pregnant or lactating woman is only able to consume a fraction of it. Resultantly, there is debate over providing cooked meals to women at the anganwadi centres. While this would require the pregnant and lactating women to go to the anganwadi every day, which might be inconvenient for many, it would ensure that the supplementary nutrition intended for them is consumed only by them, and not by other members of the family.

Acknowledging that lactating and pregnant women have enhanced nutrition requirements, on 2 May 2003, the Supreme Court ordered that lactating and pregnant women are entitled to Antyodaya cards which would entitle them to purchase 35 kilograms of grain per month, at Rs 2 per kilogram for wheat and Rs 3 per kilogram for rice. The order also stated that possession of a BPL card was not necessary for inclusion in the AAY.

Maternity Entitlements

Maternity entitlements are the rights of lactating and pregnant women to meet their own and their infants' nutrition requirements. They include additional nutritious food, nutrition education, healthcare, paid leave from work, support for breastfeeding and young infant and child feeding, and financial assistance to meet nutrition requirements.[7]

Article 11 of CEDAW and the ILO Maternity Protection Convention 2000 recognise the special requirements of women during pregnancy and the first few years of the child's life, and oblige state parties to take steps towards maternity protection and other measures required to end discrimination against women because of their maternity.[8] Article 12.2 of CEDAW obliges state parties to ensure appropriate services to women during pregnancy, confinement and the postnatal period. It also requires state parties to provide adequate nutrition to pregnant and lactating women and deliver free services where required.[9]

Legislations Entitling Maternity Entitlements to Women in the Formal Sector

India has legislated the Maternity Benefits Act, 1961, and the Employees' State Insurance Act, 1948, which entitle pregnant and lactating women to certain benefits. However, both these acts are applicable only to women in the formal sector, which accounts for a mere 7 per cent of all women workers (FOCUS survey).

The Maternity Benefits Act, 1961, prohibits employers from employing women during the six weeks immediately following the day of the delivery or miscarriage. It also prohibits employers from requiring pregnant women to do physically strenuous work which may adversely affect their health, or that of their foetus, in the last one month of their pregnancy, and makes women eligible for payment equalling their average daily wage for the period of their absence from work due to maternity. However, the act puts a ceiling on the time period for which a woman is entitled to maternity benefit at 12 weeks. The act provides other entitlements such as paid leave for illnesses arising due to pregnancy and delivery, nursing breaks, protection from dismissal during or due to maternity leave, etc.[10]

The Employees' State Insurance Act, 1948, entitles insured women to payment in case of confinement or miscarriage or sickness due to pregnancy, confinement, premature birth of child or miscarriage.[11]

Schemes Providing Financial Assistance to Pregnant/ Lactating Women

The NMBS was introduced in 1995. It was an entirely centrally sponsored scheme which provided a cash assistance of Rs 500 to pregnant BPL women for up to two live births, provided they were 19 years or above in age. The maternity benefit was to be disbursed in one instalment 8–12 weeks prior to delivery, in order to enable the women to meet at least some of their nutritional requirements during pregnancy.[12]

Ten years later, the JSY programme was introduced, which provides cash assistance to women who deliver in institutions. Apart from providing money to women to meet their nutrition requirements during lactation, the objective of this scheme is also to encourage institutional deliveries which would enable the women and the newborn babies to access better healthcare (Lim et al. 2010). The women eligible for the JSY cash assistance also have to belong to BPL households, should be 19 years or older in age, and are entitled to this support for only up to two live births. In the low-performing states (Assam, Bihar, Chhattisgarh, Jammu and Kashmir, Jharkhand, Madhya Pradesh, Odisha, Rajasthan, Uttar Pradesh and Uttarakhand), which have low in-facility birth coverage, all women, irrespective of BPL status, are entitled to the cash benefit under JSY. This is also an entirely centrally sponsored scheme, but some states such as Andhra Pradesh make their own contributions, thereby increasing the amount of cash assistance.[13]

After delivery in a government or accredited private health facility, mothers and ASHA workers are entitled to amounts as per Table 13.5.

Table 13.5. Cash assistance provided to mothers and ASHAs in JSY

Type of state	Rural areas		Urban areas	
	Cash assistance to mother	Cash assistance to ASHA	Cash assistance to mother	Cash assistance to ASHA
Low-performing	1,400	1,000	1,000	200
High-performing	700	600[a]	600	200[a]

Note: [a] With effect from 1 April 2009.
Source: *People's Union for Civil Liberties v. Union of India and Others*, IA No. 98/2010 in WP(C) No. 196/2001.

The purpose of providing cash to ASHAs is to incentivise them to bring pregnant women to health centres for deliveries, and to compensate them for the transportation costs incurred by them to bring the pregnant women to the health centre and take them back.[14]

The rationale for introducing the JSY was that the NMBS did not address the health needs of women who give birth, which entails protecting the expectant mother from avoidable morbidity and mortality due to causes related to pregnancy and neonatal care. Hence, in IA no. 37 of 2004 in the right to food case, the Government of India sought permission to modify the NMBS to introduce JSY.[15] However, in the same year, the Supreme Court had passed an order according to which the NMBS (among other schemes) could not be discontinued or restricted in any manner without prior approval of the court. Hence, in IA no. 54 of 2005, the petitioners in the right to food case questioned the legality of discontinuing the benefit under the NMBS by introducing JSY. They argued that since the cash assistance in the JSY is provided *after* the institutional delivery, it does not provide monetary support to women before delivery, i.e., during pregnancy.[16]

On 9 May 2005, the Supreme Court refused to allow the Government of India to phase out NMBS and provide maternity benefits under JSY, as it was not clear whether JSY preserved all the benefits available under NMBS (Right to Food Campaign 2008). Further, on 20 November 2007, the Supreme Court passed an order according to which the NMBS was to continue to provide Rs 500 per birth to all pregnant BPL women, irrespective of their age and number of children (ibid.). The debate on the conditionalities of age and number of children (and institutional deliveries in case of JSY) are discussed in Chapter 6, 'Conditionalities'.

Maternity Entitlements and the NFSB

Both drafts of the NFSB entitled lactating women to Rs 1,000 a month for six months immediately after the delivery of the baby. The NAC draft stated that the amount of Rs 1,000 was to be revised every year based on an appropriate price index. However, the draft introduced in the Parliament did not have any such provision, which would lead to a reduction in the real value of the maternity benefits with increase in prices over time. The Right to Food Campaign argued that since maternity entitlements were provided as wage compensation to enable women to quit work and stay at home, they should be equivalent to the minimum wage and inflation indexed. It also advocated that the maternity benefits should be provided for not just six months, but for nine months, starting from three months before the expected date of delivery. The drafts mentioned that women who are regularly employed with government, public sector undertakings, state public sector undertakings or those receiving similar benefits under any other law were not eligible for maternity benefits under the NFSB. However, some observers, such as the members of the Right to Food Campaign, were of the opinion that these benefits should be universalised and provided to all women in maternity (Right to Food Campaign 2012).

Enhancing Nutrition Security of Adolescent Girls

As adolescent girls are future mothers, the ICDS programme addresses the nutrition requirements of adolescent girls as well. The Supreme Court order of 28 November 2001 entitles every adolescent girl to 500 kcal and 20–25 grams of protein for 300 days a year (Right to Food Campaign 2008). The universalisation of supplementary nutrition for adolescent girls was reiterated in the Supreme Court order of 13 December 2006 (ibid.). However, the provision of supplementary nutrition to adolescent girls is far from universal (Commissioners of the Supreme Court 2009).

The MWCD implemented two programmes for adolescent girls which have components of nutrition—the Kishori Shakti Yojana and the National Programme for Adolescent Girls. Only 2.3 per cent of adolescent girls are being covered under the Kishori Shakti Yojana, that too assuming that there are no leakages in the programme. The objective of the National Programme for Adolescent Girls is to provide 6 kilograms of foodgrains to underweight girls (weight < 35 kilograms) in the age group of 11–19 years. However, this scheme is implemented in only 51 districts of the country (Commissioners of the Supreme Court 2009).

The NFSB drafts and the final act did not provide any legal entitlements to adolescent girls, not even those already provided by the Supreme Court under the ICDS. The drafts and act only urge the government to progressively realise universal nutritional, health and education support for all adolescent girls.

Enhancing Nutrition Security of Single Women

Public Distribution System

Acknowledging that single women have greater food insecurity, the Supreme Court in the right to food case declared in its order of 2 May 2003 that widows, other single women (and destitute women) with no regular support were entitled to Antyodaya cards which would allow them to purchase 35 kilograms of grain per month, at Rs 2 per kilogram for wheat and Rs 3 per kilogram for rice. Possession of a BPL card was not necessary for inclusion in the AAY (Right to Food Campaign 2008).

Because of the extreme deprivation of single women, the Saxena Committee recommended the automatic inclusion of single women in the priority category (GoI 2009). However, the SECC 2011 ascribed a deprivation score only to female-headed households with no adult member between 16 to 59 years of age (GoI 2011).

Widow Pension

Currently, BPL widows in the age group of 40–59 years are entitled to pensions under the Indira Gandhi National Widow Pension Scheme.[17] As different states add different amounts of funds to those received by the central government for this scheme, the amount received as pensions varies with states. For instance, the pension amount is Rs 200 a month in Bihar and Rs 500 in Gujarat and Rajasthan (Deenadayalan 2011). The state governments may also have their own eligibility criteria. Like any other scheme targeted to people who are BPL, the widow's pension scheme too suffers from exclusion errors—poor widows who have been excluded from BPL lists are unable to receive this pension. Apart from exclusion errors, the scheme also suffers from irregularities such as the pension amounts being credited in the accounts of widows after delays of several months. At many banks, middlemen help illiterate women withdraw their pension money, but in return take a percentage of the pension amount as compensation for their services,[18] despite these pensions being the only source of income for a large number of widows and hence contributing greatly to their food security (National Forum for Single Women's Rights 2011).

Unfortunately, there are no similar central government sponsored pensions for other categories of single women, such as those who are divorced, separated or abandoned, although some states provide pensions to destitute and separated women and support to single women for raising their children (National Forum for Single Women's Rights 2011).

Because of the important role played by social security pensions in contributing to the food security of widows, the Right to Food Campaign, which included groups such as the National Forum for Single Women's Rights, demanded pensions in the NFSA not just for widows, but for all single women. The campaign demanded that the pension amount should be Rs 1,300 per month (at Rs 2009–10 prices[19]). It demanded that these pensions should be provided to all single women irrespective of their age or BPL status (Right to Food Campaign 2012). However, none of the drafts of the NFSB gave legal entitlements to pensions to single women (or any other category of vulnerable people), although the NFSA, in the legally non-binding Schedule III, aims to give pensions to single women as well as senior citizens and persons with disabilities.

Swadhar Scheme

The Swadhar Scheme, which was introduced in 2001 and implemented by the MWCD, assists various types of women who are in distress. The services provided under this shelter-based scheme include provision of food, clothing and shelter, healthcare counselling, legal support, social and economic rehabilitation through education, awareness generation, skill-based training and behavioural training.[20] As of 2008, 208 shelters were operational under this scheme; in the same year, the budget for the scheme was increased from Rs 87 million to Rs 150 million. However, the budget for the scheme limits its ability to cater to more than the needs of a very small proportion of the women in need.

Access to Clean Drinking Water, Sanitation, Education and Healthcare

As discussed in Chapter 3 on the scope of the NFSA, access to clean drinking water, sanitation, education and healthcare are essential for nutrition security. In 2009, India legislated the Right of Children to Free and Compulsory Education Act, for children in the age group of 6 to 14 years. The NFSB drafts did not provide legal entitlements for healthcare, clean drinking water or sanitation, but only mentioned them in the list of goals that should be progressively realised, as is the case in the NFSA. In most households where the source of water is located away from the dwelling, it is the responsibility of the female members of the family to fetch water. Furthermore, women have greater needs of privacy and greater requirements of toilets and closed spaces for bathing, cleaning and washing; hence, absence of legal guarantees to clean drinking water and sanitation affects women much more adversely than men.

Neglect of Non-reproducing Women

Food-related schemes for single women and the Supreme Court order to include single and destitute women with no support in the AAY came years after the introduction of the ICDS in 1975, which contributes to the nutrition security of pregnant and lactating women and adolescent girls. There is a critique that the priority given to the nutrition needs of women and girls who are mothers (or potential mothers) reflects a view that instrumentally values women only in their reproductive role.

Notes

1. The bureau periodically collects nutrition-related data in the rural areas of a number of states.

2. This was discussed in greater detail in Chapter 11, 'Malnutrition'.

3. UN Women, http://www.un.org/womenwatch/daw/cedaw/text/econvention.htm (accessed 2 October 2017).

4. This draft estimated that, assuming a coverage of about 22.5 million pregnant and lactating women, the expenditure for central and state governments together would be about Rs 1.35 billion. However, actual annual expenditure would depend on the number of identified entitled beneficiaries and those actually availing of the benefit. The expenditure on maternity benefits was to be shared between the central and state governments, in accordance with the relevant scheme to be prescribed by the central government.

5. For more on the peasant women conferences that let to this recognition, see Mazumdar (1997).

6. Hindu Succession Act, 1956, http://revenue.tripura.gov.in/sites/default/files/hindu-succession-act-1956.pdf (accessed 12 March 2018).

7. 'Why Maternity Entitlements?', http://www.ibfanasia.org/PPT/Maternity-Entitlement.pdf (accessed 12 March 2018).

8. See UN Women, http://www.un.org/womenwatch/daw/cedaw/text/econvention.htm (accessed 2 October 2017); ILO, C183: Maternity Protection Convention, 2000 (no. 183), http://www.ilo.org/dyn/normlex/en/f?p=NORMLEXPUB:12100:0::NO::P12100_ILO_CODE:C183 (accessed 12 March 2018).

9. UN Women, http://www.un.org/womenwatch/daw/cedaw/text/econvention.htm#article12 (accessed 2 October 2017).

10. Maternity Benefit Act, 1951, http://www.ilo.org/dyn/travail/docs/678/maternitybenefitsact1961.pdf (accessed 12 March 2018).

11. Employees' State Insurance Act, 1948, http://www.ascconline.com/img/services/labour_welfare/ESI_Act.pdf (accessed 2 October 2017).

12. *People's Union for Civil Liberties v. Union of India and Others*, IA no. 98/2010 in WP(C) No. 196/2001.

13. *People's Union for Civil Liberties v. Union of India and Others*, WP(C) no. 196/2001, order dated 11 November 2007.

14. *People's Union for Civil Liberties v. Union of India and Others*, IA no. 98/2010 in WP(C) No. 196/2001.

15. Ibid.

16. Commissioners response on 5 April 2010.

17. Indira Gandhi National Widow Pension Scheme, https://en.wikipedia.org/wiki/National_Social_Assistance_Scheme#Indira_Gandhi_National_Widow_Pension_Scheme_(IGNWPS) (accessed 12 March 2018).

18. For more information on problems in the implementation of the widow pension scheme, see National Forum for Single Women's Rights (2011).

19. In 2009–10, Rs 100 was the minimum daily wage rate for unskilled labour. Assuming 26 working days in a month, a person working at the minimum wage rate would earn Rs 2,600 a month. The pension amount should be at least half of the minimum wage, hence Rs 1,300 a month.

20. Ministry of Women and Child Development's response to Lok Sabha Unstarred Question no. 4681 answered on 25 April 2008.

References

Afsari, Nealan. 2010. 'What do Somalia, Sudan, Iran and the U.S. have in common?' PBS, 3 December. http://www.pbs.org/wnet/need-to-know/opinion/what-do-somalia-sudan-iran-and-the-u-s-have-in-common/5518/ (accessed 12 March 2018).

Beherman, J. R. 1984. 'Intra-household Allocation of Nutrients in Rural India: Are Boys Favoured over Girls? Do Parents Exhibit Inequality Aversion?' Unpublished manuscript, Department of Economics, University of Pennsylvania, Philadelphia.

Choudhary, Neetu, and D. Parthasarathy. 2007. 'Gender, Work and Household Food Security'. *Economic and Political Weekly*, 42(6): 523–31.

Commissioners of the Supreme Court. 2008. *Eighth Report of the Commissioners of the Supreme Court: A Special Report on the Most Vulnerable Social Groups and Their Access to Food*. http://www.hrln.org/hrln/pdf/rtf/reports/Eight%20report%20.harsh%20tanveer%20final%20aug%2020%2008.pdf (accessed 2 October 2017).

———. 2009. *Ninth Report of the Commissioners*. September. http://www.sccommissioners.org/Reports/Reports/SCC9_0909.pdf (accessed 2 October 2017).

Cowan, B., and J. Dhanoa. 1983. 'The Prevention of Toddler Malnutrition by Home-Based Nutrition Education', in D. S. McLaren (ed.), *Nutrition in the Community: A Critical Look at Nutrition, Planning, Policy and Programmes*. New York: John Wiley and Sons, pp. 339–56.

Deenadayalan, Yamini. 2011. 'It's Not as Simple as Providing Us with Some Cash'. *Tehelka*, 11 October.

Dhar, Aarti. 2011. 'Low income single women demand public funds to live with dignity'. *The Hindu*, 11 October.

Ercilla, M. B., and M. P. Chikoko. 2006. 'Methodology: Nutrition and Food Security Surveillance System'. Action Against Hunger, Lilongwe.

FAO (Food and Agriculture Organization). 1996. 'FAO Celebrates International Women's Day'. Press Release 96/4. http://www.fao.org/waicent/ois/press_ne/presseng/h5f.htm (accessed 1 October 2017).

———. 2006. 'The Right to Food Guidelines: Information Papers and Case Studies'. FAO, Rome.

———. 2009. *Guide on Legislating for the Right to Food*. Rome: FAO.

GoI (Government of India). 1974. *Towards Equality: Report of the Committee on the Status of Women in India*. New Delhi: Ministry of Education and Social Welfare, Department of Social Welfare.

———. 1980. *Sixth Five-Year Plan*. New Delhi: Planning Commission.

———. 1985. *Seventh Five-Year Plan*. New Delhi: Planning Commission.

———. 1992. *Eighth Five-Year Plan*. New Delhi: Planning Commission.

———. 2009. 'Report of the Expert Group to Advise the Ministry of Rural Development on the Methodology for Conducting the Below Poverty Line (BPL) Census for Eleventh Five-Year Plan'. Ministry of Rural Development, New Delhi.

———. 2011. *Socio-economic and Caste Census 2011 in Rural India*. http://www.righttofoodindia.org/data/bpl_census_2011/August_2011_secc_2011_faq_english.pdf (accessed 2 October 2017).

Haddad, L., C. Peña, C. Nishida, A. Quisumbing and A. Slack. 1996. 'Food Security and Nutrition Implications of Intra-household Bias: A Review of Literature'. FCND Discussion Paper no. 9, International Food Policy Research Institute. http://citeseerx.ist.psu.edu/viewdoc/download?doi=10.1.1.29.9058&rep=rep1&type=pdf (accessed 2 October 2017).

Haddad, L., J. Hoddinott and H. Alderman (eds). 1997. *Intrahousehold Resource Allocation in Developing Countries*. Baltimore: Johns Hopkins University Press for IFPRI.

Harriss-White, B. 1997. 'Gender Bias in Intra-household Nutrition in South India: Unpacking Households and the Policy Process', in L. Haddad, J. Hoddinott and H. Alderman (eds), *Intrahousehold Resource Allocation in Developing Countries: Models, Methods and Policy*. Baltimore: International Food Policy Research Institute/Johns Hopkins University Press.

HUNGaMA. 2011. *HUNGaMA Survey Report*. http://hungamaforchange.org/HungamaBKDec11LR.pdf (accessed 25 September 2017).

IFPRI (International Food Policy Research Institute), Bangladesh Institute for Development Studies and the Institute of Nutrition and Food Science. 1998. 'Commercial Vegetable and Polyculture Fish Production in Bangladesh: Their Impacts on Income, Household Resource Allocation, and Nutrition'. Final report, IFPRI, Washington, D.C.

IIPS (International Institute of Population Sciences). 2007. *National Family Health Survey (NFHS-3), 2005–06: India*, vol. I. Mumbai: IIPS.

———. 2009. *National Family Health Survey (NFHS 3): India*. Mumbai: IIPS.

Khera, Reetika, and Nandini Nayak. 2009. 'Women Workers and Perceptions of the National Rural Employment Guarantee Act'. *Economic and Political Weekly*, 44(43): 49–57.

Krishnaraj, Maithreyi. 2005. 'Food Security: How and For Whom?' *Economic and Political Weekly*, 40(25): 2508–12.

Lim, Stephens S., Lalit Dandona, Joseph A. Hoisington, Spencer L. James, Margaret C. Hogan and Emmanuela Gakidou. 2010. 'India's Janani Suraksha Yojana, a Conditional Cash Transfer Programme to Increase Births in Health Facilities: An Impact Evaluation'. *Lancet*, 375(9730): 2009–23.

Mander, Harsh. 2012. 'Barefoot: Lone warriors'. *The Hindu*, 25 February.

Mazumdar, V. 1997. 'Peasant Women, Land Rights and Agricultural Development: Perspectives from the Indian Women's Movement', in N. Rao and L. Rurup (eds), *A Just Right: Women's Ownership of Natural Resources and Livelihood Security*. New Delhi: Friedrich Ebert Stiftung, pp. 375–88.

Mukherjee, Neela, and Amitabh Mukherjee. 1994. 'Rural Women and Food Insecurity: What a Food Calendar Reveals'. *Economic and Political Weekly*, 29(11): 597–99.

Nag, Moni. 1994. 'Beliefs and Practices about Food during Pregnancy: Implications for Maternal Nutrition'. *Economic and Political Weekly*, 29(37): 2427–38.

National Forum for Single Women's Rights. 2011. 'Are We Forgotten Women? A Study on the Status of Low-Income Single Women in India'. National Forum for Single Women's Rights, Udaipur.

National Nutrition Monitoring Bureau. 1999. *Report of Second Repeat Survey (Rural) (1996–97)*. Hyderabad: National Nutrition of Nutrition.

———. 2000. *Report on Food and Nutrient Intake of Individuals (Rural)*. Hyderabad: National Institute of Nutrition.

———. 2006. *Diet and Nutritional Status of Population and Prevalence of Hypertension among Adults in Rural Areas*. Hyderabad: National Institute of Nutrition.

Rao, Nitya. 2005. 'Gender Equality, Land Rights and Household Food Security: Discussion of Rice Farming Systems'. *Economic and Political Weekly*, 40(25): 2513–21.

Right to Food Campaign. 2008. *Supreme Court Orders on the Right to Food: A Tool for Action*. New Delhi: Right to Food Campaign Secretariat.

———. 2012. 'Clause by Clause Critique of the National Food Security Bill, 2011'. http://www.righttofoodindia.org/data/right_to_food_act_data/events/March_2012_clause_by_clause_critique_final_18_february_2012.pdf (accessed 2 October 2017).

Thakur, S. G. 2001. 'Women and Land Rights: Policy and Practice'. Paper presented at the workshop on Women's Empowerment, Policy and Natural Resources, 'What Progress?', New Delhi, 18–19 May.

UN (United Nations). 1980. 'Report of the World Conference of the United Nations Decade for Women: Equality, Development and Peace'. http://www.un.org/womenwatch/daw/beijing/otherconferences/Copenhagen/Copenhagen%20Full%20Optimized.pdf (accessed 12 March 2018).

World Bank. 2001. *World Development Report 2000–01: Attacking Poverty*. New York: Oxford University Press.

14

Vulnerable Groups

Saba Sharma

Even apart from the debates referred to earlier about universal and targeted programmes, there are those who believe that any food guarantee law must contain special provisioning entitlements, often distinct from those for other less vulnerable populations, which deal with special social and economic burdens and barriers to food and nutrition access created by these special disadvantages.

The Government of India under NSFA 2013 made a commitment to combating hunger and malnutrition by significantly expanding the coverage under the PDS of people entitled to subsidised grains, and ensuring access of children and mothers to a range of nutrition services in anganwadi centres and schools, combined with universal maternity entitlements. Yet, lawmakers in the Standing Committee of Parliament dropped a range of entitlements from the final act that had been discussed by policymakers and experts in the NAC in the years since the UPA government had promised to pass a comprehensive food security law, and that had been included in the NAC drafts. These entitlements were formulated with the understanding that the right to food law would be a human rights legislation to protect and ensure access to food for all, while maintaining affirmative interventions for the very poor and the most vulnerable groups, in the spirit of the constitutional right to life with dignity.

The Standing Committee of Parliament that was tasked with perusing the earlier draft of the NFSA recommended collapsing any separate entitlements for vulnerable groups, citing administrative difficulties in identification of homeless/destitute persons as a potent challenge. The Standing Committee additionally qualified the deletion of the clause on vulnerable groups by regarding it as a threat to the social fabric or to the idea of the family, wherein earning members might push old and disabled non-earning members out of their homes in order to feed themselves. During the parliamentary debates in the run-up to the passage of the final legislation, many parties demanded a reintroduction of vulnerable groups; however, no amendments were passed.

This chapter covers the issue of food security among the most vulnerable sections of the population—some might argue that it is for safeguarding their access to food that a right to food law is most important. In this regard, the chapter covers rationales for the protection of these special groups; who these groups might be, particularly in the Indian context; how they may be identified; and finally, provisions made for these groups in the NAC and government drafts as well as the final act, and their critiques.

Human Rights Perspectives on Vulnerable Groups

The FAO asserts that:

> All individuals have an equal right to food, but people's differing circumstances mean that different actions are required of the government in order for that right to be realised for all. Thus, ensuring the right to food of those who are affected by conflicts and other disasters often requires special protection and support measures. Exactly the same is true for groups who are marginalised or vulnerable because of inequalities and other structural factors within a society. A rights-based approach demands that those who are most vulnerable, for whatever reason, be identified and empowered to claim their rights. It equally demands that any discriminatory processes in governance and power structures be detected and corrected. (FAO 2006, 10)

Golay and Ozden (n.d.) equally affirm that while the right to food is universal, it must first and foremost be safeguarded for the most vulnerable populations around the world, such as those suffering from discrimination, women and children, landless rural workers, indigenous peoples and tribes, slum dwellers, the unemployed and others. The non-discrimination principle has been enshrined in identical provisions in the International Covenant on Civil and Political Rights and the ICESCR. In particular, the reports of the Human Rights Council in the context of discrimination in accessing the right to food are extremely relevant. For instance, the report on discrimination in 2010 (Human Rights Council 2010) affirms that merely addressing formal discrimination will not ensure substantive equality as envisaged and defined by Article 2(2) of the ICESCR. The effective enjoyment of covenant rights is often influenced by whether a person is a member of a group characterised by the prohibited grounds of discrimination. Therefore, strategies addressing discrimination, including with respect to the right to food, ought to take a vulnerable and disadvantaged groups approach, as the Parliamentary Standing Committee pointed out. Eliminating discrimination in practice requires paying sufficient attention to groups of individuals who suffer historical or persistent prejudice, instead of merely comparing the formal treatment of individuals in similar situations.

The report notes, for instance, that for the world in general, hunger, like poverty, is still predominantly a rural problem, and among the rural population it is the peasant farmers, small landholders, landless workers, fisherfolk, hunters and gatherers who suffer disproportionately (Human Rights Council 2010). With the size of urban population continuing to rise, the absolute number of urban poor and undernourished also continues to increase.

Other vulnerable groups, according to the study, include the millions of refugees that cross international borders to escape hunger, and are often refused entry and protection because they qualify not as refugees but as migrants. Part of the core content of the right to food with respect to people with disabilities is insufficiently addressed by governments; as underlined by comments to the Standing Committee's preliminary study, the perspective of accessibility changes with each disability and the corresponding measures for the realisation of the right to food of people with disabilities ought to be appropriately implemented. In addition, elderly people with disabilities are often subjected to multiple disadvantages, in respect of physical and economic accessibility and the particularities related to a changed nutrition regime. Indigenous people face exclusion and discrimination that have an impact on their right to food. In Guatemala, for instance, progress in lowering chronic malnutrition rates for indigenous children has been slower than for non-indigenous children. Between 1987 and 2002, the rate for non-indigenous children decreased by 13 per cent, compared to just 2 per cent for indigenous children. Other vulnerable groups in respect of the realisation of their right to food can be people living with HIV/AIDS and other chronic diseases, people deprived of their liberty and those living in conflict areas.

Defining Vulnerabilities versus Listing Vulnerable Groups

The FAO notes, in its *Guide on Legislating the Right to Food* (2009), that providing for vulnerable groups can be done in two different ways—the law can either list vulnerable groups, or it can define vulnerability. While some groups such as indigenous peoples, ethnic, linguistic or religious minorities, persons with disabilities, individuals living with HIV/AIDS, refugees and internally displaced people, women and children would unambiguously be considered vulnerable, other categories of people such as small farmers and casual workers may move in and out of poverty over time. On the other hand, vulnerabilities which may push people into food insecurity may include natural disasters, conflicts, debilitating illnesses, disabilities, unemployment, or a drastic reduction in the market value of the goods or services people produce. The FAO guide goes on to say that the advantage of listing the vulnerable groups is that 'it would force the state to think of entitlements for every single one of these categories and develop them through subsidiary legislation.' However, this method also 'risks leaving out of its coverage persons or groups that may become vulnerable' (ibid.).

The FAO's Food Insecurity and Vulnerability Mapping Systems programme (2000) has developed a methodology for assessing vulnerability. It refers to a range of factors that place people at risk of becoming food insecure. These can be external or internal. External factors include trends such as depletion of natural resources from which the population makes its living; environmental degradation or food price inflation; shocks such as natural disasters and conflict; and seasonality, such as seasonal changes in food production and food prices. Internal factors that determine the capacity of people to cope with difficulties include the characteristics of people themselves, the general conditions in which they live and the dynamics of their households.

The Eighth Report of the Commissioners of the Supreme Court (2008) in the right to food case is a special report that focuses on the most vulnerable groups of the population. In its summary of recommendations, it suggests that by way of identification:

> It is proposed to identify and list the most poor and vulnerable segments of urban populations by identifying them along objective and *verifiable criteria of vulnerability and denial of rights*. These are: a) place of residence and access to public services (shelterless, unauthorised slum dwellers, authorised slum dwellers and residents of resettlement colonies); b) social vulnerability: children without protection and child-headed households, single women and single women headed households, disabled people and family with persons with disabilities, old people without care givers, people in destitution; c) vulnerable occupational categories: such as rag pickers, casual daily wage workers, rickshaw pullers, porters, construction workers, street vendors, domestic helpers etc; and d) affirmative action categories: Scheduled Castes/Scheduled Tribes, Muslims. (Ibid., 8; emphasis added)

The report suggests that the categories be listed fully, and that the list should be revised every two years. It is also in line with the FAO's guideline of mapping vulnerabilities and the factors that lead up to vulnerability, rather than merely listing them out.

Role of the PDS in Ensuring Food Security of Vulnerable Groups

There were many who believed that universal access to the PDS must be guaranteed to ensure that the most vulnerable can achieve food security. Their contention has been that making separate legal entitlements specifically for the vulnerable sections of the population is likely to hinder their access to food security, rather than facilitate it. Targeting, which is required for providing separate entitlements for the vulnerable, is ridden with errors of inclusion and exclusion, i.e., the inclusion of rich households in a targeted PDS and exclusion of poor households from a targeted PDS, respectively. While targeting captures the status of poverty only at a given point in time, households and individuals may move in and out of poverty at any time. However, many who advocate a universal PDS—such as the left parties—are sometimes silent about the other kinds of nutritional support (such as community kitchens and pensions) required for vulnerable populations (although Brinda Karat, Politburo Member of the CPI[M], proposed an amendment to the NSFA for including community kitchens providing subsidised cooked food as entitlements). Member of Parliament Kanimozhi (Dravida Munnetra Kazhagam) asked in Parliament: 'Will the hon. Minister think of making sure that even homeless people are brought under this Bill?'

Some were of the opinion that a universal PDS, which supplied the same quantity of rations to everyone at the same price, went against the principle of affirmative action for the vulnerable sections of the population. Mander (2011a) argued that the universal principle must be combined with the idea of affirmative action:

> People vary hugely in their ability to access sufficient food. There is enormous difference in food denials and vulnerability between households, based on divergent conditions of wealth, gender, caste, community, age, disability, and a range of other factors.... What is the justification to give you and me, and for that matter a rich rural landlord subsidised grain, and that too at the same levels and price, as a destitute rural landless widow, or a homeless disabled beggar?

Some strong proponents of universal PDS went further to argue that a food security act needed to do little more than providing access to a universal PDS. All resources—financial, grains, administrative capacity—should be geared to this single goal. Small special programmes for specifically identified vulnerable groups are difficult to administer, attract disproportionate time and resources, and in many ways would become infructuous, and irrelevant, if a universal PDS were adopted. These non-PDS programmes have been considered diversionary and an excuse to move away from universalising the country's food rations programme.

On the other hand, the majority, including the Right to Food Campaign, the NAC and the government were of the opinion that apart from the PDS, additional entitlements such as cooked meals and social security transfers such as pensions and maternity benefits were also required to meet the special food requirements of vulnerable groups. However, there has been debate on the scope and importance of these entitlements.

The Role of Destitute Feeding Centres and Community Kitchens

The NAC and the government drafts entitled destitute persons to one meal every day, free of charge. They also entitled homeless persons to affordable meals through a scheme of community kitchens. The drafts did not elaborate the nature of these programmes, or on questions like whether or not there would be gatekeeping in destitute feeding centres and community kitchens to cater to these categories of people. In the act, provisions for the homeless and destitute were removed completely.

One view was that everybody should be entitled to the meals provided at the destitute feeding centres and the community kitchens. This view is based on the assumption of self-selection in such schemes. The self-selection can be aided by locating the feeding centres and community kitchens in areas which have a higher density of destitute and homeless populations. The other view propagated gatekeeping at the centres, possibly by issuing identification cards to the users. However, this is likely to be an extremely time-consuming and cumbersome process with the possibility of excluding many destitute and homeless persons.

For the purpose of destitute feeding, it was suggested that the following categories of people were to be considered as destitute:

- Persons above 60 years of age
- Persons with disabilities and infirm adults
- Persons with leprosy
- Widows
- Single-women-headed households

- Households with no able-bodied male members
- People living off alms
- Persons with no support system
- Households with no place to cook
- Any other persons identified as 'destitute' and in need

For rural areas it was suggested that the gram sabha and a village level officer would be jointly charged with verifying the eligibility of 'any other persons identified as "destitute" and in need' to be included in the scheme of destitute feeding. Categories of people expected to eat meals at the community kitchens included manual labourers such as daily wage workers, construction labourers, rickshaw pullers, load pullers, ragpickers and street vendors.

The final act, as already noted, does not provide any support for community kitchens, however. This was in conformity with the views of the Parliament Standing Committee, which noted that the bill provided that all destitute persons should be entitled to at least one meal every day free of charge, and all homeless persons should be entitled to affordable meals at community kitchens, in accordance with such schemes, including cost sharing, as may be prescribed by the central government. The committee felt that it would be difficult for the administration to identify destitute and homeless persons who may be given such benefits under the provisions of the bill. It quoted Naveen Jindal, MP: 'It is difficult for the administration to verify the claim of being homeless or being a destitute. It runs the risk of breaking the social fabric as non-earning members of the family (elderly & disabled) might be pushed out of homes to feed for themselves' (Standing Committee on Food, Consumer Affairs and Public Distribution 2013, 103).

In Parliament, MP Anto Antony from the ruling Congress Party called for reintroducing the deleted sections for destitute and vulnerable populations:

> However, the present Bill is limited in its scope if we compare with the Food Security Bill-2011 introduced in the Lok Sabha in 2011. If the Government re-insert the sub-sections 3, 4 and 9 of section 2 as well as sections 8 to 12 of the Bill to the present Bill, then the destitute, disaster affected and homeless persons respectively will entitle to get the benefits of the Food Security. (Parliament of India 2013, 297)

The next section discusses the disadvantages faced by some vulnerable groups and debates on how best to ensure their food security.

Vulnerable Groups, Their Disadvantages, and Debates on How to Ensure Their Food Security

The Aged

Due to their diminished capacity to work because of old age, the aged are able to earn less than younger generations, which not only threatens to decrease their access to food, but also makes them more vulnerable to poverty, homelessness, untreated illness and violent abuse, which in turn further undermines their right to food. Hence, any effort to ensure a population's right to food has to take special measures to ensure the food security of the elderly. The existing measures to provide support to the elderly include the National Old Age Pension Scheme. This scheme was started in 1995 to provide old age pensions to senior citizens (aged 65 years or above). Although this scheme is primarily addressed to those elderly without any assured means of subsistence, the eligibility conditions as well as the coverage of the scheme vary from state to state. The pension amounts include the contribution of the central government which is currently Rs 200 per month, which is often supplemented with a contribution from the state government.

The scheme is currently ridden with several problems, with one of the main problems being its limited coverage—there are plenty of applications for the pensions, but funds are limited. Although the Government of India gave directions to cover all BPL individuals above 60 years of age under this scheme, this order is yet to be fully implemented by all state governments. Many state governments link eligibility under this programme to those BPL individuals who are not 'supported' by other family members, further restricting the outreach of this scheme. The cash grants disbursed by the central government are often 'diverted' by state governments for other purposes, or released after long delays. While the NAC draft aimed to provide pensions for the aged, the government drafts and final act moved pensions for the aged to the legally non-binding portion.

Out-of-School Children

It is argued that all children have fundamental rights to education, health and nutrition, and these rights are so closely interlinked that they should not be seen in isolation. In the Indian context where the family is often not a reliable arrangement for ensuring these rights to children, given the circumstances in which most people live in the country, there is a case for looking at the school as the best institutional means of protecting the rights of the child (Drèze 2004). In 2009, 80.4 lakh (8.04 million) children in India (4.22 per cent of the total children in the country in this age group) were out of school. Although the proportion of out-of-school children is reducing steadily, due to legislation such as the RTE Act, the number of such children in India is still substantial (Mukul 2009). Out-of-school children are anyway deprived of their fundamental right to education, for various reasons such as their families requiring their help in household chores, raising small children or working to earn money. However, by preventing them from accessing the midday meals at schools, their right to food is also being violated.

The NAC draft of the NFSB entitled all children below the age of 14, including those who were out of school, to approach any feeding centre such as an anganwadi centre, school midday meal or destitute feeding centre for a freshly cooked nutritious meal. The draft directed that under no circumstance could any such institution deny a freshly cooked nutritious meal to any child. The Right to Food Campaign and many others advocated for including out-of-school children in midday meals as well. However, all the subsequent government drafts as well as the final act removed this entitlement.

The argument against the conditionality of school attendance for a child to access a midday meal is similar to the conditions for pregnant women to access maternity benefits. If a child is not in school, but is working, she is not to blame. The poverty and desperation of her family and the failures of state educational provisioning are responsible. It is unjust then to withhold the right to food from a child who is already denied the right to education. It is a double whammy against a child who is most deprived, and a situation of blaming the victim. There is no ethical justification to use a right as an instrumental incentive. The right to food of the child is absolute, and cannot be conditional. The state must fulfil its duty to ensure that the child is able to access her right to education, but without withholding for this purpose the child's other fundamental rights, such as the right to food.

The Right to Food Campaign demanded that the NFSA include clauses which entitled children with special needs to appropriate meals taking into account their specific dietary requirements. It also demanded appropriate measures to ensure that children who did not have access to meals through anganwadis were not denied their rights under the legislation. The campaign wanted a reinstatement of the NAC entitlement of not denying a hot cooked meal to any child up to 14 years of age. But neither the NAC draft or government draft, nor the final act, had any special provisions for extremely vulnerable children who are marginalised, such as children without adult protection, children living on the street, working children, children of migrant and homeless parents and disabled children. Marginalised children face a double deprivation of food: not only do they have greater food insecurity in their families, if they have any, but they are also often not enrolled in anganwadis or schools, and thus they are also deprived of the food provided at these institutions.

People Suffering from Starvation

The NAC draft provided comprehensive entitlements for persons suffering from starvation. These included assistance in the form of cooked meals, higher maternity benefits for pregnant and lactating women, greater quantities of foodgrains and guarantees for wage employment. The draft also obliged state governments to proactively identify persons in starvation and enquire about the extent of their food denials. According to this draft, state governments were required to establish protocols for preventing starvation, providing relief in case of starvation, investigating starvation deaths and assigning accountability. These entitlements have been excluded from the purview of the law. More information on people suffering from starvation can be found in Chapter 9 on starvation.

Those Affected during Emergency Situations, Disaster and Conflict

In Parliament, MP Bharatsinh Prabhatsinh Parmar (BJP) declared: 'There should be a provision to provide the foodgrains in the event of a natural calamity such as flood, earthquake etc.' (Parliament of India 2013, 377). Members of Parliament from the left parties shared his concern. Prabodh Panda (CPI) said:

> No food security has been ensured at the time of war, no food security has been ensured at the time of national disasters and no food security has been ensured at the time of natural calamities. If there is no security of food grains at the time of national disasters, at warlike situation and at the time of natural calamities, what is the meaning of food security? What will you do to the destitute people of our country? All these problems are there. (Ibid., 271–72)

Saidul Haque (CPI-M) similarly stated: 'Since large parts of India perennially affected in conditions of floods or droughts or sometimes both, the food security is needed most at that time.'

The government indeed has a responsibility to prevent hunger during emergency and disaster situations. The NAC draft entitled persons affected by emergency or disaster to subsidised foodgrains and 200 days of wage employment at minimum wages (or equivalent income in case wage employment is not available or the family members are unable to work for wages) for a period of one year, and two freshly cooked meals for the vulnerable sections of the population for three months after the disaster. These measures are crucial for preventing hunger and related mortality during difficult times. However, the NFSA leaves these entitlements out of its ambit.

Migrant Men, Women and Children

Migrants constitute a large number of the destitute and homeless persons in cities across India. These men, women and children often adopt circular patterns of migration, from the countryside to metropolitan cities and back, and may usually be concentrated in certain vulnerable occupations and shelterless environments. These individuals, in their struggle for survival, experience significant distress in urban cities, where they are also exploited as cheap labour, if not harassed as beggars. In such circumstances, they become vulnerable to multiple deprivations in the wake of difficulty in accessing social security programmes and basic facilities of food, health, housing and education. In the absence or denial of such rights, a large number of homeless migrants die of starvation on the streets of cities. This highlights the inadequacy and inefficiency of private and state urban feeding programmes seeking to address food insecurity among the urban poor (CES 2007, 2010).

It is crucial for the state to recognise the portability of socio-economic rights for migrants, who are seasonally on the move; they should be able to access their entitlements unconstrained by their physical location. In addition, as already noted, subsidised meal programmes supported by public funding could indeed have become an important intervention to raise the nutrition status of urban homeless women, men and children. It would have also freed up a good portion of their daily incomes which they are forced to invest in relatively expensive street food, which is low in nutrition and hygiene. However, as we noted, although such kitchens providing subsidised hot meals were part of the NAC and government drafts, they were dropped on the recommendation of the Parliament Standing Committee.

'Single Women'

There is little to no policy data and numbers available on the plight of 'single women' in the society, and their omission from the budgetary plan is pronounced. This translates into invisibilisation of the population of single women in the society at large. 'Single women' are not a homogeneous group in the marginalisation they experience. Women who are widowed, divorced, unmarried or separated differ in the degree of impoverishment and disadvantages they encounter in everyday life. A certain commonality of experience may exist; however, their experiences uniquely differ across regions, historical contexts, cultures, religion and identity. Most often, a context of differing but multiple deprivations prevents single women from exercising their rights as equal citizens. Most single women face various social restrictions and find it extremely difficult to secure gainful employment. Even when they are able to find jobs, they may be paid very low wages and forced to work for long hours. Single women are often unable to access credit, even if they have the same earning capacity as a man or a married woman. To sustain themselves and their children, single women often have to mortgage or sell their assets in distress, or resort to undignified options such as sex work, begging and sending their children to work. In the face of extreme social ostracism, sole breadwinning responsibilities, lack of dignified livelihood choices and lack of family or community support, a gamut of social security entitlements, including food security, should be prioritised for women who are single and their dependants.

The drafts mentioned single women only in the context of the legally non-binding sections, and while pensions for single women are part of the revised National Social Assistance Programme, these were not included as legal entitlements under the NSFA, in the way that old age (and disability) pensions were also excluded.

Persons with Disability

Poverty and social exclusion are crucial factors affecting the rights of persons with disabilities. Disability as a social vector of exclusion intersects with other traditionally socially excluded groups such as the Dalits, adivasis, and Muslims to create a complex matrix of vulnerability in the Indian context. Further, even though there is a high level of malnutrition in the country, the NFSA of 2013 does not link the rights of the disabled to food security in India. Given the educational, social and physical barriers faced by persons with disabilities, it is perhaps not surprising that disabled persons are at a significant disadvantage with regard to employment. From an equity lens, destitute disabled women and men, single women who are disabled, lower-caste disabled, homeless disabled, and aged disabled women need specific social protection over and above any blanket social security given to all the disabled.

Once again, although pensions for the disabled were part of initial NAC discussions, these were excluded from the final NSFA although they are part of the revised National Social Assistance Programme of the Government of India.

People with Debilitating and Stigmatised Ailments

Persons with debilitating ailments like TB and HIV/AIDS have special nutritional needs. In addition, these persons, and those with stigmatised ailments like leprosy, have difficulty securing decent livelihoods

(Debnarayan 2011). Early NAC drafts debated special nutritional support for persons with these and similar ailments, but this was dropped from later drafts, both on grounds of costs and administrative difficulties in actually implementing such entitlements.

References

CES (Centre for Equity Studies). 2007. 'Living Rough'. CES, New Delhi.

———. 2010. 'Eating Rough'. CES, New Delhi.

Commissioners of the Supreme Court. 2008. *Eighth Report of the Commissioners of the Supreme Court: A Special Report on the Most Vulnerable Social Groups and Their Access to Food*. http://www.hrln.org/hrln/pdf/rtf/reports/Eight%20report%20.harsh%20tanveer%20final%20aug%2020%2008.pdf (accessed 2 October 2017).

Debnarayan, Sarker. 2011. 'Social and Economic Implications of HIV/AIDS: Evidence from West Bengal'. MPRA Paper no. 33648. http://mpra.ub.uni-muenchen.de/33648/ (accessed 15 October 2017).

Drèze, Jean. 2004. 'Democracy and the Right to Food'. *Economic and Political Weekly*, 39(17): 1723–31.

FAO (Food and Agriculture Organization). 2000. 'Food Insecurity and Vulnerability Information and Mapping Systems (FIVIMS)'. http://www.fao.org/3/a-x8346e.pdf (accessed 12 March 2018).

———. 2006. 'The Right to Food in Practice: Implementation at the National Level'. http://www.fao.org/docrep/016/ah189e/ah189e.pdf (accessed 12 March 2018).

———. 2009. *Guide on Legislating for the Right to Food*. Rome: FAO.

Ghosh, Jayati. 2011. 'Securing Food for the People'. *Budget Track*, vol. 8, track 1, January.

Golay, Christophe, and Melik Ozden. n.d. 'The Right to Food: A Fundamental Human Right Affirmed by the United Nations and Recognized in Regional Treaties and Numerous National Constitutions'. Part of a Series on the Human Rights Programme of the Europe-Third World Centre (CETIM).

Harriss-White, Barbara. 2002. 'A Note on Destitution'. Working Paper no. 86, QEH Working Paper Series, University of Oxford.

Human Rights Council. 2010. 'Study on Discrimination in the Context of Right to Food'. Prepared by the drafting group of the Advisory Committee on the Right to Food, Sixth Session, A/HRC/AC/6/CRP.1

Khera, Reetika. 2009. 'Right to Food Act: Beyond Cheap Promises'. *Economic and Political Weekly*, 44(29): 40–44.

Mander, Harsh. 2011a. 'Ending Indifference: a Law to Exile Hunger'. *Economic and Political Weekly*, 46(25): 45–50.

———. 2011b. 'The Middle Path on the Food Security Bill?' *Infochange*, August.

Mukul, Akshaya. 2009. 'Sharp decline in out-of-school children'. *Times of India*, 13 September.

Parliament of India. 2013. *Compendium on Parliamentary Enactments: The National Food Security Act, 2013*. New Delhi: Rajya Sabha Secretariat.

Right to Food Campaign. 2011. 'Right to Food: An Open Letter'. *Economic and Political Weekly*, 46(33): 21–23.

Saxena, N. C. 2009. 'Hunger, Under-nutrition and Food Security in India'. Working Paper no. 44, Chronic Poverty Research Centre, Indian Institute of Public Administration, New Delhi.

Standing Committee on Food, Consumer Affairs and Public Distribution. 2013. *The National Food Security Bill, 2011: Twenty-Seventh Report*. http://www.prsindia.org/uploads/media/Food%20Security/SC%20Report-Food%20Security%20Bill,%202011.pdf (accessed 12 March 2018).

15

Reforming Delivery

SAJJAD HASSAN

Any attempt to deliver food and nutrition without investing adequate time and resources into considerations of programming of food security, development of standards of public service delivery, and administrative and institutional questions of planning, implementing and monitoring, among others, is most likely to be a wasted effort. In India, it is the concern around the efficient and transparent delivery of food entitlements that has been the crux of much of the debate around the NFSA, 2013. Retrospectively, it was the poor functioning of food delivery mechanisms (especially the PDS)—in a context of widespread food insecurity, including starvation—that chiefly triggered the public mobilisation around the demand for 'right to food'.

Mobilisation for 'right to food' in India—both outside and within government—thus aimed both at creating securities and rights that could be defended and enforced in courts, as well as strengthening and reforming state-led food security schemes that would deliver those rights. The chief among these programmes is the PDS, although the ICDS and MDMS are equally significant in terms of their scale and reach, as are the National Social Assistance Programme and the Destitute Feeding Programme, both of which, although comparatively small in scale, potentially have a huge impact on the security of those most in need of food.

DEBATING 'REFORM'

This evolving debate on reforms has been the outcome of extensive debate and discussion, by various stakeholders, around the national food security legislation. The first was the debate over 'cash transfer' or 'food subsidy', with supporters of the former rooting for direct fiscal transfers to beneficiaries (or the issue of redeemable food coupons), to bypass the PDS system they considered too irremediable to fix. The latter considered food, and not cash, as essential to the social protection effort, especially in a country with problematic intra-household income distribution dynamics, and a yet-to-stabilise economy with runaway inflation. On balance, the comparative benefits of food subsidy (over cash transfer) are recognised by policymakers in India, although with the condition that the PDS system—the heart of the food subsidy system in the country—be thoroughly reformed.

Conservative economists and planners, government agencies, as well as right to food campaigners all agree that reform of PDS is essential for delivery of food rights, since the inefficiencies, wastage and corruption is unsustainable. In the debates around the legislation, economists and government think tanks understood 'reform' primarily as smart cards, which could be used in any shop of one's choice (public or private), with the subsidy directly sent to the retailer's (or the purchaser's) bank accounts; end-to-end computerisation; or use of food coupons. For right to food campaigners, 'reform' implied a wider basket of interventions, leveraging the power of information technology (IT) for monitoring and oversight, but also ensuring greater community and beneficiary involvement in planning and management.

The draft NFSB of 2011 brought out by the NAC emphasised, among other things, decentralised procurement to minimise transport costs; support to local PDS models, including grain banks; giving licences to community institutions, self-help groups, local cooperatives, panchayats and NGOs; reliance on women as FPS floor managers; clarifying duties and fixing responsibilities on FPS shops for effective delivery of PDS services; and strengthening monitoring and supervision of FPS shops by local authorities, and community monitoring tools such as social audits and public grievance redressals.

Under reforms to the PDS, the NFSA 2013 creates a space for leveraging the Aadhaar system for unique identification, with biometric information of entitled beneficiaries for targeting of benefits as well as for introducing new cash transfer and food coupon schemes in place of foodgrain entitlements. The Standing Committee, to which the bill was referred after its introduction in the lower house of Parliament in 2011, however, recommended that cash transfer not be introduced till adequate banking infrastructure was made accessible throughout the country. It further suggested specific IT reforms such as CCTV cameras in godowns, use of internet, and GPS tracking of vehicles carrying foodgrains. In a welcome departure, the Standing Committee acknowledged the trend in supply of poor-quality food across many regions and suggested that consumers be given the right to refuse grains which were below the prescribed minimum quality standards.

Debates in Parliament before the passage of the act saw both opposition parties and the allies of the ruling Congress call for serious reform of the PDS. While the opposition parties used this occasion, without proffering any concrete suggestions, to attribute the failures and inadequacies of the PDS to the incumbent government, the allies,[1] all in favour of the bill, harped on the need for restructuring, revamping and modernising the PDS as a way forward.

Learning from Other States

It is true that while greater fine-tuning with respect to robust institutional delivery, implementation and monitoring of the schemes is a prerequisite to the success of the law on the ground, small and bold steps have been taken, some by central government, others by states, towards just ensuring that these programmes deliver for the citizens.

Noteworthy among these steps is the self-initiative of the Government of Chhattisgarh in passing the Chhattisgarh Food Security Act, 2012, and the state's introduction of reforms to its PDS system. Summarily, these reforms are rooted in reworking the logistics management of the schemes, relying on the use of IT to create an effective management information system as well as allowing for the participation of citizens in supervising the working of the programme. Specific interventions include process re-engineering, supply chain computerisation, digitisation of the beneficiary database, FPS automation, and robust monitoring and grievance redress mechanisms. The reformation of PDS in Chhattisgarh—focusing mainly on 'end-to-end computerisation' to improve efficiency of operations and transparency in management—is an experience that the central government is seeking to replicate in other regional states in India.

Other food security schemes too have seen their share of reform, almost all triggered by mobilisation carried out by the Right to Food

Campaign. States like Madhya Pradesh and Tamil Nadu and parts of Maharashtra have tried to reform the ICDS scheme. In Maharashtra, the reform of ICDS has relied on co-production, with self-help groups of local women as providers of hot cooked meals, as the basis for ensuring effective management of the ICDS. The engagement of self-help groups of local women meritoriously translates into community participation and monitoring, creates incentives for effective functioning and reduces leakages as well as wastage in the programme—a stark contrast to contractor-driven models of supply of ICDS meals. Hot cooked meals are locally procured and produced, ensuring both local food preferences and higher attendance. This also organises and empowers collectives of local women, many of whom go on to expand their spaces beyond production to management of nutrition programmes.

The successful MDMS school nutrition programme is another example of a national food programme where innovations and initiatives by states have resulted in significant improvements in programme performance. The MDMS success narrative has multiple variants, with some states going for centralised logistics management—large contractors providing food to schools, and others using the decentralised route. In the latter, hot cooked meals are prepared in schools by school staff and served to children directly. Both systems have their advantages; however, the common concern running through both is professional management (or in its place, direct local control); management of the supply chain; and effective monitoring and supervision to ensure compliance with quality standards.

Another example is the Destitute Feeding Programme in Tamil Nadu (or its variants, the Emergency Feeding Programme in Odisha, and Aap ki Rasoi [Your Kitchen] in Delhi), meant as a fallback for those unable to fend for themselves—the old and infirm, widows, homeless and street children. These programmes, although small in scale compared to programmes like PDS, ICDS and MDMS, all rely (or aspire to rely) on professional management and supply chain management. While the programmes in Tamil Nadu and Odisha piggyback on the ICDS or MDMS infrastructure, the Delhi programme is outsourced to private contractors providing food centrally at designated locations across the city. This has been expanded now to provide good-quality food, at reasonable price (under a scheme called Jan Aahar—'public food') through the use of professional caterers.

Lessons Learnt

What are the lessons for programme delivery emerging from these successful examples of reforms? Common to them are some recurring themes, all related to management functions, which anyone planning a large programme of public food provision would do best to bear in mind.

Awareness Generation

Without requisite information and awareness of the specific entitlements being provided, their quality and quantity, and how these entitlements might be accessed, it is difficult both for delivery agencies to deliver effectively, and for rights bearers and all those who speak on their behalf to hold the delivering agencies to account for failures of omission or commission. Successful reforms of PDS or ICDS have relied on creating awareness about entitlements (how much food or related services, at what periodicity, to whom, for example), as also about service standards around those entitlements, to push through those reforms. There are variants of this, and while the articulation might be couched in different language—rights-based 'entitlements', or more populist 'citizens' charters'—the thrust has been the efficacy, regularity and quality of service delivery for citizens.

Specifying Roles and Responsibilities

Equally important seems to be the clarity of roles and responsibilities of service providers, ensuring that every level of management of the concerned agencies within the supply chain has its specific tasks and responsibilities clearly spelt out. This is important for two related reasons: (*a*) there are multiple agencies, from national to state to local (village or town) levels, involved in management of these mega schemes, and roles and responsibilities tend to overlap; and (*b*) without clarity it is difficult to ensure that the supply chain delivers the rights promised to the bearers of those rights. Most successful food programmes have inbuilt in them clear job descriptions of all positions in the organisational structure, as also roles and responsibilities.

Last-Mile Delivery

This is a crucial change, and something on which much of the success of overall reform rests. The organisational architecture of food security programmes in India—PDS, ICDS, MDMS and the like—is complex, with central and state governments, district agencies and local authorities playing their respective roles. The 2013 NFSA mandates doorstep delivery of foodgrains as one of the crucial reforms to PDS. This last-mile, doorstep delivery to the citizen depends on the efficient and harmonious synergy among various actors in this rather complex process. Systems and roles and responsibilities, not only of agencies but also of individual positions within each of those agencies, are thus crucial for delivery, and for fixing responsibility in case of failure.

Technological Upgradation

Job descriptions are of little consequence to effective delivery of services without addressing the fundamental questions of efficiencies through supply chain reforms and process re-engineering, among others. Use of IT has been found to be a useful ally in this effort—to ensure timely flow of information up and down the chain, as well as to citizens, contributing also to effective governance of the schemes. The illustrative experience of PDS reforms in Chhattisgarh uses the IT route to effect reforms for effective performance. The change was predicated on process reforms and supply chain management, as well as end-to-end computerisation, all of which resulted in ensuring that systems and processes, and lines of communications, were reorganised and strengthened. These also helped fix the roles and responsibilities of individual actors in the supply chain architecture.

Pre-budgeting and Resourcing

Governments must naturally commit adequate resources if they want to see the programmes working effectively. Since food security schemes, in India, are funded jointly by the centre and states, it is often the case that due to less-than-adequate coordination between the two, either the funding required is not adequately budgeted, or poor and inefficient financial management systems lead to significant delays in the release of funds in an ad hoc way. This of course impacts procurement and provisioning. Budgeting right, to provide food to all those entitled to food, and making funds available on time, often in advance, is a prerequisite for effective delivery of services. This may also require creating special measures so that funds for food security programmes are ring-fenced, either through special vehicles or through clear earmarking in budgets, and are not vulnerable to weaknesses of the financial management systems in the states.

Decentralisation

Considering that delivery of food services is a complex business with national, state and local actors playing their part, overcentralisation of powers and authority and finances at higher levels would only work to create a system that is rigid and not designed to address local needs and exigencies on the ground. Moreover, fixing responsibility at the local level or on the frontline functionary, without empowering that position to make appropriate decisions, would be a futile exercise. This, of course, needs to be balanced with checks and controls so that decentralisation does not result in lack of accountability. Useful examples of decentralisation in the context of food delivery are the Emergency Feeding Programme in Odisha, and Tamil Nadu and Bihar states, among others, where a reserve stock of rice is kept at the disposal of village councils, to be used for meeting emergency food shortage needs in the village. A variant of this programme is the food bank scheme of the Government of India, which enables a stock of rice to be stored at strategic locations in areas of food shortage, and is operated by local councils to be used according to the need of villagers. The fact that this scheme is not working as well as it should be is entirely due to its poor launch and management, and not intent.

Participant Monitoring

It is how the monitoring systems are designed, and how they work to ensure effective delivery, that determines the difference between efficacy of any schemes. Traditional monitoring systems in PDS are input based, relying on information about food procured and distributed and money spent to report on programme success. Considering that all this operates in a closed loop, with all data inputted from within and no feedback on performance taken from outside the supply chain system, it is not difficult to see why monitoring in PDS has failed to provide senior managers with effective data on how the scheme is working for those who are its beneficiaries.

The Chhattisgarh reform model essentially makes two changes to this traditional monitoring system. The most significant is of course the use of IT, and it is significantly changing the monitoring environment by providing managers real-time data, for instance, on logistics movement from warehouses to point-of-sale FPS shops. And by bringing the citizen within this reporting loop, even though on the margins, by sharing logistics movement data with them, the attempt is to recruit allies, in the form of interested beneficiaries, to the monitoring cause. This creates incentives for the monitoring system to work for the benefit of the users. Bringing citizen incentives centre stage to ensure effective functioning of schemes, of course, requires going beyond just technically bringing citizens in, to making them the real drivers of monitoring programme implementation by using tools such as social audit.

Capacity Building

Food security programmes, being targeted specially at the poorest and most needy, require, besides adequate reach and efficient management, also sensitive handling by those at all levels of the delivery chain. In practice, precisely because these programmes are meant for the poorest, there is little investment of resources, talent or sensitivity in their organisation. Systems work only as well as the people working those systems. Ensuring that there is adequate strength and reach of the food delivery infrastructure to effectively deliver to all citizens must be only the first step. Ensuring that staff at all levels of the organisation are adequately trained in the technical and managerial aspects of food delivery is the next step. But this capacity effort will be incomplete without leadership positions in these organisations, as well as frontline providers, being committed to delivering entitlements as rights.

Note

1. Chiefly the CPI(M), the Telugu Desam Party and the Bahujan Samaj Party.

PART IV

ENFORCEMENT AND TRANSPARENCY SYSTEMS

16

Enforcement and Grievance Redressal

Amod Shah

The promise of any socio-economic legislation is defeated unless its enforcement can be ensured. In this regard, the right of the victims to seek redressal for violations of their rights assumes significance. Within the Indian Constitution, there is an inherent hierarchy of civil and political rights over socio-economic rights, which have mostly been relegated to the non-enforceable section on directive principles. Increasingly though, many of these socio-economic rights have been interpreted by the Supreme Court as being part of the fundamental right to life under Article 21 of the Constitution, thereby making them justiciable in a court of law. Prominent instances where the Supreme Court has played a vital role in facilitating the justiciability of socio-economic rights include a 1993 ruling that the right to education is a fundamental right (*Unnikrishnan v. State of Andhra Pradesh*), and its orders in the right to food case (*People's Union for Civil Liberties v. Union of India*), which reiterated that the government has a responsibility to ensure food security for all its citizens, and, in particular, for vulnerable persons who are denied a life of dignity as they cannot even afford three nutritious meals a day.

Partly in response to such judicial interventions, the Indian legislature in recent years has passed a number of laws formalising the provision and protection of socio-economic rights like employment and education and, most recently, in 2013, the right to food. The term *justiciability* refers to the ability to claim a remedy before an independent and impartial body when a violation of a right has occurred or is likely to occur (International Commission of Jurists 2008). Therefore, as more and more socio-economic rights move from being non-enforceable or aspirational to concrete justiciable rights, it is essential that effective mechanisms to remedy and redress violations of these rights are also put in place. Such remedies need not necessarily be judicial in nature; in some cases, an administrative approach may be cheaper, speedier and more accessible to individuals than formal court proceedings. Irrespective of the mechanism utilised, any body tasked with grievance redressal must be independent and impartial in its functioning, and have the power to effectively act on complaints.

Overview of Grievance Redressal Mechanisms for Socio-economic Rights

Traditionally, three primary mechanisms have been adopted for redressing grievances in the case of violations of socio-economic rights.

Administrative Remedies

In an administrative remedy framework, any administrative decision which violates the provisions of a right, such as the right to food, can be challenged before the higher administrative authority. This competent higher authority should have the power to impose all measures it deems necessary for redressing the violation. For example, Mexico's 2004 Law on Social Development, which guarantees a host of socio-economic rights including the right to food, provides for the 'right of any person or civil society organization to make a formal complaint for any act or omission which prejudices or could prejudice the exercise of any of the rights established by the Law', before the State Inspectorate. The draft right to food legislations in South Africa, Nicaragua and Peru also include administrative mechanisms to redress violations and punish errant officials (FAO 2009).

Administrative remedies in India typically involve the bureaucracy, made up of civil servants appointed by the central and state governments. For example, the right to public services legislation, implemented in a number of Indian states in recent years, includes a grievance redressal structure wherein appeals against non-delivery or rejection of certain public services can be made before higher administrative officials such as the subdivisional officer and the district collector.[1] Similarly, under the Supreme Court's 'interim orders' in the ongoing right to food case, district collectors are responsible for addressing complaints against agencies implementing the relevant schemes and ensuring compliance with the interim orders at the district level (Right to Food Campaign 2008).

Elected Executive Remedies

In India, there is significant decentralisation of executive power to elected bodies of local representatives, particularly in rural areas. Such local self-government bodies, called panchayati raj institutions (PRIs) in rural areas and municipal corporations or councils in urban areas, are empowered to implement and monitor a wide range of economic and social development activities. For example, panchayats have been designated as the primary authority for planning and implementation of schemes related to MNREGA; their responsibilities include the approval, implementation and monitoring of projects undertaken under the act, allocation of employment opportunities among applicants, and the sanctioning and disbursal of unemployment allowances.

In the context of socio-economic rights, PRIs also play a role in grievance redressal of violations of such rights. For example, under the RTE Act, panchayats and municipal bodies are responsible for addressing complaints regarding violations under the act. Similarly, the Supreme Court's interim orders in the right to food case empower village-level gram sabhas to conduct social audits of the food and employment schemes covered under the ruling. Gram sabhas also conduct social audits of projects undertaken under MGNREGA.

Judicial Remedies

In most countries, certain fundamental rights are protected under the constitution and citizens have access to the judiciary if these rights are violated. While the Indian Constitution does not explicitly guarantee socio-economic rights, many such rights, including the right to food, have been interpreted by the Supreme Court as being part of the fundamental right to life, thereby providing for recourse to the courts in the event of a violation.

Depending on the country's legal framework, courts operating under such a judicial remedy framework may seek to provide adequate redressal in various ways, such as restitution of the right, cessation of the violation, rehabilitation, compensation, or ordering broader, systemic remedies. Moreover, since violations of the right to food can potentially have grave consequences, it is desirable that courts have the power to issue interim relief measures before a final judgment on whether the right has been violated (FAO 2009).

The draft right to food legislation in Honduras includes provisions for a person or group to approach the judiciary for a remedy if they believe that their right to food has been violated or is at imminent risk of being violated. In certain cases where a violation has not yet been established, the legislation empowers judicial authorities to adopt interim measures to ensure fulfilment of the right and prevent a violation (FAO 2009). Interim orders aimed at ensuring food security for vulnerable groups like children, pregnant mothers, the aged, and economically and socially backward sections of society have also been issued by the Indian Supreme Court in the ongoing right to food case.

Concerns around Grievance Redressal Mechanisms for Socio-economic Rights

There exist a number of concerns around the use of grievance redressal mechanisms which rely heavily on the administrative bureaucracy, elected executive bodies or the judiciary. Key among these are the following.

Independence

A common concern with redressal mechanisms that are heavily dependent on administrative or elected executive bodies is that their independence and impartiality will be compromised, since civil servants and PRIs play a key role in implementing government policies and programmes at a local level. As mentioned earlier, PRIs are the primary implementation body for projects undertaken under MGNREGA. Similarly, the administrative bureaucracy plays a vital role in providing a wide range of social welfare services to the people, including their entitlements under various socio-economic legislations such as the RTI Act and the NFSA. In such cases, vesting the authority for both implementation and enforcement with the same body often results in a conflict of interest. It is quite unlikely that administrative and local government officials will effectively redress complaints and hold guilty parties accountable for their actions, when they or their peers are themselves responsible for such violations. Such officials are also unlikely to take effective action in the case of violations by senior bureaucrats and politicians.

Integrity, Transparency and Accountability

While devolving implementation of government programmes to civil servants and local government officials has its advantages, it also raises critical concerns about corruption. Such decentralised systems are often characterised by a high degree of discretion for implementing officials, creating ample opportunities for corruption. In addition, elected local government bodies, such as PRIs, are vulnerable to capture by local elites who benefit from such rent seeking and corruption (Prud'homme and DEC 1994). The high degree of corruption in social welfare programmes and the complicity of local officials in the process mean that there is little scope for impartial grievance redressal of socio-economic rights violations within such mechanisms.

Additionally, conventional grievance redressal mechanisms suffer from inadequate transparency and accountability. Procedures for the appointment and transfer of government officials are opaque and subject to manipulation by politicians and senior bureaucrats, rather than being solely a function of their performance. Moreover, officials are generally not required to give clear justifications for their decisions, and provisions such as Article 311 of the Indian Constitution, which protects civil servants against unfair dismissal or reduction in rank, also make it difficult to hold them accountable in cases of corruption and incompetence in redressing socio-economic rights violations. Concerns around transparency and accountability extend to the judiciary as well. Judicial appointments to the high courts and Supreme Court of India are made by a 'collegium' of senior judges, out of the public eye and based on undisclosed criteria (Shah 2012). Moreover, a criminal investigation against a judge cannot be initiated without prior approval of the chief justice of the Supreme Court, and judges can only be removed through a complicated impeachment process. The lack of transparency and accountability makes such mechanisms an unreliable alternative for redressal of socio-economic rights violations.

Accessibility

Grievance redressal for socio-economic rights legislations have to deal with a large number of small claims, primarily about non-delivery of entitlements covered under the legislation and other administrative issues, which require little legal expertise. In such cases, mechanisms that rely heavily on the judiciary suffer from significant accessibility-related concerns. The Indian court system, which extends down to the subordinate courts operating in each district, is insufficiently decentralised to address the large volume of complaints at the subdistrict and village level in an effective manner. The judiciary also has an extremely high backlog of cases, with about 55,000 cases pending with the Supreme Court, 4.2 million with high courts and 28 million cases pending before the subordinate courts, as of 2010 (Kumar 2011). Combined with the reliance on formal procedural laws and technical rules for the presentation of evidence, grievance redressal through the courts is a confusing, expensive and time-consuming process. As a result, it is unsuitable for providing the speedy, cheap and accessible grievance redressal which is essential to remedy violations of critical socio-economic rights, such as the right to food. Such concerns are less applicable to non-judicial remedies, since the bureaucracy and PRIs have a stronger presence at the local level. However, it is important that local officials responsible for remedying violations are adequately empowered to investigate and effectively redress grievances brought before them.

Additionally, vulnerable groups such as beggars, slum dwellers and commercial sex workers are often unwilling to approach conventional grievance redressal mechanisms, due to their illegal status before the courts and administration. In general, conventional grievance redressal mechanisms often remain inaccessible to the poor and marginalised, since these sections of society have a high degree of distrust of civil servants and elected representatives, who are perceived to be corrupt, insensitive and hostile to their needs.

Quasi-judicial Grievance Redressal Mechanisms for Socio-economic Rights

Within most recent socio-economic rights legislation, there is an increasing preference for quasi-judicial grievance redressal mechanisms, such as national human rights institutions, ombudspersons and tribunals. While there exists a high degree of heterogeneity among quasi-judicial bodies, they are typically characterised by similar powers to the judiciary, but with a limited mandate and jurisdiction, as defined by the law mandating their creation. For example, the grievance redressal mechanism under the proposed NFSA is meant to deal only with violations of entitlements under the various food security and welfare schemes covered under the act. Given their specific focus, quasi-judicial bodies are likely to be more efficient and effective in dealing with grievances related to their given area. At the same time, limiting their jurisdiction is meant to ensure that they do not encroach upon other legal issues which the courts are better equipped to deal with.

Well-designed quasi-judicial mechanisms can overcome a number of concerns around conventional approaches to grievance redressal discussed in the previous section. They operate independent of governmental and administrative influence, and are less prone to problems of corruption, transparency and lack of accountability. Quasi-judicial proceedings can also be more accessible, speedy and cost-effective than the judiciary, due to greater decentralisation, access to specialised technical knowledge, and a more flexible and informal nature of functioning. Moreover, unlike the courts, which can act only on cases brought before them, quasi-judicial bodies often have to the power to initiate *suo motu* action with regard to violations of rights. Rather than restricting themselves to the particular circumstances of a specific case, they can also investigate the broader systematic causes and consequences of such violations, and thereby play an important role in shaping relevant policies. In most legal systems, courts must respect the separation of powers between the different branches of the state; therefore, they tend to shy away from such policymaking responsibilities, which have traditionally been in the legislative domain. For example, with the exception of the introduction of cooked midday meals in schools, the Indian Supreme Court's interim orders in the right to food case are aimed at ensuring better enforcement of existing government food security schemes, by converting them into legal entitlements.

The FAO's *Guide to Legislating on the Right to Food* recommends the involvement of quasi-judicial bodies like national human rights institutions in the grievance redressal mechanism for right to food violations. The draft right to food legislations in Honduras, Nicaragua and Uganda provide for the National Human Rights Commissions in these countries to act in the event of any violations. In Brazil, the Public Ministry, an autonomous and quasi-judicial body, is empowered with wide-ranging powers to investigate and prosecute violations of constitutional rights, including the right to food (FAO 2009). In addition, the office of the Brazilian National Special Rapporteur for the Right to Food is another quasi-judicial monitoring body empowered to look into complaints about right to food violations and recommend appropriate action (FAO 2007).

Grievance redressal mechanisms for a number of socio-economic rights legislations in India include strong provisions for quasi-judicial monitoring and grievance redressal of violations. Some of the more prominent examples include:

a. The MGNREGA ombudsman, a quasi-judicial authority empowered to investigate and pass orders on violations under the act.
b. The RTE Act, which involves PRIs in its grievance redressal mechanism, but also provides for grievance redressal by the quasi-judicial State and National Commissions for the Protection of Child Rights.
c. The hybrid grievance redressal structure of the RTI Act, wherein violations are first addressed by a higher administrative appellate authority, and the appellate authority's decisions can be appealed before the quasi-judicial State or National Information Commissions.
d. The three-tiered quasi-judicial framework of the Consumer Protection Act, involving district consumer disputes redressal forums, state consumer disputes redressal forums and a National Consumer Disputes Redressal Commission, with a right to appeal decisions by the national commission before the Supreme Court.

The grievance redressal mechanism under the proposed NFSA is modelled along similar lines to the Consumer Protection Act. It calls for the establishment of a three-tiered quasi-judicial framework, comprising:

1. District grievance redressal officers (DGROs) in each district to investigate and redress grievances related to the distribution of foodgrain and meal entitlements under the law.
2. State Food Commissions to monitor and evaluate the implementation of the law at the state level, advise and issue guidelines to the state government, inquire into violations of entitlements provided under the law (either *suo motu* or on receipt of a complaint) and hear appeals against orders of the DGRO.
3. A National Food Commission to monitor and evaluate the implementation of the law and the various schemes under it, advise and issue guidelines to the central government, inquire into violations of entitlements provided under the bill (either *suo motu* or on receipt of a complaint) and hear appeals against orders of the state commission.[2]

To aid in the effective investigation of violations, the legislation also provides the State and National Food Commissions with the powers of a civil court; these powers include summoning and enforcing the attendance of individuals and examining them on oath, discovery and production of any document, receiving evidence on affidavits, requisitioning any public record or copy thereof from any court or office, and issuing commissions for the examination of witnesses or documents.[3] Besides the above-mentioned quasi-judicial mechanisms, grievance redressal measures in the law include provisions for periodic social audits of the various food security and welfare schemes covered under the act by local authorities and independent agencies.

Concerns around Quasi-judicial Grievance Redressal Mechanisms in India

Despite their perceived advantages over conventional grievance redressal mechanisms, a number of concerns exist around the use of quasi-judicial remedies for violations of socio-economic rights. In the context of the right to food in India, the quasi-judicial grievance redressal mechanism proposed under the NFSA faced substantial criticism on a number of counts, and important grievance redressal provisions included in the NAC draft of the food security bill were removed or watered down in the legislation passed by the government.

Some of the key concerns around quasi-judicial grievance redressal in general, and the government's provisions under NFSA, are as follows.

Selection of Officials

Ensuring diversity, expertise and independence within grievance redressal bodies is critical. While most legislations specify necessary qualifications for recruitment of officials, the actual selection process often lacks transparency and is prone to manipulation. For example, the RTI Act requires that information commissioners appointed to the state and central information commissions be persons of eminence in public life with wide knowledge and experience in law, science and technology, social service, management, journalism, mass media or administration and governance (GoI 2005). However, the selection of these officials is left to the discretion of a three-member committee consisting of members of the government and opposition. Research on the appointment of information commissioners under the law shows that 30 per cent of these posts were lying vacant, and that about two-thirds of those appointed were retired civil servants (Commonwealth Human Rights Initiative 2012). This highlights a clear bias within the selection process in favour of retired administrative officials, even when there are other qualified and competent applicants (ibid.). The lack of clearly defined selection criteria and transparency in the appointment of officials raises serious concerns about their independence and integrity, similar to those faced by conventional grievance redressal mechanisms. Similar concerns also existed with the NFSA's proposed grievance redressal structure, which also left the selection process for DGROs, state food commissioners and national food commissioners to the discretion of the government.

In the interest of greater transparency, it is also advisable that the selection of officials occurs through a public selection process. This would include adequate publicity of available positions, inviting nominations from the public, as well as adequate diversity within the committee responsible for selecting the officials and full public disclosure of the selection proceedings. A public selection process would help minimise political interference and ensure that officials are selected solely on the basis of their qualifications and competence. Provisions requiring public participation in the nomination and selection of national food commissioners were an important component of the NAC draft of the food security bill, but were removed in the government's proposal, which instead empowered the central government to make rules in this regard.

Powers and Responsibilities

The powers and responsibilities of quasi-judicial bodies vary significantly based on the legislation, and there are a number of pertinent debates around the adequacy of these powers. To aid in the effective investigation of cases where rights violations are alleged to have occurred, most rights-based legislations in India provide quasi-judicial bodies with the powers of a civil court. Besides the State and National Food Commissions under the NFSA, quasi-judicial bodies set up under other socio-economic rights legislations also have the powers of a civil court while investigating violations.

Orders and judgments by quasi-judicial bodies can be binding or advisory in nature. For example, decisions by quasi-judicial bodies under most socio-economic rights legislations are binding on the involved parties, with provisions to appeal against the decision. On the other hand, the quasi-judicial State and National Human Rights Commissions in India investigate possible human rights violations and present their findings and recommendations to the government; however, they cannot initiate action of their own accord. As there is no guarantee that their advice will be acted upon appropriately, quasi-judicial bodies with advisory powers run the risk of being perceived as ineffective and powerless.

Even if their judgments are binding in nature, quasi-judicial grievance redressal mechanisms would remain toothless if they are unable to take adequate remedial measures in cases where violations are found to have occurred, including compensating the aggrieved party and punishing errant officials. Socio-economic rights legislations in India, including the NFSA, however, give the government broad discretion to decide on the powers of quasi-judicial officials and bodies. For example, the NAC draft of the food security bill empowered the National Food Commission to monitor situations of starvation and penalise errant officials, and also specified the actions it could take once an inquiry was completed. The law excludes all of these provisions, and instead allows the central government to decide on the powers of the commission. Similarly, the law states that the powers of the DGROs and State Food Commissions will be prescribed by the central and state governments respectively. By failing to clearly define the powers of the DGROs and food commissions, and leaving them at the discretion of the government, the legislation runs the risk of setting up an ineffective grievance redressal mechanism, with little actual power to address violations and the broader concerns around food security in India.

Separation of Powers

The separation of powers between the legislature, executive and judiciary is widely regarded as an essential feature of a well-functioning state. Under such a model, each branch performs a distinct role and operates independently of the others, thereby preventing undue concentration of power with any one branch and providing for adequate checks and balances on their functioning. Quasi-judicial grievance redressal bodies, set up to remedy socio-economic rights violations, have been criticised for violating such separation of powers, since they often play an active role in formulating policies and procedures for guaranteeing these rights. However, proponents argue that such interventions are necessary, since the legislature and executive in India are widely regarded as corrupt and unaccountable to the people, and cannot be trusted to protect the rights of citizens, particularly the poor and vulnerable.

With any grievance redressal mechanism, it is also important to ensure that the same official or body does not have the power to both prosecute and adjudicate rights violations. Inadequate separation of prosecutorial and adjudicatory powers can result in conflicts of interest or an unhealthy concentration of power, thereby undermining checks and balances within the system. The Consumer Protection Act addresses such separation of power concerns by restricting the role of the district, state and national-level consumer disputes redressal forums to adjudicating complaints, and allowing for independent lawyers to prosecute cases before these bodies. Another example is the government's proposed Lokpal and Lokayuktas Bill, which aims to create an independent anti-corruption agency with separate investigation and prosecution wings, and also provides for special courts to try these cases. Brazil's Public Ministry, an independent agency empowered to prosecute a wide range of policy and rights violations before the judiciary, is a good example of how separation of powers can be achieved while protecting the right to food and other socio-economic rights (Robinson 2011).

The NFSA's grievance redressal mechanism raises two primary concerns about separation of powers. First, State and National Food Commissions are expected to advise the government on the formulation of policy and implementation of food security schemes, which may encroach upon legislative and executive powers. Second, there is a lack of separation between prosecutorial and adjudicatory powers within the grievance redressal bodies; for example, DGROs are responsible for prosecuting as well as adjudicating complaints filed before them. Moreover, in case government officials decide to appeal the DGRO's decision before the state or national commissions, it is likely that the DGRO would have to function as a prosecutor on behalf of the individual who registered the grievance. Similarly, in cases where state and national commissions inquire into violations of the right to food, the law is unclear on who will ultimately be responsible for prosecuting and adjudicating these violations. In order to ensure a fair grievance redressal process, it is desirable to have different officials responsible for prosecuting and adjudicating such violations. Possible alternatives include the use of private prosecutors, such as in the case of the Consumer Protection Act, or an independent agency for prosecuting right to food and other socio-economic rights violations, along the lines of Brazil's Public Ministry system.

Accessibility

As discussed earlier, a rights-based legislation is likely to encounter a large volume of local-level grievances, which conventional grievance redressal mechanisms such as the judiciary are ill-equipped to handle. Such accessibility-related concerns, arising from slow processing of claims and insufficient decentralisation within grievance redressal bodies, can extend to quasi-judicial mechanisms as well. Therefore, even in cases where quasi-judicial mechanisms are put in place to remedy such violations, it is important to ensure that they have a highly decentralised structure, extending down to the district or subdistrict levels. Moreover, the poor and vulnerable often suffer the most serious violations of their rights, but also have the least ability to demand redressal. Therefore, there is a need to ensure adequate support for such groups, through greater awareness of their rights and assistance in registering complaints when these rights have been violated.

An ineffective grievance redressal mechanism at the subdistrict level has been a key concern with India's food security legislation. Unlike the NAC's draft, the NFSA does not mandate the creation of block facilitation centres at the subdistrict level to provide information and assist aggrieved persons in filing complaints with the DGRO. It has been argued that a grievance redressal mechanism for the right to food which begins at the district level will remain inaccessible to a large number of people, particularly the rural population and vulnerable groups, unless significant assistance is provided at the subdistrict level. In fact, the Right to Food Campaign's recommendations go further, calling for the appointment of dedicated grievance redressal officers at lower administrative levels, namely the panchayats in rural areas and municipalities in urban areas (Right to Food Campaign 2012). The same recommendation was also seconded by the Parliamentary Standing Committee, which also desired the representation of elected women representatives on local-level grievance committees.

Integrity, Transparency and Accountability

As highlighted above, quasi-judicial mechanisms under the NFSA and other socio-economic rights legislations in India face similar shortcomings to conventional avenues for grievance redressal. The absence of clearly defined powers and selection criteria for officials raises important concerns around the independence and effectiveness of such bodies. It would also be naïve to assume that officials will always act in the best interests of citizens; in the absence of adequate checks and balances, quasi-judicial bodies are equally vulnerable to misuse of power and corruption. Therefore, adequate measures must be taken to ensure integrity, transparency and accountability in the grievance redressal process.

It is desirable that grievance redressal bodies function independently from the government departments and agencies they are expected to monitor, in terms of their funding, infrastructure and personnel. Under the NFSA, state and central governments are required to provide these resources to the DGROs and the State and National Food Commissions. This ensures that grievance redressal bodies will not be susceptible to the influence of government officials responsible for administering food security schemes and entitlements. However, it still raises questions about their independence while investigating complaints against senior government officials.

Within most quasi-judicial mechanisms, parties have the right to appeal against rulings before a higher authority. In the case of the NFSA, the DGRO's decisions can be appealed before the State Food Commission, and, in turn, judgments by these state commissions can be appealed before the National Food Commission. In addition, quasi-judicial judgments are generally subject to judicial review; for example, decisions made by the National Food Commission can be challenged before the Supreme Court. In some cases, a law may specifically exclude its judgments from the jurisdiction of the courts, such as in the case of decisions by the State and Central Information Commissions set up under the RTI Act. However, there are instances, most prominently the State Water Tribunals, where the courts have accepted appeals against decisions by quasi-judicial bodies, even though the relevant legislation explicitly bars the judicial review of such judgments (Iyer 2010). By providing a system of checks against unfair and arbitrary decisions, such a right of appeal and judicial review increases the accountability of the grievance redressal process. The power of judicial review also helps to ensure that decisions by quasi-judicial bodies do not violate fundamental rights and protections guaranteed by the Constitution. At the same time, such a system of internal and external review will reduce the expediency of the grievance redressal process.

In the interest of greater transparency and accountability, a key demand by the Right to Food Campaign was the creation of a statement of obligations and citizens' charter which clearly defined the duties and obligations of each government office and public authority towards citizens, as defined under the NFSA (Right to Food Campaign 2012). Such disclosures serve two functions; first, they make it easier to identify violations of the law and those responsible for these violations. Second, improper behaviour on the part of grievance redressal officials can also be detected easily, thereby increasing the accountability of the grievance redressal process itself. However, the creation of a citizens' charter for the right to food has not been mandated under the NFSA. However, the law mandates the establishment of local (state-, district-, block-, FPS-level) vigilance committees that may regularly supervise the implementation of the act and register written complaints with the DGRO in case of violations. The state governments are to further ensure that these vigilance committees give adequate representation to the local authorities, the SCs, STs, women and destitute persons or persons with disability.

Excessive Tribunalisation of Justice

In addition to the issues above, the NFSA's approach raises broader issues around the trend towards setting up dedicated grievance redressal mechanisms under socio-economic rights legislations, instead of relying on existing remedy mechanisms, or institutions like the State and National Human Rights Commissions. Problems associated with the excessive tribunalisation of justice include increased bureaucracy and costs, duplication and overlap of responsibilities with existing institutions, and the possibility of undermining existing redressal frameworks. At the same time, it must be acknowledged that conventional grievance redressal frameworks have been inadequate in protecting citizens, particularly the poor and vulnerable. In the case of critical socio-economic rights where the human and social cost of inaction can be extremely severe, such as the right to food, there is therefore an urgent need to have a decentralised and responsive system to address such violations. In many cases, these objectives are best met through a dedicated and independent grievance redressal mechanism, such as the one in the NFSA.

In this context, the Citizens' Right to Grievance Redressal Bill being proposed by the government and civil society organisations like the National Council for People's Right to Information (NCPRI) provides a possible middle path.[4] The bill calls for every public authority to publish a citizens' charter which clearly lists its obligations and responsibilities to the people. Moreover, the NCPRI's draft version of the bill envisions a decentralised grievance redressal mechanism similar to that in the NFSA, including subdistrict-level grievance facilitation centres, district-level grievance redress authorities, and state and national grievance redress commissions. If properly implemented, such a mechanism may represent a viable alternative to the dedicated grievance redressal frameworks under the NFSA and other similar legislations.

Notes

1. See, for instance, the provisions under the Madhya Pradesh Lok Sewaon ke Pradhan ki Guarantee Adhiniyam, 2010, www.lawsofindia.org/pdf/madhya_pradesh/2010/2010MP24.pdf (accessed 12 March 2018).

2. See the provisions under the initially proposed NFSB, 2011, http://www.prsindia.org/uploads/media/Food Security/National Food Security Bill 2011.pdf (accessed 12 March 2018).

3. Ibid.

4. 'The Right of Citizens for Time Bound Delivery of Goods and Services and Redressal of Their Grievances Bill, 2011 (Citizens Charter)', http://www.prsindia.org/billtrack/the-right-of-citizens-for-time-bound-delivery-of-goods-and-services-and-redressal-of-their-grievances-bill-2011-2125/ (accessed 12 March 2018).

References

Commonwealth Human Rights Initiative. 2012. 'A Rapid Study of Information Commissions Established under the Right to Information Laws in India', May. http://www.humanrightsinitiative.org/postoftheday/2012/ICs-compstudy-Delhi-final-May12-VenkatN&AmikarPs.pdf (accessed 16 October 2017).

FAO (Food and Agriculture Organization). 2007. *Right to Food: Lessons Learned in Brazil*. Rome: FAO.

———. 2009. *Guide on Legislating for the Right to Food*. Rome: FAO.

GoI (Government of India). 2005. Right to Information Act, 2005 (Act no. 22 of 2005). Ministry of Law and Justice, Government of India. http://www.righttoinformation.gov.in/rti-act.pdf (accessed 16 October 2017).

International Commission of Jurists. 2008. *Courts and the Legal Enforcement of Economic, Social and Cultural Rights: Comparative Experiences of Justiciability*. Human Rights and Rule of Law Series no. 2. Geneva: International Commission of Jurists. https://www.humanrights.ch/upload/pdf/080819_justiziabilitt_esc.pdf (accessed 4 November 2017).

Iyer, Ramaswamy R. 2010. 'Cauvery dispute: time for closure'. *Hindu*, 8 February. http://www.thehindu.com/opinion/lead/Cauvery-dispute-time-for-closure/article16813526.ece (accessed 4 November 2017).

Kumar, Rohit. 2011. 'Vital Stats: Pendency of Cases in Indian Courts'. *PRS Legislative Research*, 6 July. http://www.prsindia.org/parliamenttrack/vital-stats/pendency-of-cases-in-indian-courts-2011-1837/ (accessed 16 October 2017).

Prud'homme, Remy, and DEC. 1994. 'On the Dangers of Decentralization'. Policy Research Working Paper no. 1252, World Bank. https://econpapers.repec.org/paper/wbkwbrwps/1252.htm (accessed 16 October 2017).

Right to Food Campaign. 2008. *Supreme Court Orders on the Right to Food: A Tool for Action*. Right to Food Campaign Secretariat, New Delhi. http://www.righttofoodindia.org/data/scordersprimeratoolforaction.pdf (accessed 16 October 2017).

———. 2012. 'Right to Food Campaign's Critique of the National Food Security Bill, Part II: Clause by Clause Critique of the National Food Security Bill, 2011'. http://www.righttofoodindia.org/data/right_to_food_act_data/events/March_2012_clause_by_clause_critique_final_18_february_2012.pdf (accessed 28 August 2017).

Robinson, Nick. 2011. 'Closing the implementation gap'. *Hindu*, 12 March.

Shah, Ajit Prakash. 2012. 'Who should judge the judges?' *Hindu*, 26 January. http://www.thehindu.com/opinion/op-ed/article2832090.ece (accessed 16 October 2017).

17

Victim Support

Saba Sharma and Warisha Farasat

Access to Legal Redressal among Disadvantaged Groups: The Need for a Comprehensive System of Victim Support

A strong victim support component at the most accessible level is seen as important in a rights-based approach to socio-economic rights. This is because among disadvantaged communities, especially the poor, access to justice is already impeded by various social barriers. Studies of access to justice have shown repeatedly that two factors predominate in determining whether people are able to use available legal remedies. The first, and by far the most important, is access to financial resources. Hiring lawyers and using legal institutions can be very costly in themselves, but also entail opportunity costs, which for the poor usually means time away from income-generating activities. The second factor usually identified is institutional skill—the ability to understand and use the system. It is not uncommon for the poor to be wary of the official legal system.

An essential component of this wariness is a lack of respect founded on the experience of illegality. The view is commonly expressed among the urban poor that 'there must be something wrong with a law or code if it is broken daily by so many people as they go about their daily lives' (Hardoy and Satterthwaite 2014, 31). The view that the law does not serve the real needs of the poor, but serves some other set of interests is difficult to refute in such circumstances. For the most part, the poor see the law as a tool which the wealthy and well connected can use against them (Dickens 2012). To the poor, the official law is often expressed in a foreign language—English is the official language of the law in India, Kenya and the Solomon Islands, even though only a small proportion of the population can speak English in these jurisdictions. Similarly, the language of the law is French in Niger and Portuguese in Mozambique.

It is clear that for human rights to be protected, and to be effective in constraining state action, it is not sufficient that they are bestowed from above by the state. Even with a sympathetic judiciary, human and constitutional rights will only be 'parchment barriers' unless litigants in real conflicts actively invoke them. In other words, rights also need to be seized from below by active individuals and groups in civil society. Some analysts have used the term 'rights revolutions' to refer to patterns of governance in which constitutional or human rights become an important vehicle for political struggle on the part of groups seeking more egalitarian outcomes (Anderson 2003).

A well-functioning justice system has also been crucial for the effective delivery of public services and the distribution of socio-economic and political rights. In many countries, the 'rule of law' provides the fundamental constraint on executive power. Unfortunately, this ideal bears little resemblance to the reality in many countries around the world, where legal systems in fact serve to perpetuate inequitable power relations and discrimination, producing what we call 'legal inequality traps'. Formal rules that appear to protect the interests of the broader community are undermined by institutional practices, informal strategies, or conflicting rule-based systems and social norms. '*Structures of inequality affect both the creation of institutions in the justice sector and the context within which they operate*: they are embedded in the rules, practice, and norms that perpetuate these institutions' (Sage and Woolcock 2007; emphasis added).

Human Rights Perspectives

The principles of the ICESCR, to which India is a signatory, have guided right to food legislations all over the world, including provisions for effective legal redressal in the event of a violation. However, it is difficult to envisage how provisions for an effective right to remedy ('second-order' right) can be met without first of all incorporating substantive primary rights ('first-order' rights) into domestic law. Such incorporation shall provide for adequate, effective, prompt and appropriate remedies. Hence, it must enable individuals to approach domestic courts in the event of a breach of international human rights treaty norms, such as the right to food—a process that must lead to an enforceable remedy (FAO 2011).

In the context of discrimination and the right to food, the Human Rights Council (2010) observes that there is a dual disadvantage for those who are discriminated against in the context of the right to food. Not only are many of these populations vulnerable in the sense of being exposed to chronic hunger and malnutrition, but there is also a gap in their legal protection (discussed below in more detail). In the case of the urban poor in particular, the council observes that there is a gap in their legal protection at the international level. An international instrument that offers guidance on how domestic legislation and strategies ought to address the plight of the urban poor is missing. In the absence of international guiding principles, more often than not the urban poor are forgotten by national legislatives, and thus further marginalised. There exists thus a great need for international guidelines to address the specific needs of the urban poor and the ways in which their right to food should be best realised.

Defining a Victim and Victim Rights

The UN Declaration on Justice for Victims was the first international instrument to specifically focus on the rights and interests of victims in the administration of justice. This instrument is primarily concerned with the position of victims before national criminal justice systems. However, certain principles enshrined within it are general and could apply to a wide variety of human rights violations, including the violation of the right to food. Particularly in terms of the right of victims to legal redress, and also the definition of victimhood, it proves a useful guide.

The Declaration of Basic Principles of Justice for Victims of Crime and Abuse of Power (UN 1985) recognises 'victims' as not only victims of crime, but also 'persons who, individually or collectively, have suffered harm, including physical or mental injury, emotional suffering, economic loss or substantial impairment of their fundamental rights, through acts or omissions that do not yet constitute violations of national criminal laws but of internationally recognised norms relating to human rights'. A person may also be considered a victim 'regardless of whether the perpetrator has been identified, apprehended, prosecuted or convicted and regardless of the familial relationship between the perpetrator and the victim. The term "victim" also includes, where appropriate, the immediate family or dependants of the direct victim and persons who have suffered harm in intervening to assist victims in distress or to prevent victimisation.'

The UDHR also outlines the right to effective remedy to victims under Article 8: 'Everyone has the right to an effective remedy by the competent national tribunals for acts violating the fundamental rights granted him by the constitution or by law.' All principal universal human rights instruments contain provisions that establish a right to an 'effective remedy' for victims of human rights violations. The right to an effective remedy is also recognised in human rights instruments dealing with specific rights. It includes the right to investigations, prosecutions and punishment of those responsible for human rights violations, as well as the right to reparation.

Victim Support and Other Socio-economic Rights in the Context of a Human Rights Framework: The Right to Housing

The Report of the Special Rapporteur on the Right to Housing (Human Rights Council 2007) states that according to international human rights law, everyone has the right to adequate housing as a component of the right to an adequate standard of living. The right to adequate housing includes, inter alia, the right to protection against arbitrary or unlawful interference with privacy, family, home, and to legal security of tenure. Accordingly, it is the duty of the state to protect against forced evictions, and the human right to adequate housing and secure tenure are guaranteed without discrimination of any kind on the basis of race, colour, sex, language, religion or belief, political or other opinion, national, ethnic or social origin, legal or social status, age, disability, property, birth or other status.

The report is fairly clear on the elaboration of the rights of victims to redressal and reparation. In the case of the right to housing, for instance, the state must provide victims of forced evictions appropriate remedies, among them a fair hearing, access to legal counsel, legal aid, return, restitution, resettlement, rehabilitation and compensation. Compensation should be provided for any economically assessable damage, as appropriate and proportional to the gravity of the violation and the circumstances of each case, such as: loss of life or limb; physical or mental harm; lost opportunities, including employment, education and social benefits; material damages and loss of earnings, including loss of earning potential; moral damage; and costs required for legal or expert assistance, medicine and medical services, and psychological and social services.

Cash compensation should under no circumstances replace real compensation in the form of land and common property resources. Where land has been taken, the evicted should be compensated with land commensurate in quality, size and value, or better. All those evicted, irrespective of whether they hold title to their property, should be entitled to compensation for the loss, salvage and transport of their properties affected, including the original dwelling, and land lost or damaged in the process. When circumstances allow, states should prioritise these rights of all persons, groups and communities subjected to forced evictions. Persons, groups and communities shall not, however, be forced against their will to return to their homes, lands or places of origin. While all parties must give priority to the right of return, certain circumstances (including for the promotion of general welfare, or where the safety, health or enjoyment of human rights so demands) may necessitate the resettlement of particular persons, groups and communities due to development-based evictions. Such resettlement must occur in a just and equitable manner and in full accordance with international human rights law.

However, this socio-economic right is envisaged largely with regard to evictions and displacements, so that the duties of the state are seen in the context of remedies and solutions to mediate or prevent such a process from occurring. Thus, while the report is fairly detailed in its provisions for remedies due to evictions, forced evictions or displacement, it is silent on the right to redressal of those who do not have adequate housing (for any other reason except eviction or displacement), but to whom the state may be obliged to provide such a right.

Victim Support in the Indian Context

There are not many cases of victim support being actively and systematically outlined with regard to specific legislations, although a system of legal aid is established in the country under the Legal Services Authority Act of 1987, where both national and state legal authorities are provided for. The concept of legal aid is essentially derived from the dual doctrines of the legal right of fair and adequate representation, and the criminal law principle that individuals are innocent until proven guilty. However, in India, since rich persons can afford good private legal representation, it is the poor and the marginalised persons who are left at the mercy of an incompetent and often corrupt legal aid system. While the legal aid system in India is beset with problems of corruption and poor and inefficient representation for the accused, it is still the only means by which some form of free legal representation is made available for the poor and the vulnerable.

Draft Grievance Redress Bill, 2011

The draft Citizens' Right to Grievance Redress Bill, 2011, provides for a grievance redress officer at every level of administration for all public authorities—central, state, district and subdistrict levels, municipalities and panchayats. Thus, the bill aims at a comprehensive system of grievance redress for all public authorities down to the grassroots level. The grievance redress officer has several functions, among which are:

- Providing all necessary assistance to citizens in filing complaints
- Where a complainant is unable to make a complaint in writing, rendering all reasonable assistance to the person making the request orally to transcribe the same in writing
- Acknowledging complaints via receipt, in writing, through electronic means or via text messaging
- Ensuring that the complaint is addressed within 15 days
- Ensuring that appropriate action is taken
- Making observations with regard to neglect or failure of delivery
- Informing the complainant of the action taken via a report

The grievance redress officer, thus, has many duties at the grassroots level, relating to victim support but also to enforcement (this is further elaborated on in Chapter 16, 'Enforcement and Grievance Redressal'). It may be noted here that while the office of the grievance redress officer at the grassroots level is an important one, as it is necessary to make legislations like the NFSA accessible at the lowest levels, there is some cause for concern about a conflict of interest arising from the multifaceted nature of duties for the grievance redress officer, related to both the enforcement and victim support aspects of grievance redress.

The Protection of Women from Domestic Violence Act, 2005

The Protection of Women from Domestic Violence Act, 2005, while not dealing with socio-economic rights per se, is one of the only pieces of legislation in the country that outlines a framework for victim support to victims of domestic violence by way of protection officers. The role of these officers is unique in that it advocates more than mere provision of support, instead envisaging a far more *proactive* role. For instance, under Section 9 of the act, which details the duties of the protection officer, not only is the officer supposed to report the incident to the magistrate, make the necessary applications and ensure that legal aid is provided, but also to make available a safe shelter home, if the aggrieved person so requires, and forward a copy of his report of having lodged the aggrieved person in a shelter home to the police station and the magistrate having jurisdiction in the area where the shelter home is situated; to get the aggrieved person medically examined, if she has sustained bodily injuries, and forward a copy of the medical report to the police station and the magistrate having jurisdiction in the area where the domestic violence is alleged to have taken place; to ensure that the order for monetary relief under Section 20 is complied with and executed, in accordance with the procedure prescribed under the Code of Criminal Procedure, 1973 (2 of 1974).

There are, however, concerns about how well the provision functions in actual practice. For instance, a news report found that almost four years after the act was legislated, the state of Gujarat had yet to appoint protection officers at all (*Indian Express* 2009). In a separate report more than a year later, the state of Maharashtra was also found not to have appointed any full-time protection officers, and instead delegated the related work to already overworked and overburdened officers (*Daily News and Analysis* 2010). Another problem identified with the procedure under the Domestic Violence Act has been that sometimes NGOs have been designated as protection officers. While NGOs are quite proactive in supporting the women, the law enforcement agencies or the police may not take them as seriously as they would a government officer who has been designated as the protection officer.

Notwithstanding these issues, the protection officer has an important role in facilitating victims in accessing the legal justice system. The role of the protection officer is envisaged not merely as helping the woman get a divorce, but also securing her other economic rights under the Domestic Violence Act, such as maintenance or alimony. The Domestic Violence Act is perhaps one of the only legislations in India that directly provides for victim support to the women moving the court under the act, which is over and above the existing legal aid systems. Given that women who have been abused in their domestic situations are usually vulnerable, the role of the protection officer, therefore, is to facilitate the aggrieved woman at all stages in securing the rights guaranteed to her under the law. Moreover, aggrieved women who approach the courts through a competent protection officer are more likely to be able to access stronger legal support in the courts as well, since the protection officer often connects them to the more committed and capable lawyers from the pool of legal aid advocates. Perhaps this is one model of victim support that, if it were critically explored to retain the useful elements, would have served well to nuance the grievance redress structure of the NFSA.

Existing Provisions in the Act

The right to food legislation in India, or NFSA 2013, provides no specific guidelines on how a victim whose right to food has been violated may make a complaint. Its only provisions (under Chapter VII) are with regard to the establishment of an internal grievance redressal mechanism (inclusive of call centres and helplines) and appointment of district grievance redressal officers. Section 15(1) of the act provides that, for the 'expeditious and effective redressal of grievances of the aggrieved persons in matters relating to distribution of entitled foodgrains or meals under Chapters II, III and IV, a District

Grievance Redressal Officer, with requisite staff, shall be appointed by the State Government for each District, to enforce these entitlements and investigate and redress grievances.' For appeal, 'Any complainant or the officer or authority against whom any order has been passed by officer referred to in sub-section (*1*), who is not satisfied with the redressal of grievance may file an appeal against such order before the State Commission.' This two-tiered (state and district) grievance redress system outlined in the act has been critiqued for not taking remedial measures closer to the aggrieved persons at the block or panchayat level. Regrettably, debates in the Parliament as well as in the build-up to the passage of the law excluded any significant discussion on the need for a robust victim support system.

NAC Draft on Food Security

The NAC draft (2011) outlined in Section 70: '(1) Every individual has a right to make a grievance regarding entitlements under the Act to the Block Facilitation Centre orally, in writing, by using various electronic means and through the telephone helpline. (2) Every grievance shall be given a Unique Identification Number which shall be communicated to the aggrieved on the registration of the grievance.' There are provisions for online redressal in Section 65: 'An online Centralised Public Grievances Redress and Monitoring System shall be put in place at various levels for covering entitlements under this Act,' and through telephone in Section 69: 'The Central and State Government shall create a toll free telephone helpline to register grievances under this Act which will be manned at the Block Facilitation Centre' (ibid.).

In the NAC draft (2011, Section 68), there is a provision for a block facilitation coordinator, who must perform the following functions:

(a) Register all grievances of denial of entitlements under this act.
(b) Help a complainant register their grievance made through telephone, electronically, in writing or in person and issue a dated receipt.
(c) Ensure that the complaints registered under the subsection above are also entered into the centralised public grievance redress and monitoring system.

(d) Publicise the grievance redressal mechanisms under this act at the block level.
(e) Inform complainants regarding action taken in response to their complaints.
(h) Cross-check from the complainants the veracity of the action taken report and send a report to the district grievance redressal officer regarding the same.

Critique of NAC Provisions

One of the biggest critiques of the NAC draft as well as the government draft was the lack of any victim support initiatives. With the exception of certain provisions in the NAC draft, the previous drafts of the food security bill in India stopped short of outlining any comprehensive victim support mechanism, other than asserting the rights of victims to redressal. While a grievance redressal mechanism at the institutional level was detailed, a discourse about accessing justice at the ground level and mechanisms in place to aid victims of rights violations at the first point of contact continues to be missing, both from the draft and the larger debates around the law on food security.

During the debates on the NAC draft bill, there was recognition that victim support would be an important component of the law. Nonetheless, there was no consensus on the contours of victim support. Some felt that the existing legal aid lawyers/services could be drawn in to provide support to the victims when they wanted to petition the grievance redressal mechanisms. The advantage of feeding into the already existing legal aid would be that no additional financial and human resources would be required to implement victim support at the grassroots level. However, the dismal state of legal aid, and the lack of commitment and competence on the part of legal aid lawyers, was thought to be an impediment to effective victim support.

It was felt by many who joined the debate that the existing legal aid system would be ineffective in dealing with grievance redressal under the NFSA. However, dismissing the legal aid system outright was perhaps not the best way to engage with the issue of victim support, not least of all because of the extensive countrywide reach of the national and state legal aid networks. At the same time, it is only a handful of young lawyers who currently join the legal aid system with some commitment to social justice and the law. Several attempts are being made to reform the legal aid system, and if specific cadres of officers or human rights defenders or lawyers are carved out of the existing legal aid functionaries, it may be a way to energise the system, and assist in the reform processes.

Another way to incorporate legal aid is through a system of paralegals to serve as the first point of contact for victims, and to guide them through the specifics of court processes. To this end, it was suggested that a separate cadre of officers should be designated at the village and district levels to assist the victims through the grievance redressal process. These officers need not necessarily be lawyers or persons with legal training. It would be sufficient if they have some experience of working with the local administration or with food security issues at the local level. The rationale behind having such officers at the block level is linked to the skewed access that the poor and disadvantaged often have to the legal justice system. As mentioned earlier, systems and procedures in law are often set up in a way that impedes access to the most vulnerable. In particular, as we talk about victim support under socio-economic rights, those who are likely to face the violation of such rights are usually the most disadvantaged. Since the block level is the lowest level of administration in the government, it is at this level that it is easiest and most effective to facilitate support for victims.

It was further suggested in the course of discussions that, under the NFSA, a group of human rights defenders should be appointed at the district/state levels by a collegium of judges and eminent persons to provide legal aid to victims/complainants. Alternatively, it was suggested that a separate cadre of lawyers should be appointed to assist victims with the grievance redressal mechanism. Unfortunately, despite these varied suggestions, there was no consensus on the nature and extent of victim support; nor on how these human rights defenders or separate cadre of lawyers would actually be found—whether the most competent and successful lawyers would opt for such work which would have comparatively low remuneration. How would such a special cadre be trained, supported, supervised and held accountable was also not clear.

The discussion about the contours of a plan for appointment and tenure of human rights defenders was not clear even during the debates at the NAC. However, while there are serious questions about the operationalisation, the rationale behind the need for such a cadre remains intact, and more so the significance and value of having such cadres at the lowest possible level, like the block. Moreover, since the creation of a victim support unit would also require allocation of financial resources, this was eventually not agreed upon. Therefore, even the final NAC draft did not have any mechanism for victim support, and this is in fact one of the biggest lacunae in the draft given that ordinarily it would be difficult for the most marginalised persons living with acute poverty to access the grievance redressal mechanism.

Roosevelt once said to a group of reformers, 'I agree with you, I want to do it, now make me do it.' It captures succinctly the value of dissent in democracy. The state, after setting up a mechanism for the facilitation of socio-economic rights, is also obliged to set up a system where violations to such rights can be adequately addressed. Although there are no clear answers to the questions that surround victim support—about whether these processes should be facilitated from within the government itself, or supported externally by civil society organisations and individuals—there is still a need to address these problems and evolve a comprehensive system of victim support. The lack of a victim support mechanism will undoubtedly hinder access to the grievance redressal mechanisms of those individuals whose rights are violated. It must be reiterated that individuals who do not receive the entitlement under the law, would often be living with hunger and starvation, and it is naïve to expect them to access the grievance redressal system on their own.

References

Anderson, Michael R. 2003. 'Access to Justice and Legal Process: Making Legal Institutions Responsive to Poor People in LDCs'. IDS Working Paper no. 178, Institute of Development Studies, University of Sussex.

Daily News and Analysis. 2010. '"No protection officer appointed under Domestic Violence Act", says petitioner'. 3 February.

Dickens, Jonathan. 2012. *Social Work, Law and Ethics*. London: Routledge.

Draft Citizens' Right to Grievance Redress Bill. 2011. http://www.prsindia.org/uploads/media//draft/Draft%20Citizens%20Right%20to%20Grievance%20Redress%20Bill,%202011.pdf (accessed 8 November 2017).

FAO (Food and Agriculture Organization). 2011. *Constitutional and Legal Protection of the Right to Food around the World*. Rome: FAO.

Hardoy, Jorge E., and David Satterthwaite. 2014. *Squatter Citizen: Life in the Urban Third World*. Routledge.

Human Rights Council. 2007. 'Report of the Special Rapporteur on Adequate Housing as a Component of the Right to an Adequate Standard of Living, Miloon Kothari'. A/HRC/4/18, 5 February.

———. 2010. 'Study on Discrimination in the Context of Right to Food'. Prepared by the Drafting Group of the Advisory Committee on the Right to Food, Sixth Session, A/HRC/AC/6/CRP.1.

The Indian Express. 2009. 'Domestic violence: wait on for protection officer'. 22 August.

NAC (National Advisory Council). 2011. 'National Food Security Bill, 2011'. http://www.indiawaterportal.org/sites/indiawaterportal.org/files/National_Food_Security_Bill_approved_by_the_National_Advisory_Council_2011.pdf (accessed 12 March 2018).

Protection of Women from Domestic Violence Act. 2005. http://ncw.nic.in/acts/TheProtectionofWomenfromDomesticViolenceAct2005.pdf (accessed 12 March 2018).

Sage, Caroline, and Michael Woolcock. 2007. 'Breaking Legal Inequality Traps: New Approaches to Building Justice Systems for the Poor in Developing Countries'. Working Paper no. 17, Brooks World Policy Institute.

UN (United Nations). 1985. Declaration of Basic Principles of Justice for Victims of Crime and Abuse of Power. http://www.un.org/documents/ga/res/40/a40r034.htm (accessed 8 November 2017).

18

Fines, Penalties and Compensation

WARISHA FARASAT

The most significant contribution of the Indian Constitution is the Bill of Rights, known as fundamental rights, which are enshrined in Part III of the Constitution of India. Out of all the substantive guarantees, Article 32 provides that if any of the fundamental rights are violated, they could be enforced by citizens in a court of law, ensuring that these rights are not merely guaranteed but also made enforceable. Given that the Indian Constitution was promulgated in the 1950s, there was a separation between civil and political rights that were made enforceable, and socio-economic rights that were mostly non-justiciable in nature, and contained in Part IV of the Constitution as directive principles. However, through the enunciations of the Supreme Court and high courts as well as legislations of the Indian Parliament, this distinction between these two sets of rights has been blurred, and several of the socio-economic guarantees have metamorphosed into concrete rights that individuals could demand in a court of law.

In this context, activists filed *People's Union for Civil Liberties v. Union of India* before the Indian Supreme Court in the year 2001, popularly referred to as the right to food case. The interim orders passed by the Supreme Court in this case bolstered the demand for enacting a legislation that would guarantee a universal right to food to persons living in India. The urgency of promulgating such legislation was prompted by reports of widespread deaths of persons due to hunger and starvation, farmer suicides and acute child malnutrition. However, the NFSA would be rendered useless if an individual, whose entitlements under the act have been denied, does not receive any compensation, or the errant official is not censured for acts of commission or omission for ensuring that the relevant food entitlements reach the persons concerned. There was no significant debate and discussion in the Indian Parliament on fines, penalties and compensation. However, a concern about effective implementation and enforcement of the legislation was articulated by Sonia Gandhi, the president of the ruling (Indian

National Congress) party which enacted the NFSA, delivering on its electoral promise.

The present chapter examines the concept of punitive measures and compensation which would ensure that there is a mechanism that secures the rights guaranteed to individuals under the act.

Legal Concepts of Punitive Measures and Compensation

Foremost, the concepts of fines and penalties differ significantly from the legal concept of compensation. While the former are in the nature of punitive measures imposed for the failure to perform an act or adhere to the law, the latter provides redressal for a harm done to an aggrieved person. If the right of an individual as prescribed in the Bill of Rights or another statute is violated, they would ordinarily be entitled to receive compensation from the state. Therefore, these two legal concepts have been discussed separately to identify the international best practices and domestic Indian jurisprudence on each of these two areas.

Compensation

With the development of a rights-based approach, the term 'compensation' is beginning to be replaced by 'reparations', which symbolises a more holistic, dignified and just way to address harm suffered for violations. Financial or monetary compensations are just one of the components of a comprehensive reparations policy. Reparations initiatives can be designed in many ways. They may include a range of measures such as financial compensation, social services such as healthcare or education, and symbolic measures such as formal apologies or public commemorations.[1] Within this scheme, reparations are viewed more as a matter of right for the victims rather than charity on the part of the government.

While most of the legal jurisprudence of reparations internationally has developed around the violation of civil and political rights, there is increasing realisation that a similar set of reparative principles would be applicable to violation of socio-economic rights. The UN Basic Principles on the Right to Remedy and Reparations provide for compensation, restitution, rehabilitation and guarantees of non-repetition.[2] Under these principles, effective remedy includes: (*a*) equal and effective access to justice; (*b*) adequate, effective and prompt reparation for harm suffered; and (*c*) access to relevant information concerning violations and reparation mechanisms.[3]

Gradually, domestic courts have also begun to rule on enforcing socio-economic guarantees. In the landmark decision of the South African Constitutional Court in the petition filed by Treatment Action Centre, the court ruled on the right of access to healthcare services, including reproductive healthcare. The directions in this decision paved the way for effective measures to be adopted by government for the provision of medical services to combat mother-to-child transmission of HIV.[4] Similarly, in *Government of the Republic of South Africa v. Grootboom*,[5] the court ruled upon the socio-economic right to adequate housing, and opined that the state's housing programme was inconsistent with the constitutional obligations.

Even within Indian legal jurisprudence, there is a growing recognition that violation of socio-economic rights should be compensated. Under Indian law, the law on compensation has evolved through cases of custodial killings, torture and communal violence. With the decisions in *D. K. Basu v. State of West Bengal*[6] and *Nilabati Behera v. State of Orissa*,[7] the right to receive compensation for violation of Article 21 of the Indian Constitution or the right to life has been crystallised. Alongside cases of communal violence, government has been ordered by the Supreme Court to pay compensation when a mother of three children with an infant in her womb died of gas leakage in a godown of UP Cooperative Union Ltd,[8] and also when a fire broke out in *jhuggis* (slums) in Delhi due to short-circuiting of illegal electricity connections.[9]

Recent socio-economic legislations have recognised that compensation forms an important part of ensuring that these rights reach the beneficiaries in a timely manner. Most recently, MGNREGA, which assures 100 days' work each year for one person per family, provides for compensation to workers if their wages are delayed by more than 15 days under Section 3(3) of the act. Schedule II of MGNREGA further provides that the compensation for delay in payment of wages should be paid in accordance with the provisions of the Payment of Wages Act, 1936.

Punitive Measures

Under Indian law, punitive measures such as fines and penalties mostly fall in the domain of administrative law. Competent authorities have been created in various government departments, sometimes in the nature of disciplinary committees that have the power to impose fines and penalties for failure to perform functions incumbent on the officers. The appellate administrative tribunals that have been established under various statutes and rules can review the orders passed by the disciplinary committees.

In general, the courts can review matters related to fines and penalties imposed by administrative or executive authorities, but these powers are circumscribed by certain principles, most notably the Wednesbury principle. The Supreme Court has held that to

> judge the validity of any administrative order or statutory discretion, normally the Wednesbury test is to be applied to find out if the decision was illegal or suffered from procedural improprieties or was one which no sensible decision-maker could, on the material before him and within the framework of the law, have arrived at. The court would consider whether relevant matters had not been taken into account or whether irrelevant matters had been taken into account. The court would also consider whether the decision was absurd or perverse. The court would not however go into the correctness of the choice made by the administrator amongst the various alternatives open to him. Nor could the court substitute its decision to that of the administrator.[10]

The court further observed:

> whether in the case of administrative or executive action affecting fundamental freedoms, the courts in our country will apply the principle of 'proportionality' and assume a primary role, is left open, to be decided in an appropriate case where such action is alleged to offend fundamental freedoms. It will be then necessary to decide whether the courts will have a primary role only if the freedoms under Articles 19, 21 etc. are involved and not for Article 14.[11]

Debates around the NFSB Draft of the NAC

As part of the grievance redressal mechanism, the contours of compensation for persons whose rights are affected, and the imposition of fines and penalties for non-performing public officials, were discussed at great length. The NAC draft addressed the offence of dereliction of duty on the part of public servants in implementing the provisions of the act (Sections 85, 86). It further provided for penalties (Section 87). The power to ask the relevant authority to grant compensation is perhaps the most important immediate relief that an individual can seek under this act. We must remember that often these may be persons living with hunger and starvation. Here, it must be noted that the word 'compensation' is not even mentioned once in the NFSA of the Government of India. No provision for compensation has been outlined.

Compensation

The principle of providing compensation for harm done due to violations by a public official or the failure of the state to enforce socio-economic rights is slowly being recognised in India. The first socio-economic legislation that provided for compensation for failing to provide employment or for delay in payment of the wages was MGNREGA, but it has largely been a weak provision.

Under MGNREGA, compensation is provided for if the wages to be paid under the act are delayed by more than 15 days. However, largely, this provision has remained on paper, and not been implemented on the ground. Most of the states have not made any rules regarding the payment of compensation, even though MGNREGA prescribes that the quantum of payment should be in accordance with the provisions of the Payment of Wages Act, 1936. While the raison d'être of the drafters in tying the amount of compensation to the Payment of Wages Act, 1936, was limited to calculating the quantum only, unfortunately, it has been interpreted to mean that labour courts should step in to decide questions of compensation under MGNREGA.

It was envisaged by the drafters that the payment of compensation under MGNREGA should be automatic in each and every case of delay, and that the Payment of Wages Act should be consulted only for the limited purpose of calculating the compensation.[12] But now due to the interference of the labour courts, which are in themselves inefficient and corrupt, the issue of compensation under MGNREGA is unnecessarily getting stuck within this system.

With the exception of Andhra Pradesh, where compensation is paid directly into the accounts of the workers whenever there is a delay in payment, none of the other states has provided for automatic payment of compensation. In the states of Jharkhand, Rajasthan and Karnataka, compensation has been paid in a handful of cases in which pressure was brought to bear on the state government by organisations and activists. Perhaps the most important learning from the consistent failure of payment of compensation across states under MGNREGA is that the quantum of compensation is not as significant as ensuring that it is paid in an efficient and timely manner. If compensation is paid in a timely manner, only then can the provision be useful to beneficiaries of socio-economic rights based legislation. Obviously, the quantum should not be merely symbolic or a pittance, but rather mindful of the harm that has occurred to the individual.

In this context, the discussion on compensation under the NFSA centred around two major issues: first, who should pay the compensation, and second, what should be the quantum of compensation granted to individuals whose rights were affected under the act? While it was largely felt it was the duty of the state to provide compensation, another viewpoint that was articulated was that the negligent or errant public servant should also contribute to the total compensation amount. However, the consensus was that the government should be made responsible for paying compensation to the victims. On the question of who should pay, if the liability is the public official's, only then would it serve as a deterrent. But the problem is that this would amount to a punishment and cause delay in disbursement, as the liability would then have to be imposed after the due process of fixing responsibility. On the other hand, if the liability is the state's, then all we need to establish is the failure of the right, and not who was responsible.

There were varying views on the quantum of compensation to be paid. It was felt that three times the cash amount of the entitlements that were denied would not be adequate compensation given that the victims may often be quite impoverished, and the delay in receiving entitlement would cause them extreme hardship. In this regard, a few members of the drafting committee felt that the compensation should be 10 times the cash amount of the entitlements that were denied. In the end, it was decided that the actual amount of compensation should be left to the discretion of the granting authority, but in no case should it be less than three times the amount of entitlements that were denied. This would ensure that if the facts of a particular case demanded exemplary compensation or damages, the grievance redressal authorities could exercise their discretionary powers. But it also threw up some difficult questions: what would be the cost of a starvation death or a child's malnutrition, which causes permanent harm?

While there was no resolution for difficult questions such as what amount of compensation could be paid for starvation deaths or other serious violations under the act, compensation in and of itself was felt to be very important. In its final version, Section 88 of the NFSB draft of the NAC provided that when it is found that any person has been deprived of his/her entitlements under this act, he/she shall be entitled to compensation to be payable by the appropriate government or local body whose duty it was to provide such entitlement as the case may be, of a minimum of three times the cash equivalent of the entitlements that have been denied. In addition, several other details for compensation were outlined in the draft. Furthermore, the National Food Commission was granted the power to direct the appropriate government to grant compensation or damages to the complainant or to the members of his or her family as the commission may consider necessary in accordance with Section 88 of the draft act.

Punitive Measures

Punitive measures, such as imposition of fines or penalties for the violation of rights, is recognised under Indian law. Under the RTI Act, if a public information officer fails to provide information without any reasonable cause, the central and state information commissions can impose a penalty of 250 rupees each day till the application is received or the information is furnished, up to a maximum amount of Rs 25,000.[13] The burden of proving that the officer acted reasonably and diligently lies upon the erring official. But mostly, as in the RTI Act, these penalties are civil in nature, and do not entail a jail sentence for wrongdoing.

Nonetheless, the penalty provision in the RTI Act has ensured that the public information officer fulfils the duty imposed on him/her under the act. The penalty provision has helped in ensuring compliance of public officials with the provisions of the RTI Act. Furthermore, the RTI Act provides that the central and state commissions shall also recommend disciplinary action against the central public information officer or the state public information officer, as the case may be, under the service rules applicable to them. However, state or governmental institutions are still resistant to the idea of accountability, and this provision has not been used as effectively as possible. But the threat of the penalty has surely acted as a deterrent for public officials.[14]

At the same time, if a public official commits a criminal offence, then he or she would be liable to criminal action under the provisions of the Indian Penal Code or some other statute prescribing criminal punishment. But the prosecution of public officials runs up against the bar provided in Section 197 of the Code of Civil Procedure that provides for compulsory sanction before civil servants can be prosecuted for criminal offences, which has often impeded criminal trials.

Even during discussions on the NFSA, these questions were debated at length. Should there be fines or penalties for violations under the act? Should these be civil or criminal in nature? While there was near-consensus that these penalties should be civil in nature (also because criminal penalties can usually only be levied by judicial bodies), it was considered difficult to affix responsibility on the individuals responsible. It was thought that detailed job charts containing the responsibility of the various officials should be drawn up, which would make it easier to identify the official responsible for the entitlements under the act not reaching the intended beneficiaries. This would assist in fixing liability upon erring officials and imposing penalties. But the question remained of how to ensure that liability is fixed at senior and command levels and not merely at the junior-most peg in the wheel.

In the final draft presented by the NAC, the district, state and national grievance redressal authorities were given the power to impose penalties. More specifically, Section 87 of the draft NFSB stated that if, while deciding a complaint or appeal, the DGRO or the State Food Commission or the National Food Commission is of the opinion that any public servant discharging duties under the act is, without any reasonable cause, guilty of the dereliction of duties in accordance with this act, a penalty not exceeding Rs 5,000, at 2010–11 prices, shall be imposed in the first instance, and Rs 100 for every day thereafter until the relief, as directed by the DGRO, is granted. It was also provided that the penalty could be deducted from the salary of the public servant. It was suggested that the pool of money collected from the officers as fines could then be used for general development and welfare of the village or of several villages in a block.

There was a lot of internal discussion and debate on whether or not these grievance redressal bodies should have the power to impose fines and penalties. It was felt that government officers responsible for the implementation of the act, particularly at the block and village levels, might themselves be receiving modest salaries, and therefore it would be too harsh to impose high fines on them for non-performance of their duties. At the same time, it was felt that there should be some monetary censure of non-performing officials, as their negligence could lead to hunger and starvation deaths. Therefore, it was felt that at least some penalty should be imposed on errant officials to ensure compliance with provisions of the act.

Moreover, it was felt that an element of *mens rea*, or some form of intent, should be introduced in sections that imposed fines or penalties on erring public officials. This could be couched in language such as 'reasonably detects violations', or 'deliberately fails to detect violations', as otherwise many if not most officers in India would be guilty of violating the act no matter how diligently or efficiently they performed their duties. However, the 2013 legislation fails to acknowledge that for effective implementation of the act, it is necessary that the district grievance redressal authorities should have the power to impose fines and penalties as well as disburse compensation. Instead, the power to impose fines and penalties (not exceeding Rs 5,000) on erring officials rests only with the state and national commissions (and their chosen adjudicating officers).

General Comments on the Nature and Role of Enforcing Institutions

In addition to the specific discussions on punitive measures and compensation, there were also general discussions on the overall framework of grievance redressal and implementing mechanisms, which would naturally have an impact on the issues of penalties and compensation. What is the role of these institutions, particularly the national commission? Differing opinions were put forward. In some senses, they were seen to have a dual role, as they can formulate guidelines as well as provide redressal if these guidelines are violated. A general comment is that in its current form, the pyramidical structure of the officers and commissioners created by the bill presents some separation of powers concerns. At each level, prosecutorial and adjudicatory powers are combined, and at some levels policymaking powers are also given to the same group of individuals. This would mean that they would have to punish themselves if they do not properly implement the act.

Beyond conflict of interest, this structure also creates pragmatic difficulties. For example, compensation should be given to individuals at the level of the DGRO so as to incentivise complaints with timely redressal. However, officers who are convicted by the DGRO will certainly appeal. A poor individual will not appeal or continue to argue the case. Therefore, either the DGRO will have to argue the case at the state level (and their other prosecutorial and adjudicatory responsibilities under the act will be nullified), or a separate prosecutor needs to be created to carry the appeal forward on behalf of the poor person.

Notes

1. Quote from the ICTJ website, www.ictj.org.

2. UN, 'Basic Principles and Guidelines on the Right to a Remedy and Reparation for Victims of Gross Violations of International Human Rights Law and Serious Violations of International Humanitarian Law', http://www.ohchr.org/EN/ProfessionalInterest/Pages/RemedyAndReparation.aspx (accessed 12 March 2018).

3. Ibid.

4. *Minister of Health v. Treatment Action Campaign* (no. 2) 2002 (5) SA 721 (CC) (S. Afr.), http://www.tac.org.za/Documents/MTCTCourtCase/ConCourtJudgmentOrderingMTCTP-5July2002.pdf (accessed January 2012).

5. *Government of the Republic of South Africa v. Grootboom* 2001 (1) SA 46 (CC).

6. *D. K. Basu v. State of West Bengal*, AIR 1997 SC 610.

7. *Nilabati Behera v. State of Orissa* (1993) Crl. LJ 2899.

8. *Naresh Dutt Tyagi v. State of UP*, 1995 Supp (3) SCC 144.

9. *Harvinder Chaudhry Srivastava v. Union of India* (1996) 8 SCC 80.

10. *V. Ramana v. APSRTC & Ors* (2005) 7 SCC 338.

11. Ibid.

12. The following discussion on the issue of compensation under MGNREGA draws heavily on comments received from Mr Nikhil Dey.

13. Section 20(1), RTI Act, 2005, http://rti.gov.in/rti-act.pdf (accessed 11 November 2017).

14. The following discussion on the issue of penalties under the RTI draws heavily upon comments received from Mr Nikhil Dey.

19

Accountability and Transparency

Sajjad Hassan

Experience shows that despite the best of designs of programmes—especially those for the poor and marginalised—performance and efficacy are a problem. This can be overcome only if we confront directly the issue of accountability, ensuring that, using a concept borrowed from economics, the 'agents' (meaning here, officials and providers) are responsible to the 'principals' (standing here for the rights bearers) for effective working of the schemes and delivery of the latter's entitlements. Accountability, unless approached explicitly and directly, tends quickly to regress into meaningless ritualism, at the loss of citizens.

Explicit in NFSA 2013 are clauses making the supply side more transparent—through IT reforms such as end-to-end computerisation; 'leveraging Aadhaar' or biometric information for proper targeting of beneficiaries (however, this is hotly contested because of its potentialities for misuse); but more importantly, through public disclosure of records pertaining to the TPDS, although the manner in which this information will be disclosed is unspecified and left to the discretion of the state governments. The act also seeks to make the demand side more proactive by: (*a*) establishing adequately represented vigilance committees at the local level who regularly supervise the implementation of the act and hold the DGRO accountable; and (*b*) conducting necessary social audits through local bodies or independent agencies.

Delivering Food Rights: The Supply and Demand of Accountability

It is helpful to approach the design of accountability systems as comprising a dyad, with a *supply-side* set of interventions, and a *demand-side* set. Supply-side interventions are about creating systems, processes and practices favourable to the establishment of robust systems of answerability between different parts of the provider organisation, and between the organisation and the citizen, together ensuring that the entitlement of citizens is really delivered. Citizen charters, job descriptions, work plans all contribute to that larger objective. Demand-side interventions, on the other hand, are those systems, practices and spaces that enable rights bearers to make further and continuing demands on the system/organisation to obtain their rights. Transparency measures and community participation in planning and monitoring are examples of demand-side tools contributing to effective accountability.

What is important here is to see the two sides of the dyadic relationship working together to produce the right results for the poor. Often, in government reform programmes, all efforts are made at ensuring refinement of supply-side interventions, creating major design changes, elaborating systems of command and control between different levels of the organisation, including through use of IT, and creating spaces for frequent reviews and periodic monitoring. But in the absence of citizen participation (demand-side effort), exclusive investment in internal accountability measures (vertical or horizontal), however well meaning, very often reduce to box ticking and number crunching exercises, with little relevance to the working of programmes on the ground.

It is equally fallacious to assume that mere demand making and citizen activism will, by itself, solve the problems of programme failure and poor delivery. While this 'people approach' is admittedly a reaction to the utter failure of food programmes for citizens, and the complete absence of citizens' voices in their management and oversight, by themselves they can do little to solve the problem. Citizens' activism helps create awareness and a push for performance and results, as transparency measures through the use of RTI and 'public hearings' and 'social audits' of food programmes have done in India in the context of the right to food movement. But for awareness and activism to result in effective delivery of services in a systematic and institutionalised manner—rather than episodically in response to specific cases of public agitation—requires activism (demand side) to link up to reforms within the organisation (supply side) for effective management of the programmes. Successful cases of reforms working for the poor have seen the two sides of the supply–demand dyad working together: a sensitive and efficient delivery structure focusing on results while creating spaces for citizens to engage; and citizens' groups making use of those spaces to inform, empower and correct state actors to provide results in the interest of citizens.

It is important in reform exercises to talk about the demand side (as against only the supply side), so that rights bearers have the realisation that their 'rights' have been denied; when this occurs, they decide to do something about it, take the trouble of approaching a grievance redress authority or mobilise a collective to do that, and keep up the pressure (at various levels of administration) until they obtain justice. Demand-side measures are important especially in poorer countries where education levels are low; where there is a large section of the marginalised; where civil society is weak; and where public systems mostly need to be forced into action. It should not be assumed in those contexts that once a 'right' has been created, through legislation or otherwise, rights bearers (the weakest here) will automatically use 'voice' to their benefit. They mostly neither have the awareness nor the means or organisational capacity to make use of the opportunity. To enable this to happen requires creating systems and processes promoting 'voice', building the individual and organisational capacity of rights bearers, and creating networks among others—exactly the sort of things that the right to food movement does in India the context of food security, or the national campaign for people's right to information in the context of the transparency/RTI movement in the country. So what are the transparency and accountability issues that have been thrown up by these movements?

For one, taking a leaf out of the experience of RTI and MGNREGA mobilisation as well as the experience of their implementation over the past half decade or so, civil society organisations have, in the context of the right to food campaign and legislation, pushed for building into the food law specific and enabling provisions that would tilt the scale in favour of citizen, to hold providers of food entitlements accountable to the law and rules. These provisions and spaces are mostly about transparency, including in the decision-making process; rights of stakeholders to demand, obtain and examine official documents and other proof; and institutionalising stakeholder roles in decision making, such as in mandatory social audits and in vigilance and oversight committees. With the proposed food security act seen as encompassing the working of the largest pro-poor programmes in the country, pro-reform citizens' groups have seen this debate and consultation as an opportunity to raise the bar on transparency and accountability measures in the provision of public services for the poor, sometimes learning from the lessons of the RTI Act and MGNREGA, sometimes drawing fresh lines and building fresh yardsticks. The NAC's draft NFSB, which came out of a process of intense consultations with civil society groups, especially the Right to Food Campaign, thus pulled together all these ideas and thoughts, when it proposed the following principle for transparency—proactive mandatory disclosure; inspection of all documents, of sites of delivery of entitlements, and offices; making accessible copies of documents, records and samples of material ensuring the transparency of the decision-making processes, through social audits and implementation of RTI, 2005.

The premise is that if all documents are freely available to all interested parties, especially those directly affected by a public scheme, and that the bases of decisions arrived at by public servants are also open, offering spaces and opportunities to demand answers from the providers on the working of schemes, there would be enough incentive in the system to ensure that public oversight led to checks on arbitrary decision making and improvements in management and implementation. The NAC draft sought to make 'proactive disclosure' by the providers, already present in the RTI law of 2005, into a mandatory requirement. It also sought to give powers to ordinary citizens to examine related public documents and inspect stocks and sites of point of delivery. But most importantly, it sought to consolidate the gains of social audit and vigilance committees, already introduced in recent laws (MGNREGA, for example) into institutional spaces for those aggrieved to demand and obtain from their providers not only answers but actual relief and entitlements. And it sought to give legal authority to those provisions and drill in accountability by making it mandatory for public authorities to provide that relief within a specified time frame, on pain of punitive action.

Government accepted these provisions of the NAC when it included in the NFSB, laid before Parliament, provisions asking for all records to be placed in the public domain and open for inspection; making social audits by public agencies, or by independent agencies, mandatory for PDS and other food-related schemes, making their findings public and taking appropriate action on them; and making it mandatory for states to set up vigilance committees and ensure they act as effective watchdogs. After the passage of the legislation, a greater test is its rigorous implementation, as to the degree to which intent (of building accountability through a mix of supply- and demand-side innovations) is able to be translated into on-the-ground reality for the poor.

The supply–demand dyadic structure elaborated above raises two issues that are important for this partnership to work. One concerns the responsiveness of the delivery system to demands thrown up as a result of engagement with citizens' groups. Unless the system is designed and willing to act in a sustained manner in response to information and 'voice' created as a result of engagement with citizens, the net result will be naught, in most cases creating expectations that are then not met, resulting in further anomie. A useful example here comes from the significant innovation of 'social audits'. Most pro-poor programmes—and definitely those for food security—now have an element of social audit built into them almost by default. The principle is simple: internal systems of monitoring and evaluation and auditing of public programmes are clearly not working, as their performance continues to be below par. But these schemes exist for citizens, who must have a say in ensuring their working. One way to create spaces for citizens is to give them a role in auditing the working of those programmes so that, given the incentives affected citizens have to ensure sound working of the programmes, they will act as better drivers of change than 'agents' themselves.

'Social audit' requires agents to ensure that, as part of their periodic reporting exercise, they will also ensure the scheme is audited in public with all stakeholders being informed of the details of the project or scheme, and hold a public hearing on performance with inputs obtained from members of the public. Reports of this public exercise, in what was until now only an internal audit, are necessary documentation today. The assumption is of course that agents will be forced into taking corrective measures due to the evidently public nature of the auditing exercise. What is happening in most situations, however, is that social audits and reports are written out ritualistically, without adequate preparation for ensuring that the hearings are effective, representing the range of opinions on the working of the specific project/scheme; worse, there is little follow-up on the findings of the audits, thus little action taken on the findings and recommendations. The result is that people are slowly losing faith in the social audit innovation. Where implemented with due preparation and regard to impact on programme performance, as in attempts in Rajasthan and Andhra Pradesh, social audit seems to be having the expected results.

The lesson coming out of Rajasthan and Andhra Pradesh also links up to a debate under way on social audits, specifically their manner of implementation. The debate is really about whether government agencies—service providers—should be taking on the conduct of social audit, or whether they should be enshrined as legal processes with non-state and preferably community organisations and groups conducting them, and the providers being duty bound to respond to the findings. But as is evident, it's a bit of a non-debate really: unless state agents—the providers—assume ownership of accountability measures, including social audit, their ability to respond to the findings of an independent audit might be limited, thus defeating the very purpose of the exercise. On the other hand, even to be able to conduct social audits in an effective manner and on a scale large enough to be of more than occasional interest, community organisations would need the support of the state for capacity building, organising, and documentation and data collection. All this is possible if state providers work in partnership with community organisations to plan, manage and make use of social audits as an institutionalised tool to manage their programmes effectively. This is precisely what the governments of Rajasthan and Andhra Pradesh have done with their new directorates of social audit—community organisations helping state departments plan and conduct, and then the departments taking ownership of the findings and acting upon those resolutely.

Finally, the supply–demand framework points to the necessity of state agents working in close and meaningful collaboration with non-state actors to strengthen accountability and improve programme performance. Demand-side measures naturally belong to the realm of the non-state. These include articulating demands, public activism, people's movements, development NGOs and the like. Yet, it is for the state to create the spaces and opportunities and nurture and empower citizens' groups to be able to strengthen people's voices to engage effectively with state agents. The state itself cannot act as a proxy for citizen groups. This is as it should be. Citizens' groups have the range and reach in local communities that state agents do not. They aggregate and articulate public opinion, especially those of specifically targeted groups, something that even political parties are unable to do.

For any meaningful drive towards accountability, the state must work with non-state actors at the policy and design level, but also critically at the cutting edge, in blocks and villages and hamlets, to ensure that pro-poor policies are effectively translated into results on the ground. This is where citizen voices and concerns can be fed back directly into the working of schemes and programmes—helping empower providers to become learning organisations so as to maximise performance. This is also the level where providers can partner with non-state agents to reach out to citizens, mobilising communities and expanding the reach and depth of schemes to cover more and more of the excluded and the voiceless, thus further enhancing impact.

Voicing Demand: Transparency, Participation and People's Organisation

Evidence shows that in India, and globally, information and transparency work wonders when it comes to creating 'voice' for demand making by all, but particularly the poor, for services and entitlements. Other powerful demand-side interventions are creating organisations of the poor and mobilising them around pro-poor agendas such as the right to food, right to information, or indeed the right to basic services. The following sections look briefly at helpful interventions creating transparency and enabling community participation in the monitoring of food schemes.

Any food security regime relies on an effective, appropriate and citizen-friendly transparency environment, accessible to all. The transparency regime for provision of information to any person should include proactive mandatory disclosure; inspection of all documents, of offices, and of sites of delivery of entitlements; making accessible copies of documents, records and samples of material; and ensuring transparency of the decision-making processes. Clear timelines need to be devised for responding to requests for information within specified durations. This should include details of the type of information to be recorded, the manner and pro forma in which it is to be recorded and made public; the authority by and at the level of which this is to be recorded; the mode and the process by which the public can ask for the information; and other relevant details related to each entitlement, in rules to be specified by governments. Cost to be paid by a person obtaining information should not only be reasonable but should in no case exceed the market cost of making copies of documents.

Planning, Implementation, Monitoring and Evaluation

The process of planning, implementing, monitoring, evaluating, and the consequent corrective processes undertaken for all interventions carried out to deliver guaranteed entitlements should be transparent. The monitoring and evaluation process should include:

1. Ensuring access to entitlements.
2. Specifically ensuring that every potential beneficiary has information card(s) that provide appropriate information about food entitlements and details of the cardholder. These cards could also contain information regarding entitlements, status, all transactions including measurements, weights and statistics relevant to the act, and with a clear timeline of delivery.
3. Creating universal awareness of entitlements, of the process of access to these entitlements, about the officials/functionaries responsible for ensuring access to entitlements, and information about the process by which grievances/appeals/complaints are to be registered, and manner of processing and disposal of grievances/appeals/complaints.
4. Mechanism for speedy and effective redressal of grievances, appeals, corrective/punitive action on complaints.
5. Clear parameters and definite time frames for delivery of entitlements, and clearly laid out process of evaluation of performance.
6. Mechanism to fix responsibility and corrective action on the basis of the evaluation.
7. Budgets, accounts and statements of expenditure, and stock registers along with other material relevant for monitoring expenditure, as further specified in the schedule and/or rules of the act.

Proactive Disclosure

Agencies implementing food security schemes should make information available to citizens without their seeking it, or needing to apply for it. All important documents related to food security legislations/programmes/schemes should be proactively disclosed to the public. Details of content and other information required to be put out proactively, and their periodicity, should also be made public. Proactive disclosure should include the dissemination of records in a functional form, so that a layperson is able to make sense of the information and use it to advance their rights. This would include the dissemination of information in a consolidated and summarised form wherever appropriate. Proactive disclosure could include:

1. Reading out aloud of essential information in a useful, easy-to-understand format.
2. Displaying information at prominent public locations, including offices of local bodies in villages and districts, again in helpful, easy-to-understand formats.
3. Putting out information through newspaper reports/advertisements, press releases or information sheets/leaflets and reports, and through radio and TV announcements, among others.
4. Making available key records on the internet. A web portal of food security related schemes and programmes would be a helpful step to provide and make information accessible. The portal should be a gateway for information on schemes, guidelines as also infrastructure and staffing, and budgets, funds and material releases for various components of the schemes. These ought to be user-friendly, with user interfaces allowing citizens to make complaints and seek information.

Vigilance Committees

Equally useful is involving citizens and local communities in delivery as well as monitoring of food schemes. Successful examples from India of community monitoring include those of FPSs through establishment of local-level vigilance committees, made up also of beneficiaries of food schemes and weaker sections of society, with powers to inspect and report violations of guidelines. Similar institutions directly giving beneficiaries a say in monitoring food schemes exist also in the ICDS programme of nutrition, with the establishment of mothers' committees for each feeding centre (anganwadi).

Social Audits

Statutory social audits should be conducted of all food security schemes, to further provide opportunities for community-based monitoring, and participatory auditing of quality, performance, financial expenditures, and services and entitlements. These could be of two types:

1. First, a process of continuous public vigilance, where any person can, at any point of the planning, implementation or evaluation process, ask questions and demand answers about the functioning of the act.
2. Second, a mandatory joint audit of all food schemes by rural and urban local bodies every six months, on pre-announced dates. This should include auditing of receipts and distribution of foodgrain at each FPS, and would include reading aloud in public of a summary of transactions in the previous six months.

The following principles of social audit ought to be ensured:

1. The process should be universal and open.
2. It should allow for ongoing social audit (concurrent) as well as post facto audit.
3. The geographical unit of social audit ought to be decided according to the appropriate unit for delivery, e.g., FPSs, anganwadi centres, primary health centres, etc. Social audits may also be held at the panchayat, block and district levels. While the focus ought to be on participation of residents and beneficiaries, social audits should be public events, with free access to anyone interested.
4. Governments should develop clear standard operating procedures on social audit, clearly identifying methods and procedures for conducting social audits, as also how findings of such audits would be used by agencies implementing food schemes as well as grievance redressal authorities/oversight bodies and watchdogs in their judgments and awards.

5. Governments would need to develop capacities for social audit among stakeholders. It would also require to facilitate conduct of social audits. Institutional mechanisms and capacities would need to be developed to drive the social audit agenda, such as a separate social audit unit or directorate, as in the case of the state governments of Andhra Pradesh and Rajasthan. These would act as capacity builders and facilitators, handholding agencies on the ground conducting social audits.
6. Independent audits, carried out by civil society or citizens' groups, of the working of any scheme, or any aspect of any scheme, ought to be encouraged. These should be taken cognisance of both by agencies implementing various schemes, and grievance redressal authorities set up under food security regimes.
7. There should be necessary administrative and legal outcomes of social audits. Their findings cannot be directly actionable for punishment, because this could rapidly deteriorate into kangaroo courts. But likewise, they should not be ignored—their findings should have the same status as a preliminary investigation, and should mandate detailed investigation into the findings of the social audit, and suitable administrative, civil or criminal consequences should follow.

Telephone Helpline

Other participation tools include a special toll-free telephone-based helpline in each district to help citizens secure their entitlements. This is a helpline number, displayed prominently, that people could access to ask questions and clarify doubts; make complaints about the working of schemes; and obtain updates on their complaints.

Experience shows that despite the most citizen-friendly laws and programmes, it is only when robust accountability mechanisms are built into these, and where citizen groups and pro-poor platforms exercise 'voice', that the desired results can be realised. The right to food, especially for those on the margins, can be realised if accountability mechanisms embedded in the law are practised by duty bearers and rights holders exercise the opportunities these provide. The catalogue of helpful mechanisms and tools listed in this chapter, if applied robustly, can have meaningful impact on the ability of the poor to access their right to food.

20

Duty Bearers

SABA SHARMA

In any rights-based legislation, there are five necessary elements: (*a*) the content of the legal right which is being created; (*b*) the rights holders, or individuals or groups to whom these legal rights accrue; (*c*) the duty bearers or agency (or agencies) who have the legal duty to ensure that the rights holders access the designated rights; (*d*) consequences for failures to ensure that these rights accrue, which would necessarily fall upon the legal duty bearers; and (*e*) the institutional mechanisms to ensure that these consequences for failures actually accrue to the defaulting duty bearers. The question this chapter will examine is: who should be the appropriate duty bearers to ensure socio-economic rights such as the right to food? Who should bear the legal duty to ensure the right to food of all? Should it be individuals, families, local communities, private charities, the private sector or the state?

I shall elaborate the broad consensus in various official drafts of the food legislation in India, that whereas each of these groups has a moral duty to ensure the right to food of all, the legally enforceable duty for the right to food must lie with the state alone. The chapter will examine the legal, moral and philosophical rationale behind the state being held responsible for food security, and for ensuring that people have adequate access to food and other socio-economic rights. While it is now widely accepted that civil and political rights must be protected and upheld by the state, holding the state accountable for socio-economic rights is a relatively recent idea.

The responsibility for implementation of the NFSA lies with the regional states. Given the federal nature of the Indian polity, the state government has the predominant duty to set the eligibility criteria and distribute the foodgrains, while the central government has the duty to procure the foodgrains and make sure there is adequate stock. This arrangement emerged over time against the backdrop of fierce demands by regional parties for the centre to bear greater cost-sharing responsibilities and give support to state governments, while simultaneously wanting a more decentralised legislation that would give individual states a greater ability to tailor the act to their respective needs. However, in this context, the NFSA excluded the recommendations of the Parliamentary Standing Committee on greater cost sharing between centre and states and further decentralisation, such that states voluntarily disclose information to the gram sabhas at least twice a year regarding availability of foodgrains; inclusion or exclusion of beneficiaries; appointment and termination of FPSs; quantity of foodgrains received and distributed; and other relevant information. In the discussions around the various draft bills on food security, a loud emphasis remained on decentralisation of powers to local bodies and self-governance units.

Moreover, the act allows room for the entry of private contractors to provision food at anganwadis and for midday meals at schools without explicitly defining their obligations.

A Rights-based Approach

In recent times, the rights-based approach to development has gained more and more prominence, and has also led to a discourse about the need for socio-economic rights. A paper from the Danish Institute for Human Rights differentiates a rights-based approach from both charity and needs-based approaches, which have so far guided aid and development policy in most nations, particularly developing countries, although this is beginning to change. In a charity-based approach, for instance, the emphasis is on the inputs, and recipients of charity are often seen as victims. In a needs-based approach, people are seen more as objects of intervention. In both approaches, it is agreed that the individuals in question deserve assistance. In a rights-based approach, however, such assistance is seen as an *entitlement*. The focus is as much on the structural causes of the deprivation and the nature of the intervention as it is about providing assistance; it is about empowering individuals, and not merely providing services. In this sense, socio-economic rights are not about acts of charity or need-based interventions, but recognition that something has been denied, and that the situation has to be changed (Boesen and Martin 2007). The recognition of denial must necessarily also address further questions about the structural nature of the problem, as discussed earlier, and entitlement must then transcend notions of deserving and non-deserving recipients, since rights must be applicable to everyone.

While a rights-based approach necessarily creates a set of rights holders, it also identifies a set of duty bearers, i.e., those who will be responsible for upholding and protecting these rights, and who take responsibility when the system fails. Until socio-economic rights began to be legislated, the state, while deemed responsible for many of these 'services', was conceived of as providing 'welfare', and could not be held legally accountable when the welfare system failed. (It could, of course, be held morally accountable, and also politically accountable during elections for these failures.) In addition, because socio-economic rights were not seen as entitlements, the size of the welfare state could shrink or expand in different periods of time, often dependent on the health of the economy or the ideology of those in power. In more recent times, the advent of neoliberal policies has meant that 'structural readjustment' has taken place in many nations, particularly developing nations, in some cases leading to a progressive retreat of the state from the area of socio-economic development. The importance of a legal regime of rights is that these programmes graduate from welfare to legally enforceable entitlements.

The Role of the Individual and the Family

Development economist Jean Drèze explains that individuals and families, as well as other institutions, cannot be exempt from the duty to provide food. He points out that while the state holds the primary responsibility for upholding the constitutional rights of individuals, and that it has the resources to do this best, it is not the only one responsible:

> However, the right to food is not the responsibility of the state alone. To illustrate, suppose that I come across someone who is dying of starvation on the street. If I am able to do something about it, and if I recognise that every citizen has a right to be free from hunger, it would clearly not be right for me to wash my hands off the situation and say that it is the responsibility of the state. The fact that the state bears the primary responsibility for letting this happen does not absolve me from the duty of intervening, if I am

> in a position to do so. In other words, in some circumstances at least, the responsibility for protecting the right to food is a shared responsibility, involving not only the state but also other institutions or individuals. (Drèze 2004, 1726)

Drèze takes the example of a girl who is undernourished. At one level, it is the state that is responsible, but at another, if the parents control the distribution of food within the family, then theirs is also an important role in the larger context of her health, and it is difficult to say who should be held responsible to what extent.

However, despite this, there is no argument for the *legal responsibility* of the family in the case of the right to food. None of the drafts of the food security bill, whether from the NAC or the government, accorded any legal responsibility to the family where access to food is denied. This is not the view held by the Right to Food Campaign either, which also argues that it is the state, first and foremost, that is responsible. Nor is the legal duty of individuals and the family upheld in any of the human rights instruments, like the UDHR or the ICESCR. Even Drèze, while saying that the dynamics of inequity within the family cannot be overlooked, argues that the primary legal responsibility must lie with the state.

The reason for this is, first, that the family may be too impoverished and deprived, or discriminated against, to provide food, and it would be doubly unjust to blame and punish the victim for their oppressed and deprived condition. There are also, admittedly, possibilities of discrimination within the family, as pointed out by Drèze. This may be against girls or women, or the aged or disabled. There are even graver instances of irresponsible or alcoholic and violent parents, and of child abuse. But the belief is that both the protection of vulnerable children and action against child abuse and domestic violence should occur under dedicated laws such as the Juvenile Justice Act and laws against child sexual abuse and domestic violence. The provision of food in a modern society should remain the legal duty of the state.

It is interesting and instructive to observe parallel discussions in the context of the Right to Free and Compulsory Education Act, 2009, or RTE. When the RTE was passed in India, much of the debate revolved around who the primary duty bearer should be. One position was that while the term 'right' implied that there were children whose parents wanted to send them to school but could not, and that the state became primarily responsible in this case, the notion of 'compulsory' education also placed some responsibility on parents, implying that there may also be parents unwilling to send their children to school, and that they must be compelled to do so. It was argued, furthermore, that in many cases children's employment is deemed by very poor families as essential to the family's survival, and in this case, instead of penalising the family, strategies should be devised so that children are able to study as well, and do not suffer because of the adverse conditions in which their families find themselves. Therefore, it is necessary for the state to intervene (Mehta and Kapoor 2010).

In the final RTE law passed by Parliament, families are mentioned as duty bearers as well. The act states that it shall be 'the duty of every parent or guardian to admit or cause to be admitted his or her child or ward, as the case may be, to an elementary education in the neighbourhood school'. But what is significant is that the law envisages no consequences for the failure of families to fulfil this duty. The reasons for this are the same as mentioned earlier in the context of the right to food: that impoverished families which do not send their children to school are victims themselves, and it would be unjust to punish them doubly for their suffering. The principal duty bearer in the RTE is the 'appropriate government', which is in various contexts the central, state or local government. However, the act does not really specify the extent of the family's responsibility, or how it is placed in relation to the responsibility of other actors, such as the state, civil society or the private sector.

In the draft right to food bills of the NAC and the government, the responsibility of the family or parents in the context of children is not stated anywhere. Instead, the NAC's draft says, in the context of child malnutrition, that 'Parents or guardians of every malnourished child shall be provided support through nutrition counselling for appropriate therapeutic foods, health checkups and referral services, free of charge' (NAC 2011). Thus, here, it is seen as the state's responsibility to help out impoverished parents instead of making them responsible for the situation in which they find themselves. While abuse and neglect are a reality in many families, it can be argued that there are other laws (such as the Juvenile Justice Act, or the Prevention of Domestic Violence Act) to deal with these, and that they need not be part of socio-economic legislations like the right to food or RTE.

The Role of the Private Sector

In the debates on the right to education, the law places primary duty on appropriate governments to secure the right to education for all, but also includes some duties for the private sector. For instance, there is a provision in the RTE that there will be 25 per cent reservation in private schools for children from the marginalised and weaker sections. The thinking behind this was to use the existing infrastructure of many leading private schools, infrastructure that is perhaps among the best in the country, towards the end of making quality education widely accessible not merely to those among the rich and middle classes, but also to the poor and disadvantaged. It was also felt that children on both ends of the spectrum would benefit from interaction with one another. The resistance of many elite and leading private schools to this particular provision has raised pertinent questions about the nature of entitlement to education in India.

Private schools allege that the provision is unconstitutional, and that it is the duty of the state to provide free and compulsory education, and that this responsibility cannot be transferred to the private sector by making the 25 per cent provision mandatory. There are also concerns that while the government has provided for a reimbursement mechanism for the additional costs incurred, these will not be at market rate, and that the result of this would be increased costs for other children, or a deterioration in the quality of education. On its part, the government claims that it is not shifting the entire burden of providing compulsory education, but asking for 25 per cent of the seats to be reserved for children from weaker sections. Moreover, the Supreme Court upheld the provision as constitutional, saying that while the state is obliged to provide free and compulsory education, it is free to decide how this is to be done.[1] The court also added that while private schools have the right to run their institutions as desired, they cannot be exempt from reasonable restriction in the interest of the public, since education is recognised not as a business but as a charitable activity. A dissenting judgment by J. Radhakrishnan states that Article 21A, which obliges the state to provide free and compulsory education, says that the state shall 'provide' and not 'provide for'. To this end, he says, the state must fulfil this duty on its own, and cannot place this duty on non-state actors.

There are diverse, even polarised views, about the role of the private sector as duty bearers in education. One view, held by educationist Anil Sadgopal and others, is that only a state-run common school system should supply education as a right. The current law, according to this view, is equivocal with measures like 25 per cent reservations, and leaves open opportunities for the privatisation of education (Sadgopal 2011). Thus, neighbourhood schools run by the state are the only solution. However, it can also be argued that this will not do away with privilege, since neighbourhoods are also likely to cluster around class, and even real-estate values have been known to fluctuate due to the presence of certain schools in certain neighbourhoods.

There are others who believe the private sector creates models of high quality in education, and even poorer parents are opting for fee-charging private schools in preference to public schools. Here the view is that while the primary duty lies with the state, there is a recognition of the private sector, and certain standards and regulations are laid down for it (such as the 25 per cent provision). However, it is recognised that universal education cannot be attained through private schools, and it has never happened anywhere in the world until now (Raina 2010). Yet another camp believes that the education sector should be opened up much more to the private sector, and that there are also many private schools that do not run on a for-profit basis. It is believed that parents should have a choice, and if they choose to send their children to private schools, more of these should be allowed to come up.[2]

Privatising education would create a new set of duty bearers in the private sector. An editorial in *The Times of India* asserts that the state has the primary responsibility, and acknowledges that passing the buck entirely to the private sector does not help the cause of universalising education. Nevertheless, it suggests that the corporate sector should be allowed to set up schools as 'brands', on private sector principles. Further, the editorial suggests that the government should provide land, tax breaks and scholarships but leave the schooling up to the private schools (*The Times of India* 2008). However, it does not make clear what accountability these schools would have as far as the right to education is concerned, and whether the state would direct the private sector to set up schools in certain parts of the country; if not, how far would this move go in providing education in areas where the most disadvantaged live? Moreover, it leaves us with an interesting question—to what *extent* is the state a duty bearer? Is the duty of the state merely to provide resources and let the market handle the rest, or is the state accountable in a broader sense?

There is an equally passionate debate about the role of the private sector in the provision of the right to food as well. Issues such as cash transfers and the role of commercial interests in preparing ready-to-eat food for children are hotly contested. On one side, corruption in the government and inefficiency of state delivery systems are cited by some as a reason for the entry of the private sector. On the other hand, others have argued that corruption is often as rampant where private contractors are involved, and certainly the ethical dilemma of leaving food security hostage to the profit motive cannot be sidestepped. In the case of both cash transfers and commercial interests (both of which are discussed at length in other chapters in this volume), it has been argued that by delegating responsibility to the private sector, the state is abdicating its own duties towards food security.

The position of many in the government, including the erstwhile Planning Commission and former prime minister Manmohan Singh, was that the role of the private sector must be strengthened in areas such as the PDS; claims to greater efficiency of the market and reduced corruption were offered as the justification for this position (*Hindu* 2011; Planning Commission 2007). It remains a matter of debate whether this is in fact true. More importantly, however, a greater role for the private sector does not necessarily result in greater access to food and nutrition security, and this concern must trump any gains to be made from efficiency. In the framework of rights-based access to food, what must in fact be of the greatest priority is accountability, and how rights holders can claim entitlements. While the logic of the market may provide the most efficient solution, it does not necessarily provide the most just or appropriate one, one that is in tune with treating all rights holders as legitimate.

In recent times, the growth of transnational corporations has made them huge players in the market for food provisions, and evidence of this can be seen globally. For instance, the UN Special Rapporteur for the Right to Food notes that 'Many transnational corporations have revenues far exceeding the revenues of the Governments of the countries in which they operate. Concentration has produced huge transnational corporations that monopolise the food chain from the production, trade, processing to the marketing and retailing of food, narrowing choices for farmers and consumers' (Ziegler 2008, 17). Addressing the unaccountability of such corporations, the report says that there are still relatively few mechanisms in place to ensure that those that often control the supply and distribution of food do not violate human rights. In addition, however, the report stresses that it is the job of the state to mediate between consumers, producers and transnational corporations. Quoting Rousseau, Special Rapporteur Jean Ziegler adds, 'As Jean Jacques Rousseau wrote 246 years ago in the Social Contract: "Between the rich and the poor, it is freedom which oppresses and it is law which liberates"' (ibid., 26).

In another note, the secretary general reiterates this position, stating that both as producers and as lobbyists engaging with policies, transnational corporations exert a great influence over the price, quality, health and safety of food. Corporations have no 'social contract' with individuals, but while human rights laws are supposed to ensure governments do not abuse their powers, how can we hold accountable corporations that are sometimes more powerful than governments? (UNGA 2008). The secretary general elaborates that there are two ways of doing this, either through indirect obligations, in that states have to monitor transnational corporations, or through direct obligations imposed on the corporations themselves. The right to food imposes three levels of obligations on the state: to respect, protect and fulfil the right to food, and in this context the right to protect may be invoked. Quoting the Maastricht Guidelines, the report states:

> The obligation to protect includes the State's responsibility to ensure that private entities or individuals, including transnational corporations over which they exercise jurisdiction, do not deprive individuals of their economic, social and cultural rights. States are responsible for violations of economic, social and cultural rights that result from their failure to exercise due diligence in controlling the behaviour of such non-state actors (para. 18). (Ibid., 13)

Since states have a primary responsibility to protect human rights, this responsibility cannot be delegated.

Corporations must be held accountable under national law. The report goes further to state that transnational corporations also have obligations of their own to fulfil. According to General Comment 12 of the ICESCR: 'While only States are parties to the Covenant and are thus ultimately accountable for compliance with it, all members of society—individuals, families, local communities, non-governmental organizations, civil society organisations, as well as the private business sector—have responsibilities in the realisation of the right to adequate food (para. 20)' (UN Economic and Social Council 1999). Some companies have their own policies on human rights, such as Shell, for instance, with its commitments detailed on its website.[3] However, these voluntary commitments have not prevented Shell from violating these policies time and again. Its notorious record in environmental damage in the Niger Delta is well known—there was an oil spill as recently as December 2011 (Vidal 2011)—as is its collusion in human rights abuses, such as through the funding of armed gangs and other violent activities (Reuters 2011). Thus, asking corporations to adopt voluntary guidelines for the protection of human rights is not enough, and the state is still required to ensure compliance.

Civil Society and Private Charities

The role of NGOs and private and religious charities is another aspect that must be considered when thinking about duty bearers and rights holders, particularly in the context of feeding the most vulnerable and marginalised populations.[4] A study done by the Centre for Equity Studies (CES) focuses exclusively on the role of charities and feeding programmes in the context of urban impoverished persons (CES 2010). The study examines many religious and non-religious charities as well as organisations running subsidised feeding programmes, and examines their impact on the overall situation of hunger on the city streets. For instance, the study estimates that religious charities possibly contribute to less than 4 per cent of the homeless population in the city (ibid., 50).

In addition to this, there are many problems with religious charities, primarily, of course, their sporadic nature, save for a few places like gurudwaras and a few other religious centres. In addition, religious charities tend to focus on donations and more on the goodwill of the giver, as opposed to the receiver. Often, a consequence of this is that the food served is not particularly nutritious or hygienic, and is also often served in fairly degrading conditions without a sense of dignity. In certain places, the most destitute and vulnerable sections of the population may be turned away or served separately if they do not look 'clean' enough. The lack of consistency and accountability is difficult to remedy, although one exception is in Tamil Nadu, where the government has made it mandatory for some 360 temple trusts to feed urban destitute people (CES 2010, 90). As the CES study also points out however, while this is a welcome measure as a supplementary initiative, it cannot function as a substitute for state action. Additionally, if it is to be implemented elsewhere, it cannot focus on trusts of any one religion (ibid.).

Non-religious charities and organisations also run feeding programmes in cities. The CES study finds, yet again, that while there are some impressive models present, they operate at too small a scale to make much of a dent in the hunger situation. Many operate with some form of state support, whether in the form of subsidised rates for land or subsidised ingredients for the preparation of food. The most effective systems are those where an active state role is evident, such as destitute feeding in Tamil Nadu, where the system is linked to the already existing ICDS programme. This not only greatly widens the scale of coverage, it also significantly lowers the cost since it uses existing infrastructure.

Both religious and non-religious institutions can have an important role to play in feeding the most vulnerable sections of society, but the effects will only really be seen if the state acts to facilitate, enforce and monitor these enterprises. Even if the state decides to leave it up to the organisations to implement and run the programmes, it must still provide crucial resources like land or space, and other budgetary support depending on the kind of model being used. This is applicable whether it is the provision of free meals, or destitute feeding, or to set up community kitchens to provide affordable, nutritious meals for homeless populations in urban areas. In both cases, a marginalised and vulnerable population's food needs are being met, and for this it is imperative that they be able to claim their right to food from the state. Moreover, a rights-based approach fundamentally moves away from a charity-based approach to an entitlement-oriented one, and charity is by nature a voluntary phenomenon and cannot substitute for the accountability of the state.

The NAC draft bill on food security provided for community kitchens for the homeless to be set up by the state governments, and allowed for the running of such kitchens to be undertaken by any agency that is seen as appropriate by the government; but it did not provide any specific model or way to do this. This draft thus envisaged a possible role for civil society, including women's self-help groups and trade unions, while providing that the state should be held accountable finally to ensure that these systems work such that the rights are delivered to the rights holders. The government draft of the bill also mentioned community kitchens but without any indication as to their running, except that they must be in accordance with any such scheme of the central government.

Rationale for the State as Duty Bearer

The role of the state as a duty bearer stems first from the enshrinement of the fundamental right to life in the Constitution of India. The Supreme Court of India has asserted that the right to food can be seen as an extension of this right, which should be interpreted as the right to live with human dignity. In addition to this, Article 39(a) (which says that the state has a duty to assure livelihood), and Article 47 (the duty of the state to raise the standard of living and ensure adequate nutrition), also deal with the right to food. And while these are part of the directive principles, and so not enforceable in court, they should be seen as enforceable given that they are an extension of the right to life (Right to Food Campaign 2008, 2–3). This is the view also taken by the National Human Rights Commission in its hearing on starvation deaths in Odisha, where it says:

> The citizen's right to be free from hunger enshrined in Article 21 is to be ensured by the fulfillment of the obligations of the State set out in Articles 39(a) and 47. The reading of Article 21 together with the obligations of the State under Articles 39(a) and 47, places the issue of food security in the correct perspective. The Right to Food is thus a guaranteed Fundamental Right which is enforceable by virtue of the constitutional remedy provided under Article 32 of the Constitution. (NHRC 2003, 199)

Further, the principles in the Constitution are in tandem with the ICESCR, 1976, to which India is also a signatory. The ICESCR is the international instrument that pertains most directly to socio-economic rights. It is stated clearly in the preamble to the covenant: 'Considering the obligation of States under the Charter of the United Nations to promote universal respect for, and observance of, human rights and freedoms', and further, in Article 2: 'Each State Party to the present Covenant undertakes to take steps, individually and through international assistance and co-operation, especially economic and technical, to the maximum of its available resources, with a view to achieving progressively the full realisation of the rights recognised in the present Covenant by all appropriate means, including particularly the adoption of legislative measures.' In each of the articles, it is the state parties whose duties and responsibilities are invoked; thus, this agreement is much more explicit about where accountability lies.

The Committee on Economic and Social Rights of the UN, moreover, adds in General Comment 12 that with regard to the right to adequate food, states have three obligations: 'the obligations to *respect*, to *protect* and to *fulfil*. In turn, the obligation to *fulfil* incorporates both an obligation to *facilitate* and an obligation to *provide*' (UN Economic and Social Council 1999). Furthermore, these duties are elaborated thus:

> The obligation to *respect* existing access to adequate food requires States parties not to take any measures that result in preventing such access. The obligation to *protect* requires measures by the State to ensure that enterprises or individuals do not deprive individuals of their access to adequate food. The obligation to *fulfil* (*facilitate*) means the State must proactively engage in activities intended to strengthen people's access to and utilisation of resources and means to ensure their livelihood, including food security. Finally, whenever an individual or group is unable, for reasons beyond their control, to enjoy the right to adequate food by the means at their disposal, States have the obligation to *fulfil* (*provide*) that right directly. This obligation also applies for persons who are victims of natural or other disasters. (Ibid.)

This envisages a broad and fairly interventionist role for the state.

In the UDHR of 1948, to which India is a signatory, states are never directly implied as duty bearers in the text of the agreement, although the primary responsibility of the state is presumed, particularly in that it is states that are signatories, while individuals and groups are addressed as rights holders. Where the question of socio-economic rights is addressed, for instance, the state is held accountable primarily through its control of resources. Article 22 states: 'Everyone, as a member of society, has the right to social security and is entitled to realisation, through national effort and international cooperation and in accordance with the organisation and resources of each State, of the economic, social and cultural rights indispensable for his dignity and the free development of his personality.'[5]

In the twentieth century, the perception of what constitutes the role of the state has undergone many changes. During the Great Depression, when western economies were at their lowest in terms of unemployment and wage deflation, the state was actively interventionist, injecting money into the economy and thus raising the level of demand, and slowly lifting economies out of deep recession. In the postwar boom of the 1950s and 1960s, when these economies saw a great expansion in their wealth, the welfare state also expanded and provided social security and safety nets more widely. The philosophy here was that the market could take care of the economy and the accumulation and distribution of wealth, while the state would be present to provide a cushion against any individual or temporary setbacks. At the same time, an alternative model of governance was envisaged in the former Soviet Union, where the state took the responsibility not merely of ensuring social security for all, but in the process also controlling economic resources and curtailing certain freedoms.

The latter half of the twentieth century also saw the emergence of independent post-colonial governments, which sought to intervene based largely on the idea that their duty was to correct for the long-term disadvantages that their countries had suffered because of colonial rule. Since the 1990s, however, the world over, there has been an increased shift towards a greater role for the market, and a smaller and smaller role for the state. This has taken the form of structural adjustment programmes in many developing countries, often with disastrous consequences, and a general shift towards the neoliberal framework in others, where social security and socio-economic rights are increasingly compromised in the name of austerity or fiscal discipline. In recent times, the ascent of neoliberal thinking in mainstream discourse about government has meant that it is more important to secure socio-economic rights like the right to food, education, health, etc., through legislation, meaning that the state is accountable regardless of what kind of government is in power.

Surabhi Chopra makes the argument that socio-economic rights like the right to food must be seen as inextricably linked to civil and political rights. She says,

> The right to food entails an entitlement on the part of the citizen, an obligation on the part of the state, and the ability to sue the State when it fails to meet its obligations. This, in turn, implies the existence of certain civil and political rights such as freedom of expression and association, and institutional structures such as an independent judiciary and freedom of the press. (Chopra 2009, 9)

Furthermore, Chopra makes the valid point that it is not enough for regimes to merely provide food, or even access to food, but that the very idea behind the democratic state being a duty bearer is that such a state is accountable to the people, and that is what is crucial about rights-based approaches.

Citing studies by Drèze and Sen, Chopra says:

> It is important to clarify that illiberal regimes which do not guarantee basic civil rights and lack an independent judiciary can and have successfully tackled chronic hunger.... Their successful implementation of social security measures may be instrumental, intended to improve workforce productivity and stimulate growth, for instance, or benevolent and paternalistic. However, where social security measures are discretionary rather than obligatory, where citizens cannot demand these measures, or publicly critique state policy, the rights to food, to health, shelter, or education are not being respected, protected and fulfilled. (Chopra 2009, 9)

Similarly, the colonial British establishment in India had a Famine Code, but its lack of accountability towards its subjects meant that the code could be implemented at its discretion.

This chapter has argued that while there are multiple means of providing food security—namely, through the individual and the family, the private sector, and civil society—the ultimate responsibility must lie with the state. The rights-based approach to socio-economic rights must, at its core, hold the democratic state accountable as a primary duty bearer.

Notes

1. *Society for Unaided Private Schools of Rajasthan v. Union of India & Anr.*, Writ Petition (C) no. 95 of 2010, http://courtnic.nic.in/supremecourt/temp/wc%209510p.txt (accessed 20 December 2017).

2. This, for instance, is the view of Parth J. Shah, president of the Centre for Civil Society (see Shah 2010).

3. Shell Global, 'Human Rights', https://www.shell.com/sustainability/transparency/human-rights.html (accessed 12 March 2018).

4. For a longer discussion on civil society and private charities in the context of destitute feeding and community kitchens, see Chapter 14, 'Vulnerable Groups'.

5. Universal Declaration of Human Rights, http://www.un.org/en/universal-declaration-human-rights/ (accessed 12 March 2018).

References

Boesen, Jakob Kirkemann, and Tomas Martin. 2007. 'Applying a Rights-Based Approach: An Inspirational Guide for Civil Society'. Danish Institute for Human Rights.

CES (Centre for Equity Studies). 2010. *Eating Rough: Hidden Hunger on the City Streets—Searching for Solutions*. http://centreforequitystudies.org/wp-content/uploads/2017/07/Gender_REPORTS_AND_BRIEFS_community_kitchens_edit_Final_Report_4_Dec_2010-1.pdf (accessed 22 December 2017).

Chopra, Surabhi. 2009. 'Holding the State Accountable for Hunger'. *Economic and Political Weekly*, 46(33): 8–12.

Drèze, Jean. 2004. 'Democracy and Right to Food'. *Economic and Political Weekly*, 39(17): 1723–31.

The Hindu. 2011. 'Centre will table Right to Food Act in Parliament, says Manmohan', 11 February.

Mehta, B. C., and Kranti Kapoor. 2010. 'Implementing Right to Education Act'. *Mainstream*, 48(10).

NAC (National Advisory Committee). 2011. 'National Food Security Bill, 2011'. http://www.indiawaterportal.org/sites/indiawaterportal.org/files/National_Food_Security_Bill_approved_by_the_National_Advisory_Council_2011.pdf (accessed 12 March 2018).

NHRC (National Human Rights Commission). 2003. *Annual Report 2002–2003*. http://nhrc.nic.in/Documents/AR/AR02-03ENG.pdf (accessed 22 December 2017).

Planning Commission. 2007. 'Public Distribution System'. *Tenth Five Year Plan*, vol. 2, chapter 3. New Delhi: Government of India.

Raina, Vinod. 2010. 'Are private schools better?' *Business Standard*, 3 February.

Reuters. 2011. 'Shell fuelled human rights abuses in Nigeria—NGO', 3 October. https://www.reuters.com/article/nigeria-shell/shell-fuelled-human-rights-abuses-in-nigeria-ngo-idUSL5E7L33Q720111003 (accessed 23 December 2017).

Right to Food Campaign. 2008. 'Supreme Court Orders on the Right to Food: A Tool for Action'. Right to Food Campaign Secretariat, New Delhi.

Sadgopal, Anil. 2011. 'Neoliberal Act'. *Frontline*, 28(14). http://www.frontline.in/static/html/fl2814/stories/20110715281400400.htm (accessed 21 December 2017).

Shah, Parth J. 2010. 'Are private schools better?' *Business Standard*, 3 February.

The Times of India. 2008. 'School for all'. Editorial, 3 November.

UN Economic and Social Council. 1999. 'Substantive Issues Arising in the Implementation of the International Covenant on Economic, Social and Cultural Rights: General Comment 12'. E/C.12/1999/5, 12 May.

UNGA (United Nations General Assembly). 2008. 'The Right to Food: Note by the Secretary General'. A/58/330, 28 August.

Vidal, John. 2011. 'Nigeria on alert as Shell announces worst oil spill in a decade'. *The Guardian*, 22 December. https://www.theguardian.com/environment/2011/dec/22/nigerian-shell-oil-spill (accessed 22 December 2017).

Ziegler, Jean. 2008. 'Promotion and Protection of All Human Rights, Civil, Political, Economic, Social and Cultural Rights, including the Right to Food: Report of the Special Rapporteur on the Right to Food'. Human Rights Council, A/HRC/7/5, 10 January.

21

Financing the Right to Food

AMOD SHAH

Government spending on social sectors like education, health and nutrition remains far below globally accepted norms of the ratio of this investment as part of GDP in India. Some believe that this is on account of misplaced priorities and fiscal constraints. With respect to the NFSA, many, including members of the government proposing it, have voiced concerns about the implications of binding commitments to food security on the government's financial situation, at a time when it is already running a large fiscal deficit. This chapter discusses the financing of the NFSA in the context of such debates. The first section provides a brief background on the history of public spending on social sectors in India, along with international comparisons. The second section discusses the costs of the proposed NFSA. The third section summarises the key debates around the use of subsidies in facilitating public social spending. The fourth section looks at the major debates around government financing of the NFSA, including fiscal concerns, the role of taxation, ethical justifications and the economic impact of such spending.

PUBLIC SPENDING ON SOCIAL SECTORS IN INDIA AND OTHER DEVELOPING COUNTRIES

India initiated large-scale economic reforms after the balance of payments crisis in 1991, with the aim of transitioning from a socialist economy to a liberalised, market-based one. Measures adopted as part of the reforms process include increased fiscal discipline, public expenditure cuts and tax reforms. This has also resulted in deep cuts in government spending on social sectors and poverty alleviation programmes, with important implications for the poor and vulnerable sections of society.

Social sector spending in India experienced a significant decline in the post-reform period, but has since recovered some of these losses.

Total central government expenditure on social services and rural development was 1.5 per cent of GDP in 2005–06, similar to peak levels seen in 1989–90. For states, such spending stood at 5.3 per cent of GDP in 2005–06, compared to a high of about 6 per cent of GDP in the late 1980s (Ramkumar 2008). Despite improvements in recent years, social sector spending remains significantly below required levels. For example, total central and state government expenditure in 2009–10 on education was 3.7 per cent of GDP (significantly below the 6 per cent benchmark recommended 40 years ago) (CBGA 2012), whereas spending on health and family welfare was 1 per cent of GDP. In the same year, total social expenditure (including the food subsidy) in India stood at only about 7.9 per cent of GDP. This compares poorly to the Organisation for Economic Co-operation and Development (OECD) countries, where average social spending is around 14 per cent of GDP (CBGA 2011a). Among developing countries, South Africa's 2012–13 budget set aside approximately 16 per cent of its GDP for social sectors (*Sowetan Live* 2012); social spending by Brazil in 2009 was almost 16 per cent of GDP (*Portal Brasil* 2011).

Brazil, South Africa and a number of other developing countries have entrenched the duty to progressively realise economic and social rights in their constitutions, and this has significant positive implications for actual public expenditure on social sectors. In South Africa's case, a whole range of socio-economic rights are explicitly enshrined and protected in the Constitution, whereas the Brazilian Constitution mandates minimum levels of educational spending as a share of tax revenues (Nolan and Dutschke 2010, 281). Brazil's Nutritional Security Framework Law, passed in 2006, mandated the creation of a set of institutions specifically for monitoring the right to food (Patnaik 2009). In contrast, socio-economic rights are not included in the Constitution of India, but government action in these matters is guided by the directive principles of state policy. The lack of clearly defined constitutional protections may explain the low levels of social spending in India, particularly in the post-1991 reform period, where increased fiscal tightening has curtailed overall public expenditure.

Comparing public spending on food security across countries is more difficult, given the broad range of programmes covered under this category and varying levels of poverty. The flagship of Brazil's Zero

Hunger programme is the Bolsa Família, a system of conditional cash transfers with a budget of more than US$8 billion in 2010; there are also large allocations to the National Programme for Strengthening Family Farming and the National School Feeding Programme (Souza and Chmielewska 2011). South Africa's Integrated Food Security Strategy allocated US$12.4 billion to grants under its flagship National Comprehensive Social Security Scheme in 2010–11 (Koch 2011). For India, spending in 2011–12 on the PDS, the primary food security programme, was Rs 72,823 crores (or about US$13 billion) (*Economic Times* 2012), with additional allocations to related programmes like the ICDS, MDMS and MGNREGA.

However, the scale of poverty and food insecurity is significantly higher in India than in other countries. Bolsa Família in Brazil benefits about 12.4 million households, while South Africa's National Comprehensive Social Security Scheme reaches about 14 million individuals. In contrast, the PDS in India aims to target about 65.2 million BPL households (*Economic Times* 2010). As discussed in the next section, this number was expected to rise sharply with the passage of the NFSA. As a result, per household spending under the PDS in India is only about US$200 per year, compared to about US$645 and US$885 under the flagship food security schemes in Brazil and South Africa respectively. Given the much higher degree of need in India, the outlay of resources required to provide for food security is significantly higher than these countries, as well as current budgetary allocations.

Estimating the Costs of the NFSA

Under the NFSA, the TPDS was expected to provide subsidised foodgrains to 75 per cent of the rural and 50 per cent of the urban population of India. This was a significant increase from the existing scope of the PDS before NFSA, which aimed to reach about 37.2 per cent of the country's population classified as BPL. Additionally, the act includes special provisions, outside the purview of the TPDS, for certain groups, including nutritional support and maternity benefits for pregnant and lactating women and nutritional support for children (through the ICDS and MDMS). The costs associated with providing these entitlements are as follows:

1. According to official estimates, the foodgrain requirements to meet the increased scope of the TPDS and nutritional support entitlements under the NFSA were expected to result in an annual food subsidy expenditure of about Rs. 1.3 lakh crore in 2013–14.[1] This was significantly higher than previous government estimates, as well as costs estimated in the NAC's draft NFSB and the Report of the Expert Committee on the National Food Security Bill, and compared to a food subsidy bill of Rs 85,000 crores (or 0.85 per cent of GDP) in 2012–13.

   The food subsidy is the difference between the economic cost of foodgrains (total cost of procurement, storage and distribution) borne by the government and the subsidised prices at which these are provided through the TPDS and for nutritional support programmes. These estimates assume 100 per cent offtake by beneficiary households, though in reality the offtake and hence the actual subsidy incurred is likely to be lower. At the same time, the amount of food subsidy will vary annually depending on the difference between the economic cost of foodgrains and the issue price guaranteed under the act. There exists a considerable degree of uncertainty about the actual food subsidy bill in future years; by one estimate, this would rise to Rs 1.4 lakh crore and Rs 1.58 lakh crore in 2014–15 and 2015–16, respectively.
2. The cost of providing maternity benefits of Rs 6,000 to an estimated 20 million mothers is Rs 12,000 crores, which will be shared between the central and state governments.
3. Costs associated with setting up the grievance redressal mechanisms mandated under the act are estimated to be Rs 460 crores.
4. The act also requires that governments strive to progressively realise certain provisions for advancing food security, including the revitalisation of agriculture, interventions related to the procurement, storage and transportation of foodgrains, and other measures for increasing access in areas like healthcare, drinking water, sanitation, pensions, etc. While such measures are undoubtedly important for the effective realisation of the goals of this act, they have not been specifically mandated under the act. Moreover, their costs are quite difficult to estimate, and since most of them have linkages to other areas besides food security, such expenditure would be incurred irrespective of the passage of the act.

In analysing the financing implications of the act, it is important to differentiate between new allocations being made for the purposes of food security, and those which were already mandated (such as the ICDS and MDMS) or have linkages to other areas besides food security (such as funding for revitalisation of agriculture and MGNREGA).

Accounting only for new allocations related to expanded PDS coverage and maternity benefits mandated under the NFSA, the government's estimate for the food subsidy in 2012–13 was Rs 112,205 crore, representing an increase of Rs 23,228 crore over the estimated food subsidy bill for the same year (GoI 2012). At the same time, there is considerable uncertainty over the final subsidy cost, which will vary depending on leakages, offtake by beneficiaries, changes in the economic cost of foodgrains, and expenditures on increasing system capacity and PDS reforms.

The financing of the NFSA was the chief cause of divisive debates in Parliament, with opposition MPs not only criticising the potential costs of the act but also singularly maligning the incumbent Congress government for perpetuating poverty in India. While the demand for greater decentralisation is a lever that the regional parties (or state governments) have repeatedly voiced to pressurise the centre to evolve a more context-specific legislation, the states are reluctant to bear the majority of the costs and demand that the centre bear the bulk of the finances on social sector spending on food.

Debates around Public Social Spending and the Role of Subsidies

The nature of government spending, and in particular the impact of subsidies, remains a topic of considerable debate. Since markets are often imperfect in their functioning and allocation of benefits, governments must step in to ensure socially optimal and equitable results. Some instances of 'market failure', such as monopolistic competition and the impact of externalities, can be remedied by improved government regulation. In other cases, however, government intervention in the form of direct public expenditure is required, for example:

1. When the level of output or welfare derived in a free market is lower than desirable. For example, in the case of health and education,

free market prices are likely to be too high for poorer sections of society. Therefore, the government has to step in to ensure adequate provision of such essential services.

2. When markets often lead to unequal distribution of income and other benefits. In such cases, government spending on social sectors can help in reducing inequalities through income redistribution. Ethical arguments in support of such redistribution are based on the principle that the overall societal good is best served by ensuring a minimal level of welfare for the poor and marginalised.

There is little disagreement on the principle that a basic level of public spending is necessary to provide for socially optimal levels of essential services and to reduce inequality. A number of studies also show that public expenditure in areas such as education and healthcare can have positive impacts on incomes and economic growth (Chu et al. 1995). However, high levels of public expenditure face criticism for being an inefficient means of improving societal welfare. Due to a number of factors, including lack of expertise and accountability within the political class and civil service, perverse political incentives, lobbying by interest groups and corruption, government spending can sometimes be diverted into unproductive areas, which provide little economic or social benefit. Such wasteful expenditure is also harmful to a country's future fiscal health if it is funded by government borrowing. Opponents of increased government spending argue that many social services can be provided in a more efficient and cost-effective manner by the private sector.

In the context of such opposition to high levels of public expenditure, subsidies, particularly targeted ones, have been criticised for a wide range of reasons, including their fairness, effectiveness, distortionary nature and inefficiency. In India's case, much of the spending on food security is through the provision of subsidised foodgrains to poor households. In this context, it is necessary to analyse if subsidies are the best way to provide for food security. Some of these major debates about the impact of subsidies are summarised in what follows.

Fairness of Subsidies

While public expenditure on social sectors benefits the entire population, targeted subsidies are often criticised for being unfair, since they cater to specific sections of society. Since such subsidies are funded through government tax revenues, it is argued that this places an unfair burden on taxpaying members of society, who bear the costs without receiving any benefit. Subsidy programmes are also vulnerable to the influence of special interest groups, who benefit from economic distortions created by the subsidy. As a result, subsidy programmes are often subverted to serve such vested interests, or it becomes difficult to discontinue them once they have fulfilled their purpose, at a cost to intended beneficiaries and honest taxpayers. In the context of food subsidies in India, critics point out that powerful farmer and trading lobby groups have been able to push for higher grain procurement prices, causing rapid increases in the government's food subsidy without any benefit to poor households.

Such concerns of fairness, while sometimes valid, can often be addressed through better management and design of subsidy programmes. For example, most taxpayers are unlikely to take exception to cost-effective food subsidies which benefit genuinely food-insecure households. Alternatively, a programme with universal entitlements, such as the PDS before the introduction of targeting, may also satisfy such concerns (but it would create other concerns of high subsidy costs on undeserving segments of the population). In either case, it is necessary to ensure checks on the lobbying power of special interest groups.

Subsidies as a Means of Addressing Inequalities

Through the transfer of benefits, in cash or kind, to poor households, subsidies serve as a means of income transfer and reduction in inequalities. At an economic level, inequality can also undermine a country's growth and development through a variety of channels, including insufficient investment and educational opportunities for poorer sections of society, higher and more persistent poverty, and increased socio-political conflict (Galor 2009). Therefore, there are strong economic justifications for addressing such inequalities through subsidies and other means of income redistribution.

Both cash subsidies and direct support for spending that is important for poorer individuals, such as subsidised food and housing, have a large impact on income distribution. At the same time, public expenditure on social sectors can also have indirect but significant effects. For example, public spending on education, health and job training programmes leads to improvements in human capital for the poor, thereby improving their economic status and incomes. Since poorer sections of society are likely to benefit more from public spending on basic services (for example, primary rather than tertiary education, and local clinics rather than central hospitals), such expenditure may be considered progressive in nature. In addition, strong institutional arrangements enforcing the rule of law and access to justice can also contribute to a fairer distribution of income by reducing abuses and corruption (Afonso et al. 2007). Unlike subsidies, such measures do not have an immediate, direct impact on inequality. However, appropriate public spending and institutions can serve as an important complement to the role played by subsidies.

In India's case, rapid economic growth in recent years has been highly unequal. The Gini coefficient, a measure of income inequality, increased from 0.30 in 1993–94 to 0.36 in 2009–10, indicating that inequality actually increased during the reforms period. Employment growth has been abysmal; the period between 2004–05 and 2009–10 resulted in the creation of only one million jobs, against a projected estimate of 60 million (Himanshu 2012). Public social expenditure also remains well below required levels. In this context, subsidies can play an important role in facilitating income transfers and reducing inequality.

The Impact of Subsidies on Work Incentives

Subsidies are criticised for creating disincentives to work and increasing dependency among beneficiaries. If a subsidy is tied to employment or income levels, people may reduce work efforts in order to qualify, or may choose to drop out of the workforce altogether, creating a culture of dependency among beneficiaries. Evidence from industrialised countries shows some evidence of such labour disincentives in the case of unemployment transfers (Vodopivec 2009). There is limited empirical evidence about the impact of subsidies on work incentives in developing countries.

Such concerns are relevant with respect to unemployment and other similar cash transfers, but their impact on in-kind transfers such as food subsidies is likely to be minimal; it is highly unlikely that people will reduce their work effort in order to qualify for a small quota of free foodgrains each month. Moreover, in cases where perverse incentives exist, these can be minimised through better design of subsidy programmes. For example, in the case of unemployment transfers, measures to reduce work disincentives can include assistance to beneficiaries in terms of job services and training, complementary supports such as childcare and transport assistance, and time limits on the duration of benefits.

Inefficiency of Subsidies

Subsidies face severe criticism for their high degree of inefficiency in fulfilling their intended objectives. Critics argue that a large governmental role in the provision of subsidies leaves them open to the bureaucracy and corruption that often characterise government interventions. These are most severe in the case of targeted subsidies, which suffer from large exclusion and inclusion errors. Targeting also gives rise to rent-seeking activities in the process of identification and certification of beneficiaries. Besides this, there are high costs associated with the execution of such programmes on account of mismanagement and administrative inefficiencies. Therefore, it is argued that subsidies are a costly and inefficient way of providing such benefits.

Many of these concerns about cost-effectiveness and efficiency are relevant to food subsidies in India. As per the Planning Commission's evaluation of the TPDS in 2005, about 58 per cent of subsidised foodgrains did not reach BPL families because of identification errors, non-transparent operation and unethical practices in the implementation of the system. The report also estimated that for every one rupee of income transfer to the poor through the system, the government spent Rs 3.65; this equates to an efficiency rate of just 27 per cent (Planning Commission 2005). At the same time, the recent experience with the PDS in a number of states suggests that with some much-needed interventions, such as expanded coverage, reduced prices, diversification of product mix and the use of technology, the efficiency of the system can be improved significantly (Khera 2011). Reforms to address shortcomings in the PDS, most prominently the move towards universalisation of entitlements and the possible use of cash transfers, have been the focus of much debate; these are discussed in Chapter 4, 'Universal or Targeted Entitlements', and Chapter 7, 'Cash Transfers or Transfers in Kind', respectively.

Subsidies as a Means of Boosting Growth

By providing cash or freeing up spending from other areas, subsidies effectively provide additional income to their beneficiaries. Some proponents of subsidies argue that since the poor have a higher propensity to consume than the rich, they are more likely to spend this additional income, thereby boosting economic demand. However, since subsidies are often financed by taxation, removal of subsidies may also lead to lower taxation, which itself would stimulate the economy and also support growth. Moreover, in cases where subsidies are funded by an increase in the government's budget deficit, their demand-boosting impact is counteracted by higher interest rates and other negative impacts of budget deficits, which have an adverse effect on economic performance.

Debates around Financing of the NFSA

The financing of the NFSA must be analysed in the context of the ability and willingness of the government to commit public expenditure to social sectors, as well as the various benefits arising from such spending. This section summarises some of the key debates in this regard.

Government's Fiscal Position and Subsidy Expenditure

Sustainable public finances are essential for balanced and sustainable economic growth. When governments choose to spend beyond their means, such spending must be funded through increased borrowings, which is unsustainable in the long run. Over time, countries running budget deficits build up huge public debt and large interest payment obligations, which prevent them from spending adequately on social sectors, infrastructure and other key areas; this in turn results in weak and inequitable economic growth. Moreover, high levels of public debt can lead to inflation, if the government chooses to print money to pay off its obligations. Lastly, large government borrowing reduces the availability of credit to the private sector, hurting investment and economic growth.

India's 2011–12 fiscal deficit stood at a fairly high 5.9 per cent of GDP, and critics pointed to the large negative impact that high subsidies had had on the government's financial position. The government's total subsidy bill in 2001–12 was Rs 2,08,503 crores (approximately 2.5 per cent of GDP), of which food subsidies stood at Rs 72,823 crores. The budget's allocation of Rs 75,000 for food subsidy costs in this year was significantly lower than the government's own estimates, making it highly likely that the actual food subsidy would overshoot budgeted amounts. Citing the impact of high subsidy costs on the fiscal deficit, the budget aimed to restrict total subsidies to 2 per cent of GDP in 2012–13 and reduce it further in later years (*Economic Times* 2012).

Previous attempts to reduce or rein in the food subsidy have included a move away from universal entitlements to the TPDS in 1997, and an increase in foodgrain prices in 1999–2000. However, the food subsidy bill has been difficult to contain in recent years, accounting for about 0.9 per cent of GDP in 2011–12; this compares to an average of about 0.5 per cent of GDP for the period between 1990 and 2000 (CBGA 2011b). While it is clear that food subsidies represent a substantial fiscal burden on the government, much of the recent increase can be attributed to higher procurement prices of grains, as well as leakages and inefficiencies in the distribution system, which have increased after the introduction of the TPDS. There has been little benefit to beneficiary households, either in terms of improved coverage or efficiency of distribution.

Similarly, inefficiency and misuse are rampant in the provision of fuel and fertiliser subsidies. For example, diesel car sales have risen rapidly in recent years, as private car owners also benefit from the diesel subsidy (Centre for Science and Environment 2012). Increasing taxes on these vehicles can help to curtail such misuse and ensure that the subsidy benefits the transportation sector. Similarly, the fertiliser subsidy is channelised through fertiliser companies and retailers, who often do not pass on the full subsidy to farmers. Therefore, instead of simply capping subsidy costs as a share of GDP, it would be much more preferable to tackle such issues, since they can contribute substantially to the lowering of subsidy costs.

Funding for subsidies and other social expenditure is also determined by the government's spending priorities. Given limited resources and competing priorities, securing adequate funds for social programmes is a contentious and difficult process. Poor and marginalised sections often lack the societal support and political influence necessary to lobby for their causes, thereby losing out to competing interests in the budget-making process. For example, India's defence budget for 2012–13 was increased by more than 17 per cent over the previous year, to Rs 19.3 million; in 2011–12, there was an 11 per cent increase. However, such outlays are considered non-negotiable, with little debate about their impact on budget deficits. Rather, various studies show that pro-poor expenditures are the most affected during budget consolidation (Ortiz et al. 2011). Therefore, creating fiscal space for the increased food subsidy under the NFSA is likely to be more a question of the government's willingness, rather than its ability to fund social spending.

As discussed earlier, fiscal considerations are important in order to ensure robust and sustainable growth. At the same time, social infrastructure in India remains extremely poor, and it would be dangerous to allow spending in critical areas like food security, health and education to be determined solely by budgetary considerations. Therefore, even in times of fiscal constraints, government needs to ensure adequate allocation of funds for social spending, either through rationalisation and reprioritisation of current expenditures or through an increase in revenues.

The Role of Taxes

Among the instruments available to the government to fund public expenditure, taxation is the most important. Under progressive taxation systems, tax rates increase with increases in the taxable amount, so that wealthier sections of the population bear a larger share of the tax burden. Through higher taxation of the wealthy and channelising these revenues into public spending on social programmes, progressive taxation aims to facilitate income redistribution and reduction in societal inequalities.

Taxes can be direct or indirect; direct taxes are those which are paid directly by an individual or organisation, whereas indirect taxes are imposed on goods and services, rather than on a specific entity. Direct taxes, such as personal and corporate income tax, are more progressive than indirect taxes, which are generally applied equally, irrespective of the income of the taxpayer.

There are two major impacts of taxation on growth: First, higher taxes may hurt investment returns and the incentive to work, consequently lowering growth and government revenues. Second, taxation provides the means to finance public expenditure, which if wisely invested enhances productivity and growth. Therefore, in setting tax rates, policymakers need to ensure that tax rates are neither so high that they destroy the incentive to work and encourage evasion (as was the case in India in the mid-1970s, when the highest income tax bracket in the country was 97.5 per cent, leading to large-scale tax evasion and significant damage to the government's ability to raise revenues from taxation), or so low that they stifle the government's ability to provide for worthwhile public expenditure.

There is considerable debate over the optimal level of taxation and the relative preference between direct and indirect taxes. The neoliberal economic policy framework adopted by multilateral agencies like the IMF and World Bank calls for a 'broad-based, low-rate' approach to taxation, through the adoption of broad-based, uniform indirect taxes like VAT and a flatter, more uniform income tax (Bird 2010). Proponents of such an approach argue that broad-based indirect taxes increase government revenues since they are cheaper to implement and more difficult to evade than direct taxes, whereas lower income taxes for individuals and corporations help stimulate economic growth. However, such policy frameworks face criticism because of their highly regressive nature and inability to facilitate adequate income redistribution. In fact, indirect, consumption-based taxes, such as VAT and sales tax, are likely to hit poorer sections of society harder, since they spend a larger share of their income on consumption of goods and services. Moreover, it is argued that raising direct tax revenues by tackling evasion and reducing tax exemptions and concessions would be preferable to raising indirect taxes.

Level of Taxation

India's tax reforms since the early 1990s have been broadly successful in fulfilling their objective of reducing tax rates without hurting government revenues. The lowering of tax rates initially led to a decline in the overall tax to GDP ratio for India, from 15.7 per cent in 1991–92 to a low of 13.8 per cent in 2001–02. However, strong economic growth, improved compliance and a broadening of the tax base have resulted in improvements in tax revenues since, with the tax to GDP ratio rising to about 17.6 per cent in 2007–08 (GoI 2010). In 2010–11, India's overall tax to GDP ratio was about 16 per cent (*The Economic Times* 2011). Personal income tax in India remains progressive, with the rich paying higher taxes.

However, the increase in tax revenues since 2001–02 has not been matched by a corresponding improvement in social spending, which in fact was on a declining trend until a few years back. Therefore, it is important to ensure that increased tax revenues are channelised to social sectors. The absence of adequate income redistribution and investment into essential social sectors is likely to result in greater inequality and unsustainable, low-quality growth. Moreover, international comparisons show that India's tax to GDP ratio in 2010–11 was less than half the 2008 OECD average of 38.4 per cent (*The Economic Times* 2011), and Brazil's 2009 tax to GDP ratio of 32.6 per cent (OECD 2011). This suggests that there exists significant scope to raise tax revenues, thereby increasing availability of public resources for social sector spending.

Type of Taxation

In 1990–91, indirect taxes accounted for a huge 86 per cent of total government tax revenues. While the post-1991 tax reforms have been successful in decreasing the relative importance of indirect taxes, India's taxation system remains highly regressive. In 2009–10, direct taxes were estimated to account for 37.3 per cent of total government tax revenues, while the share of indirect taxes was 62.7 per cent. There has been an improvement in the direct tax share of central government revenues, which rose from only 19.2 per cent in 1990–91 to 56.8 per cent in 2009–10. However, states continue to rely heavily on indirect taxation, and the share of direct taxes in state government revenues in 2009–10 was just 2.3 per cent (GoI 2010). This is particularly concerning, since states contribute the vast majority of public social sector spending in India .

In an international context, most developed countries and a number of developing countries exhibit a much higher reliance on direct taxes; on average, direct taxes accounted for 68 per cent of government tax revenues for OECD countries in 2005 (OECD 2007). Therefore, it is necessary to adopt measures to make India's tax system less regressive, by increasing the share of direct taxes in total tax revenues. Failure to do so impairs the government's ability to facilitate income transfers and reduce inequalities through progressive taxation.

Tax Compliance and Efficiency

Besides the level and type of taxation, the system's efficiency and compliance play a vital role in determining government tax revenues. India's post-1991 tax reforms have led to some impressive improvements in these areas. Tax collection charges have declined steadily and accounted for about 0.4 per cent of total revenues in 2008, indicating the increased efficiency of the system. In addition, India's tax base (number of persons and companies being taxed) broadened, from 23 million in 2000–01 to 32.6 million in 2008–09 (GoI 2011).

At the same time, there exists significant scope for improvements in tax compliance. The rise in the tax base between 2000–01 and 2008–09 was largely due to an increase in personal income tax payers, from 21.2 million in 2000–01 to 30.8 million in 2008–09 (GoI 2011). Corporate tax assessments actually fell marginally during this period, which appears inconsistent with strong economic growth.

Additionally, a huge magnitude of tax revenue is foregone each year due to tax exemptions and incentives; in 2010–11, total central government tax revenue foregone was estimated to be Rs 5.12 lakh crore.[2] In contrast, India's fiscal deficit for that year stood at Rs 3.69 lakh crore (Reuters 2011). While tax benefits are useful in incentivising much-needed investment and savings, many of them are also prone to misuse. For example, certain states in India offer corporate tax holidays to incentivise corporate investment. However, companies often book the majority of their profits through subsidiaries based in these states, thereby lowering their tax liabilities. Moreover, there are numerous instances of companies shutting down operations as soon as their relevant tax incentives have expired. In such cases, tax concessions do little to promote their objective, but rather result in revenue being foregone by the government. Similarly, personal tax exemptions offered for individual taxpayers to promote savings and charitable contributions are often misused. Checking such tax avoidance and misuse can play an important role in raising government revenues.

In aggregate, it is estimated that India loses Rs 14 lakh crore annually due to tax evasion (*Bloomberg Businessweek* 2011). A wide range of activities are classified under evasion, including under-reporting of income, illegal or unreported transactions and the transfer of assets to overseas tax havens where they can be concealed from domestic tax authorities. Besides exacerbating already high income inequality and placing an unfair burden on honest taxpayers, such large-scale evasion severely impairs the government's ability to fund critical infrastructure and social sector spending. In the context of high fiscal deficits and limited government resources, tackling tax evasion is extremely important. Measures being adopted by the government to address evasion include greater international cooperation in order to obtain information about assets being held abroad by Indian residents, sanctions against non-cooperative tax havens, greater vigilance against money laundering and other illegal transactions, and strengthening of laws dealing with tax evasion.

The Role of Other Revenues in Social Spending

About 14 per cent of total government revenues come from non-tax sources, such as profits from state-owned enterprises, interest payments, disinvestment proceeds and grants (Sinha 2010). However, since many of these payments are uncertain or erratic in nature, they are not a viable alternative to a well-functioning tax system as a source of funds for public expenditure on social sectors.

An innovative use of taxation for social spending is the education cess, introduced in July 2004. This 2 per cent cess is imposed on income tax, corporation tax, excise and customs duties and service tax, and revenues raised are earmarked for education. While such measures are important, they cannot be a substitute for government budgetary allocations. In 2006–07 and 2007–08, the education cess funded 38 per cent and 49 per cent of total expenditure on education respectively, reflecting the reduced contribution of other budgetary resources. Thus, the education cess is being used as a substitute for budgetary allocations and not as an additional source of revenue to boost the existing allocation to education, as was originally intended (CUTS International 2009). Care needs to be taken to ensure that in exploring avenues for supplementary funding, the government does not shirk its obligations to allocate adequate budgetary funds to social programmes.

The Economic Impact of Hunger and Malnutrition

India's high economic growth rate has not led to any significant improvement in the country's social and human development indicators. The failure to ensure food security for the population means that levels of hunger and malnutrition remain extremely high even today. Almost one in two Indian children are stunted and 40 per cent are underweight. One-third of all Indian women are underweight, while micronutrient deficiencies are extremely high in both groups; 80 per cent of children and 56 per cent of women are anaemic (Kadiyala et al. 2012). India is one of only three countries where the IFPRI's Global Hunger Index has risen, from 22.9 to 23.7, between 1996 and 2011 (Himanshu 2012).

Improvements in the nutritional status of the population have significant positive impacts on health, education and productivity, and consequently on incomes and economic growth. Malnutrition affects educational outcomes through a reduced capacity to learn (as a result of early cognitive deficits or lowered current attention spans) and fewer total years of schooling for malnourished children (Alderman 2005). In terms of productivity losses, the economic impact of reduced employment resulting from childhood malnutrition in India is estimated at Rs 10,300 crore (Bhandari and Zaidi 2004). Other studies suggest that the total productivity loss from malnutrition in India is almost 3 per cent of GDP (Horton 1999). Thus, there are strong economic justifications for the government to adopt a more proactive approach in tackling hunger and malnutrition.

Ethical Justifications for Public Spending on Food Security

The philosophy of universal human rights recognises all human beings as having equal intrinsic worth simply because they are human (Mander 2012). Therefore, at an ethical level, everyone, irrespective of their economic worth, must be assured a floor of dignified survival, whereby their fundamental needs, such as food, shelter and access to health and education, are satisfied adequately and in a dignified manner. In the context of food security, the right to food is a basic human right. Therefore, societies are obligated to provide the able-bodied with the means to obtain food and enable those who are unable to feed themselves to receive food directly (FAO 2001). The existence of hunger and malnutrition is ethically unacceptable under any circumstances, and cannot be justified on the grounds of inadequate resources.

The Role of Public Expenditure in Human Development and Inclusive Growth

Government attempts to raise the level of human development of its people are best achieved through public spending on social sectors. A wide body of research indicates that there are strong linkages between public expenditure in areas like education, health and nutrition, and indicators of human development. Increases in social spending can have a stronger impact on human development indicators than growth in per capita income. This suggests that when the aim is social upliftment of the population, direct interventions are preferable to indirect attempts such as income and employment generation (though these have an important role to play in ensuring economic security and stable livelihoods). Baldacci et al. (2004), using data for 120 developing countries between 1975 and 2000, find that public spending on education and health has a significant positive impact on improvements in these areas. Using data for nearly 200 countries between 1980 and 2006, Conceição and Kim (2009) find a negative impact on indicators of health and education during periods of economic deceleration, which are generally associated with decreased social sector spending; such negative effects were most severe for developing countries. Therefore, government spending on social programmes is critical for ensuring strong improvements in human development indicators.

More broadly, higher public expenditure on social sectors facilitates a move towards a more inclusive growth model which benefits all sections of society, including the poor and marginalised. It is argued that a reliance on high private profits to fund economic growth has resulted in largely unproductive investment and inequitable distribution of wealth. Instead, developing countries are better off adopting an employment-intensive growth process, which in turn supports stronger consumption and domestic demand (ILO 2011). Besides ensuring sustainable economic growth, wage-led growth is effective in reducing inequality and poverty levels. Under such wage-led growth models, increased social spending in areas like health, education, nutrition and basic infrastructure is extremely important, because in addition to improving societal welfare and labour productivity, such spending also facilitates rapid economic growth through increased employment and demand.

Notes

1. 1 lakh = 100,000; 1 crore = 10 million.

2. 'Revenue Foregone under the Central Tax System: Financial Years 2009–10 and 2010–11', http://indiabudget.nic.in/ub2011-12/statrevfor/annex12.pdf (accessed 4 January 2018).

References

Afonso, A., L. Schuknecht and V. Tanzi. 2007. 'Income Distribution Determinants and Public Spending Efficiency'. *Journal of Economic Inequality*, 8(3): 367–89.

Alderman, H. 2005. 'Linkages between Poverty Reduction Strategies and Child Nutrition: An Asian Perspective'. *Economic and Political Weekly*, 40(46): 4837–42.

Baldacci, Emanuele, Benedict Clements, Sanjeev Gupta and Qiang Cui. 2004. 'Social Spending, Human Capital, and Growth in Developing Countries: Implications for Achieving the MDGs'. IMF Working Paper WP/04/217, International Monetary Fund, Washington, D.C.

Bhandari, L. and L. Zaidi. 2004. 'Reviewing the Costs of Malnutrition in India'. Indicus Analytics, November.

Bird, R. 2010. 'Taxation and Development'. *Economic Premise*, no. 34, October.

Bloomberg Businessweek. 2011. 'In India, tax evasion is a national sport', 28 July.

CBGA (Centre for Budget and Governance Accountability). 2011a. 'UPA's Promises and Priorities: Is There a Mismatch? Response to Union Budget 2011–12'. http://www.cbgaindia.org/publication/upas-promises-priorities-is-there-a-mismatch/ (accessed 4 January 2018).

CBGA (Centre for Budget and Governance Accountability). 2011b. CBGA Budget Track vol. 8, track 1, January. http://www.cbgaindia.org/budget-tracks/vol-8-track-1-2011/ (accessed 12 March 2018).

———. 2012. 'Charter of Demands for Union Budget 2012–13'. People's Budget Initiative, CBGA, New Delhi. http://www.cbgaindia.org/wp-content/uploads/2015/01/Charter-of-Demands-For-Union-Budget-2012-13.pdf (accessed 12 March 2018).

Centre for Science and Environment. 2012. 'Budget 2012 Is a Mixed Bag, Says CSE'. 16 March. http://cseindia.org/content/budget-2012-a-mixed-bag-says-cse (accessed 4 January 2018).

Chu, Ke-young, S. Gupta, B. Clements, D. Hewitt, S. Lugaresi et al. 1995. 'Unproductive Public Expenditures: A Pragmatic Approach to Policy Analysis'. Pamphlet Series no. 48, Fiscal Affairs Department, International Monetary Fund, Washington, D.C.

Conceição, Pedro, and Namsuk Kim. 2009. 'The Asymmetric Impact of Growth Fluctuation on Human Development: Evidence from Correlates of Growth Decelerations and Accelerations'. UNDP/ODS Working Paper, United Nations Development Programme, New York.

CUTS International. 2009. 'Social Sector Budgeting: Higher Allocations Mask Poor Utilisation'. http://www.cafindia.org/pages/Social%20Sector%20Budgeting.pdf (accessed 4 January 2018).

Economic Times. 2010. 'PDS revamp comes a day ahead of food law meet'. https://economictimes.indiatimes.com/news/economy/policy/pds-revamp-comes-a-day-ahead-of-food-law-meet/articleshow/6165247.cms (accessed 12 March 2018).

———. 2011. 'India's tax-GDP ratio still less than half OECD's', 21 April.

———. 2012. 'Union budget 2012–13 pegs FY'13 subsidy bill 14% lower than current fiscal', 16 March.

FAO (Food and Agriculture Organization). 2001. 'Ethical Issues in Food and Agriculture', FAO Ethics Series. FAO, Rome.

Galor, O. 2009. 'Inequality and Economic Development: An Overview'. Working Paper no. 2009-3, Department of Economics, Brown University. https://EconPapers.repec.org/RePEc:bro:econwp:2009-3 (accessed 12 March 2018).

GoI (Government of India). 2010. *Indian Public Finance Statistics 2009–10*. Department of Economic Affairs, Ministry of Finance, Government of India. July, 2010. https://dea.gov.in/indian-public-finance-statistics (accessed 12 March 2018).

———. 2011. 'Chapter 6: Direct and Indirect Taxes', *Statistical Year Book India 2011*. New Delhi: Ministry of Statistics and Programme Implementation, Government of India.

———. 2012. 'National Food Security Bill, a Historic Initiative by Central Government'. Press Information Bureau, Ministry of Consumer Affairs, Food and Public Distribution, 11 February. http://pib.nic.in/newsite/PrintRelease.aspx?relid=80259 (accessed 12 March 2018).

Himanshu. 2012. 'Food Insecurity'. *Caravan Magazine*, 4(2), February.

Horton, S. 1999. 'Opportunities for Investments in Nutrition in Low-income Asia'. *Asian Development Review*, 17: 246–73.

ILO (International Labour Organization). 2011. 'An Interview with Jayati Ghosh, Winner of the ILO Decent Work Research Prize', 11 November. http://www.ilo.org/global/publications/world-of-work-magazine/articles/WCMS_167557/lang--en/index.htm (accessed 5 January 2018).

Kadiyala, S., P. K. Joshi, S. Mahendra Dev, T. N. Kumar and V. Vyas. 2012. 'A Nutrition Secure India: Role of Agriculture'. *Economic and Political Weekly*, 47(8): 21–25.

Khera, R. 2011. 'Revival of the Public Distribution System: Evidence and Explanations'. *Economic and Political Weekly*, 46(44): 36–50.

Koch, J. 2011. *The Food Security Policy Context in South Africa*. Country Study no. 21. Brasilia: International Policy Centre for Inclusive Growth.

Mander, H. 2012. 'In whose welfare?' *Hindustan Times*, 25 March.

Nolan, A., and M. Dutschke. 2010. 'Article 2(1) ICESCR and States Parties' Obligations: Wither the Budget'. *European Human Rights Law Review*, 3: 280–89.

OECD (Organisation for Economic Co-operation and Development). 2007. *Revenue Statistics 1965–2006*. Paris: OECD Publishing.

———. 2011. *Revenue Statistics in Latin America*. Paris: OECD Publishing

Ortiz, I., J. Chai and M. Cummins. 2011. 'Identifying Fiscal Space: Options for Social and Economic Development for Children and Poor Households in 184 Countries'. UNICEF Social and Economic Policy Working Paper, UNICEF, New York, December.

Patnaik, B. 2009. 'How to tackle India's hunger'. *Mint*, 2 January.

Planning Commission. 2005. *Performance Evaluation of Targeted Public Distribution System (TPDS)*. New Delhi: Programme Evaluation Organisation, Government of India, March.

Portal Brasil. 2011. 'Social investments grew 140% in 15 years', 12 July.

Ramkumar, R. 2008. 'Levels and Composition of Public Social and Economic Expenditures in India, 1950–51 to 2005–06'. *Social Scientist*, 36(9–10): 48–94.

Reuters. 2011. 'India's 2010/11 fiscal deficit at 4.7 pct of GDP', 31 March.

Sinha, R. 2010. 'An International Comparison of Tax Regimes'. Discussion paper, Centre for Budget and Governance Accountability, New Delhi. http://www.cbgaindia.org/working-paper/an-international-comparison-of-tax-regimes/ (accessed 12 March 2018).

Souza, D., and D. Chmielewska. 2011. 'Public Support for Food Policy in India, Brazil and South Africa: Elements for a Policy Dialogue'. Working Paper no. 80, International Policy Centre for Inclusive Growth, April.

Sowetan Live. 2012. 'South Africa's first R1 trillion budget'. 22 February.

Vodopivec, M. 2009. 'Introducing Unemployment Insurance to Developing Countries'. SP Discussion Paper 907, World Bank, Washington, D.C., May.

Notes on the Contributors

Ankita Aggarwal is Coordinator, Right to Food Campaign, New Delhi. She was previously a research scholar at Centre for Equity Studies, New Delhi.

Warisha Farasat is an alumnus of the National Law School, Bangalore, and Columbia University. She practises law in New Delhi, and is co-author of *Splintered Justice: Living the Horror of Mass Communal Violence in Bhagalpur and Gujarat* (2016).

Sajjad Hassan is a researcher on governance, pro-poor development and social inclusion. He has served in the Indian Administrative Service (IAS), and supported grassroots organisations, policy research and human rights advocacy. In 2015, he founded Misaal, a platform for minority rights capacity, research and advocacy.

Harsh Mander, writer, human rights worker, columnist, researcher and teacher, is Director, Centre for Equity Studies, New Delhi, an independent organisation engaged in research and advocacy on a range of social and economic justice issues in India. Some of his books include *Looking Away: Inequality, Prejudice and Indifference in New India (2015); Fatal Accidents of Birth: Stories of Suffering, Oppression and Resistance* (2016); *Living with Hunger and Ash in the Belly: India's Unfinished Battle against Hunger (2012).*

Ashwin Parulkar is Senior Researcher, Centre for Policy Research, New Delhi. He was previously with Centre for Equity Studies, New Delhi.

Amod Shah is a Ph.D. candidate, International Institute of Social Studies in the Hague, The Netherlands. He was previously a researcher with the S. R. Sankaran Unit on Hunger and Social Exclusion at the Centre for Equity Studies, New Delhi.

Saba Sharma is currently a Ph.D. student, Department of Geography, University of Cambridge, studying the state and ethnic conflict in Assam. She was a former research associate at the Centre for Equity Studies, New Delhi.

Index